AF702600

RT-PCR Protocols

METHODS IN MOLECULAR BIOLOGY™

John M. Walker, Series Editor

220. **Cancer Cytogenetics:** *Methods and Protocols,* edited by *John Swansbury, 2003*
219. **Cardiac Cell and Gene Transfer:** *Principles, Protocols, and Applications,* edited by *Joseph M. Metzger, 2003*
218. **Cancer Cell Signaling:** *Methods and Protocols,* edited by *David M. Terrian, 2003*
217. **Neurogenetics:** *Methods and Protocols,* edited by *Nicholas T. Potter, 2003*
216. **PCR Detection of Microbial Pathogens:** *Methods and Protocols,* edited by *Konrad Sachse and Joachim Frey, 2003*
215. **Cytokines and Colony Stimulating Factors:** *Methods and Protocols,* edited by *Dieter Körholz and Wieland Kiess, 2003*
214. **Superantigen Protocols,** edited by *Teresa Krakauer, 2003*
213. **Capillary Electrophoresis of Carbohydrates,** edited by *Pierre Thibault and Susumu Honda, 2003*
212. **Single Nucleotide Polymorphisms:** *Methods and Protocols,* edited by *Piu-Yan Kwok, 2003*
211. **Protein Sequencing Protocols, 2nd ed.,** edited by *Bryan John Smith, 2003*
210. **MHC Protocols,** edited by *Stephen H. Powis and Robert W. Vaughan, 2003*
209. **Transgenic Mouse Methods and Protocols,** edited by *Marten Hofker and Jan van Deursen, 2002*
208. **Peptide Nucleic Acids:** *Methods and Protocols,* edited by *Peter E. Nielsen, 2002*
207. **Recombinant Antibodies for Cancer Therapy:** *Methods and Protocols.* edited by *Martin Welschof and Jürgen Krauss, 2002*
206. **Endothelin Protocols,** edited by *Janet J. Maguire and Anthony P. Davenport, 2002*
205. ***E. coli* Gene Expression Protocols,** edited by *Peter E. Vaillancourt, 2002*
204. **Molecular Cytogenetics:** *Protocols and Applications,* edited by *Yao-Shan Fan, 2002*
203. ***In Situ* Detection of DNA Damage:** *Methods and Protocols,* edited by *Vladimir V. Didenko, 2002*
202. **Thyroid Hormone Receptors:** *Methods and Protocols,* edited by *Aria Baniahmad, 2002*
201. **Combinatorial Library Methods and Protocols,** edited by *Lisa B. English, 2002*
200. **DNA Methylation Protocols,** edited by *Ken I. Mills and Bernie H, Ramsahoye, 2002*
199. **Liposome Methods and Protocols,** edited by *Subhash C. Basu and Manju Basu, 2002*
198. **Neural Stem Cells:** *Methods and Protocols,* edited by *Tanja Zigova, Juan R. Sanchez-Ramos, and Paul R. Sanberg, 2002*
197. **Mitochondrial DNA:** *Methods and Protocols,* edited by *William C. Copeland, 2002*
196. **Oxidants and Antioxidants:** *Ultrastructure and Molecular Biology Protocols,* edited by *Donald Armstrong, 2002*
195. **Quantitative Trait Loci:** *Methods and Protocols***,** edited by *Nicola J. Camp and Angela Cox, 2002*
194. **Posttranslational Modifications of Proteins:** *Tools for Functional Proteomics***,** edited by *Christoph Kannicht, 2002*
193. **RT-PCR Protocols,** edited by *Joe O'Connell, 2002*
192. **PCR Cloning Protocols, 2nd ed.,** edited by *Bing-Yuan Chen and Harry W. Janes, 2002*
191. **Telomeres and Telomerase:** *Methods and Protocols,* edited by *John A. Double and Michael J. Thompson, 2002*
190. **High Throughput Screening:** *Methods and Protocols,* edited by *William P. Janzen, 2002*
189. **GTPase Protocols:** *The RAS Superfamily,* edited by *Edward J. Manser and Thomas Leung, 2002*
188. **Epithelial Cell Culture Protocols,** edited by *Clare Wise, 2002*
187. **PCR Mutation Detection Protocols,** edited by *Bimal D. M. Theophilus and Ralph Rapley, 2002*
186. **Oxidative Stress Biomarkers and Antioxidant Protocols,** edited by *Donald Armstrong, 2002*
185. **Embryonic Stem Cells:** *Methods and Protocols,* edited by *Kursad Turksen, 2002*
184. **Biostatistical Methods,** edited by *Stephen W. Looney, 2002*
183. **Green Fluorescent Protein:** *Applications and Protocols,* edited by *Barry W. Hicks, 2002*
182. **In Vitro Mutagenesis Protocols, 2nd ed.**, edited by *Jeff Braman, 2002*
181. **Genomic Imprinting:** *Methods and Protocols,* edited by *Andrew Ward, 2002*
180. **Transgenesis Techniques, 2nd ed.:** *Principles and Protocols,* edited by *Alan R. Clarke, 2002*
179. **Gene Probes:** *Principles and Protocols,* edited by *Marilena Aquino de Muro and Ralph Rapley, 2002*
178. **Antibody Phage Display:** *Methods and Protocols,* edited by *Philippa M. O'Brien and Robert Aitken, 2001*
177. **Two-Hybrid Systems:** *Methods and Protocols,* edited by *Paul N. MacDonald, 2001*
176. **Steroid Receptor Methods:** *Protocols and Assays,* edited by *Benjamin A. Lieberman, 2001*
175. **Genomics Protocols**, edited by *Michael P. Starkey and Ramnath Elaswarapu, 2001*
174. **Epstein-Barr Virus Protocols,** edited by *Joanna B. Wilson and Gerhard H. W. May, 2001*
173. **Calcium-Binding Protein Protocols, Volume 2:** *Methods and Techniques,* edited by *Hans J. Vogel, 2001*
172. **Calcium-Binding Protein Protocols, Volume 1:** *Reviews and Case Histories,* edited by *Hans J. Vogel, 2001*
171. **Proteoglycan Protocols,** edited by *Renato V. Iozzo, 2001*
170. **DNA Arrays:** *Methods and Protocols,* edited by *Jang B. Rampal, 2001*
169. **Neurotrophin Protocols,** edited by *Robert A. Rush, 2001*
168. **Protein Structure, Stability, and Folding,** edited by *Kenneth P. Murphy, 2001*
167. **DNA Sequencing Protocols,** *Second Edition,* edited by *Colin A. Graham and Alison J. M. Hill, 2001*
166. **Immunotoxin Methods and Protocols**, edited by *Walter A. Hall, 2001*
165. **SV40 Protocols**, edited by *Leda Raptis, 2001*
164. **Kinesin Protocols,** edited by *Isabelle Vernos, 2001*
163. **Capillary Electrophoresis of Nucleic Acids, Volume 2:** *Practical Applications of Capillary Electrophoresis,* edited by *Keith R. Mitchelson and Jing Cheng, 2001*
162. **Capillary Electrophoresis of Nucleic Acids, Volume 1:** *Introduction to the Capillary Electrophoresis of Nucleic Acids,* edited by *Keith R. Mitchelson and Jing Cheng, 2001*
161. **Cytoskeleton Methods and Protocols**, edited by *Ray H. Gavin, 2001*
160. **Nuclease Methods and Protocols,** edited by *Catherine H. Schein, 2001*

METHODS IN MOLECULAR BIOLOGY™

RT-PCR Protocols

Edited by

Joe O'Connell

Department of Medicine,
National University of Ireland,
Cork, Ireland

WITHDRAWN

Humana Press Totowa, New Jersey

©2002 Humana Press Inc.
999 Riverview Drive, Suite 208
Totowa, New Jersey 07512

www.humanapress.com

All rights reserved. No part of this book may be reproduced, stored in a retrieval system, or transmitted in any form or by any means, electronic, mechanical, photocopying, microfilming, recording, or otherwise without written permission from the Publisher. Methods in Molecular Biology™ is a trademark of The Humana Press Inc.

All authored papers, comments, opinions, conclusions, or recommendations are those of the author(s), and do not necessarily reflect the views of the publisher.

This publication is printed on acid-free paper. ∞
ANSI Z39.48-1984 (American Standards Institute)
Permanence of Paper for Printed Library Materials.

Production Editor: Kim Hoather-Potter.

Cover design by Patricia F. Cleary.
Cover illustration: Background from Fig. 3A in Chapter 17 "RT-PCR-Based Approaches to Generate Probes for mRNA Detection by *In Situ* Hybridization" by Joe O'Connell; foreground from Fig. 2 in Chapter 18 "Amplified RNA for Gene Array Hybridizations" by Valentina I. Shustova and Stephen J. Meltzer.

For additional copies, pricing for bulk purchases, and/or information about other Humana titles, contact Humana at the above address or at any of the following numbers: Tel.: 973-256-1699; Fax: 973-256-8341; E-mail: humana@humanapr.com or visit our Website: http://humanapress.com

Photocopy Authorization Policy:
Authorization to photocopy items for internal or personal use, or the internal or personal use of specific clients, is granted by Humana Press Inc., provided that the base fee of US $10.00 per copy, plus US $00.25 per page, is paid directly to the Copyright Clearance Center at 222 Rosewood Drive, Danvers, MA 01923. For those organizations that have been granted a photocopy license from the CCC, a separate system of payment has been arranged and is acceptable to Humana Press Inc. The fee code for users of the Transactional Reporting Service is: [0-89603-875-0/02 $10.00 + $00.25].

Printed in the United States of America. 10 9 8 7 6 5 4 3 2 1

Library of Congress Cataloging in Publication Data

RT-PCR Protocols/edited by Joseph O'Connell
p.cm.-- (Methods in molecular biology)
Includes bibliographical references and index.
ISBN 0-89603-875-0 (alk. paper)
1. Polymerase chain reaction--Laboratory manuals. I. O'Connell, Joseph. II. Methods in molecular biology (Clifton, N.J.); v. 193.

QP606.D46 R8 2002
572'.43--dc21

2002190221

Preface

Until the mid 1980s, the detection and quantification of a specific mRNA was a difficult task, usually only undertaken by a skilled molecular biologist. With the advent of PCR, it became possible to amplify specific mRNA, after first converting the mRNA to cDNA via reverse transcriptase. The arrival of this technique—termed reverse transcription-PCR (RT-PCR)—meant that mRNA suddenly became amenable to rapid and sensitive analysis, without the need for advanced training in molecular biology. This new accessibility of mRNA, which has been facilitated by the rapid accumulation of sequence data for human mRNAs, means that every biomedical researcher can now include measurement of specific mRNA expression as a routine component of his/her research plans.

In view of the ubiquity of the use of standard RT-PCR, the main objective of *RT-PCR Protocols* is essentially to provide novel, useful applications of RT-PCR. These include some useful adaptations and applications that could be relevant to the wider research community who are already familiar with the basic RT-PCR protocol. For example, a variety of different adaptations are described that have been employed to obtain quantitative data from RT-PCR. Quantitative RT-PCR provides the ability to accurately measure changes/imbalances in specific mRNA expression between normal and diseased tissues. Because of its remarkable sensitivity, RT-PCR enables the detection of low-abundance mRNAs even at the level of individual cells. RT-PCR has afforded many opportunities in diagnostics, allowing sensitive detection of RNA viruses such as HIV and HCV. RT-PCR facilitates many diverse techniques in research, including *in situ* localization of mRNA, antibody engineering, and cDNA cloning. In particular, the present work highlights how RT-PCR complements other technological advances, such as laser-capture microdissection (LCM), real-time PCR, microarray technology, HPLC, and time-resolved fluorimetry.

RT-PCR has become one of the most widely applied techniques in biomedical research, and has been a major boon to the molecular investigation of disease pathogenesis. Determination of the pathogenesis of diseases at the molecular level is already beginning to inform the design of new therapeutic strategies. It is our hope that *RT-PCR Protocols* will stimulate the reader to explore diverse new ways in which this remarkable technique can facilitate the molecular aspects of their biomedical research.

Joe O'Connell

Contents

Contributors

SIMON AARONS • *Cork Cancer Research Centre, Mercy Hospital, Cork, Ireland*

CLAIRE ADAMS • *Department of Microbiology, National University of Ireland, Cork, Ireland*

FRANZ BACHMAIR • *Department of Obstetrics and Gynecology, University Hospital, University of Innsbruck, Austria*

BRIGITTE BALCZ • *Department of Pediatrics, Division of Pediatric Cardiology, University of Vienna, Austria*

MICHAEL W. BENNETT • *Department of Medicine, National University of Ireland, Cork, Ireland*

MARIE BEUZARD • *Laboratoire de Biotechnologie des Anticorps and INSERM U255, Institut Curie, Paris, France*

SANDRA BUSTEED • *Department of Medicine, National University of Ireland, Cork, Ireland*

YI CAO • *Division of Cellular Immunology, German Cancer Research Center, Heidelberg, Germany*

PIERRE CARAYON • *Sanofi-Synthélabo, Montpellier, France*

TATJANA CRNOGORAC-JURCEVIC • *Molecular Oncology Unit, Faculty of Medicine, Hammersmith Hospital, London, UK*

GUENTER DAXENBICHLER • *Department of Obstetrics and Gynecology, University Hospital, University of Innsbruck, Austria*

STEVEN D. DOUGLAS • *Division of Immunologic and Infectious Diseases, Joseph Stokes Jr. Research Institute at the Children's Hospital of Philadelphia, Philadelphia, PA*

LIAM J. FANNING • *Hepatitis C Unit, Department of Medicine, Cork University Hospital, Ireland*

CHRISTINE FLEISCHMANN • *Division of Pediatric Cardiology, Department of Pediatrics, University of Vienna, Austria*

TRIONA GOODE • *Huffington Center on Aging, Baylor College of Medicine, Houston, TX*

SUSANNE GREBER-PLATZER • *Division of Pediatric Cardiology, Department of Pediatrics, University of Vienna, Austria*

NADÈGE GRUEL • *Laboratoire de Biotechnologie des Anticorps, Institut Curie, Paris, France*

MARKUS HALMINEN • *Department of Virology, University of Turku, Finland*

CHRISTOPH HANSKI • *Department of Gastroenterology, Universitätsklinikum Benjamin Franklin, Freie Universität Berlin, Germany*

VINCENT HEALY • *Wellcome Trust Cellular Physiology Research Unit, Department of Physiology, National University of Ireland, Cork, Ireland*

WEN-ZHE HO • *Division of Immunologic and Infectious Diseases, Joseph Stokes Jr. Research Institute at the Children's Hospital of Philadelphia, PA*

AILEEN HOUSTON • *Department of Medicine, National University of Ireland, Cork, Ireland*

CHRISTIAN G. HUBER • *Institute of Analytical Chemistry and Radiochemistry, University of Innsbruck, Austria*

JORMA ILONEN • *Department of Virology, University of Turku, Finland*

CAROLINE JOYCE • *Department of Biochemistry, University Hospital, Cork, Ireland*

TOMOHIRO KATO • *Rheumatology, Immunology and Genetics Program, Institute of Medical Science, St. Marianna University School of Medicine, Kawasaki, Japan*

RAYMOND KELLY • *Department of Medicine, National University of Ireland, Cork, Ireland*

BEATRIX KOTLAN • *National Institute of Hæmatology and Immunology, Budapest, Hungary*

TSUTOMU KUMAZAKI • *Research Institute for Radiation Biology and Medicine, Department of Biochemistry and Biophysics, Hiroshima University, Japan*

MANAE SUZUKI KUROKAWA • *Rheumatology, Immunology and Genetics Program, Institute of Medical Science, St. Marianna University School of Medicine, Kawasaki, Japan*

JIAN-PING LAI • *Division of Immunologic and Infectious Diseases, Joseph Stokes Jr. Research Institute at the Children's Hospital of Philadelphia, PA*

ALLAN S. LAU • *Department of Pediatrics, University of Hong Kong, Queen Mary Hospital, Hong Kong*

GARY LEE • *Department of Pathology, Mercy Hospital, Cork, Ireland*

NICK R. LEMOINE • *Molecular Oncology Unit, Faculty of Medicine, Hammersmith Hospital, London, UK*

GERT LUBEC • *Division of Pediatric Cardiology, Department of Pediatrics, University of Vienna, Austria*

MIKA MÄKELÄ • *Department of Clinical Allergology and Pulmonary Diseases, University of Turku, Finland*

BENNO MANN • *Department of Surgery, Universitätsklinikum Benjamin Franklin, Freie Universität Berlin, Germany*

JEAN MARCHAND • *Sanofi-Synthélabo, Montpellier, France*

STEPHEN J. MELTZER • *University of Maryland School of Medicine, Baltimore, MD*

MICHAEL G. MOLLOY • *Department of Rheumatology, Cork University Hospital, Ireland*

SANDRINE MOUTEL • *Laboratoire de Biotechnologie des Anticorps, Institut Curie, Paris, France*

BRIAN MULLAN • *Hepatitis C Unit, Department of Medicine, National University of Ireland, Cork, Ireland*

KENNETH NALLY • *Department of Medicine, National University of Ireland, Cork, Ireland*

CHANG HOON NAM • *Laboratoire de Biotechnologie des Anticorps and INSERM U255, Institut Curie, Paris, France*

TORSTEN O. NIELSEN • *Molecular Oncology Unit, Faculty of Medicine, Hammersmith Hospital, London, UK*

KAISA NIEMINEN • *Department of Clinical Allergology and Pulmonary Diseases, University of Turku, Finland*

KUSUKI NISHIOKA • *Rheumatology, Immunology and Genetics Program, Institute of Medical Science, St. Marianna University School of Medicine, Kawasaki, Japan*

DARREN O'BRIEN • *Department of Medicine, National University of Ireland, Cork, Ireland*

JOE O'CONNELL • *Department of Medicine, National University of Ireland, Cork, Ireland*

TERRY O'CONNOR • *Department of Medicine, National University of Ireland, Cork, Ireland*

FERGAL O'GARA • *Department of Microbiology, National University of Ireland, Cork, Ireland*

GERALD C. O'SULLIVAN • *Department of Surgery, National University of Ireland, Cork, Ireland*

AIDEEN RYAN • *Department of Medicine, National University of Ireland, Cork, Ireland*

PAUL RYAN • *Cork Cancer Research Centre, Mercy Hospital, Cork, Ireland*

JOHANNES SAVOLAINEN • *Department of Clinical Allergology and Pulmonary Diseases, University of Turku, Finland*

JOERG F. SCHLAAK • *St. Mary's Hospital, London, UK*

FERGUS SHANAHAN • *Department of Medicine, National University of Ireland, Cork, Ireland*

VALENTINA I. SHUSTOVA • *University of Maryland School of Medicine, Baltimore, MD*
MINNA SJÖROOS • *Department of Virology, University of Turku, Finland*
DANIEL G. SULLIVAN • *Department of Laboratory Medicine, University of Washington, Seattle, WA*
JEAN-LUC TEILLAUD • *Laboratoire de Biotechnologie des Anticorps and INSERM U255, Institut Curie, Paris, France*
FRANK WALDRON-LYNCH • *University of Cambridge Clinical School, Addenbrooke's Hospital, Cambridge, UK*
MATTI WARIS • *Laboratory of Biophysics, University of Turku, Finland*
MICHAEL C. YEUNG • *Snyder Research Foundation, Winfield, KS*

I

INTRODUCTION

1

RT-PCR in Biomedicine

Opportunities Arising from the New Accessibility of mRNA

Joe O'Connell

1. Introduction

Reverse-transcriptase-polymerase chain reaction has become one of the most widely applied techniques in biomedical research. The ease with which the technique permits specific mRNA to be detected and quantified has been a major asset in the molecular investigation of disease pathogenesis. Disease-related imbalances in the expression of specific mRNAs can be sensitively and quantitatively determined by RT-PCR. RT-PCR also offers many opportunities in diagnostics, allowing sensitive detection of RNA viruses such as Human Immunodeficiency Virus (HIV) and Hepatitis C Virus (HCV). RT-PCR is an integral component of many methodologies that are essential to biomedical research, including *in situ* localization of mRNA, antibody engineering, and cDNA cloning. This chapter provides an overview of some of the ways in which RT-PCR can be utilized in biomedical science, and summarizes the importance and applicability of the protocols described in this volume. These protocols include some useful adaptations and applications that may have significance for those in the wider research community who are already familiar with the basic RT-PCR protocol. Each individual chapter in this volume contains complete experimental detail for the protocols described, so that even a newcomer to RT-PCR should be able to perform the techniques. In particular, this volume demonstrates how RT-PCR complements other technologies, such as laser-capture microdissection (LCM), real-time PCR, microarray analysis, high-pressure liquid chromatography (HPLC) and time-resolved fluorometry.

From: *Methods in Molecular Biology, vol. 193: RT-PCR Protocols*
Edited by: J. O'Connell © Humana Press Inc., Totowa, NJ

2. Highly Sensitive Detection and Analysis of mRNA

The greatest advantage of RT-PCR in the analysis of mRNA is its extraordinary sensitivity. Using nested RT-PCR, mRNA can essentially be detected at the level of single copies. Many of the chapters in this volume demonstrate the highly sensitive detection of mRNA. In Chapter 4, nested RT-PCR is used to analyze mRNA expression in a single cell; a single cell is lysed and placed directly in the RT-PCR reaction. Using appropriate, intron-spanning primers, differential mRNA splicing may also be analyzed by this technique at the single-cell level ***(1)***. Although the technique was demonstrated using a single cell in suspension, RT-PCR detection of mRNA in small numbers of cells in solid tissues is also made possible by use of laser-microdissection ***(2)*** (*see* Chapter 15). The capability for such sensitive detection and analysis of mRNA is an enormous asset in many research areas, such as developmental biology. For example, mRNA expression can now be analyzed from the earliest phases of embryogenesis, at the level of only a few embryonic cells. Another area that requires analysis of small local populations of cells is brain research. Understanding the function of the brain is one of the greatest challenges in biology today. Research in brain biology will benefit significantly from RT-PCR; the expression of low-abundance mRNAs can now be measured in small tissue samples from specific areas of the brain ***(3,4)***. In Chapter 3, for example, RT-PCR is applied to detect and quantify specific mRNA at the femto/attogram level in minute amounts of brain tissue. Although the extraordinary sensitivity of nested RT-PCR is a huge advantage for the detection of low-abundance mRNA, this level of sensitivity also presents some risks in the interpretation of results obtained using this technique. Chapter 5 highlights a caveat pertaining to the use and interpretation of data from nested RT-PCR; unless a quantitative approach is employed, sensitivity controls should be adopted to estimate the level of mRNA detected by the nested RT-PCR assay. Otherwise, the amount of detected mRNA can be overestimated ***(5)***.

3. Quantitative RT-PCR: Approaches and Applications

The sensitivity of RT-PCR makes it particularly useful for detecting low-abundance mRNA, especially in small amounts of tissue such as biopsy specimens. A disadvantage of standard RT-PCR with respect to less sensitive techniques such as Northern blot is that it is only semi-quantitative. This is because of the "plateau" in the kinetics of PCR product accumulation, in which linearity in the relationship between product and initial template tapers off with increasing cycles. Many strategies have been developed to enable quantitative data to be obtained from RT-PCR. Some of the most commonly used approaches are reviewed in Chapter 6, and methodologies and applications of several of these approaches are explored in this volume.

Many quantitative RT-PCR approaches are based on competitive PCR *(6)*. Essentially, a control PCR template is constructed that has identical primer sites to the target template, but has a difference—for example, in size—which allows amplification products from this control template to be distinguished from those of the target template. This control template is spiked in at known concentration as an internal standard prior to amplification of the target template. The standard will compete directly with the target template during PCR amplification, so that if the internal competitive standard template is present in equal amount to the target template, equivalent PCR products are obtained from both. In practice, multiple PCR reactions (usually 5–7) are set up containing serially increasing amounts of the internal standard. Following PCR amplification, the equivalence point—where there is equal yield of target and competitive standard PCR products—is determined. The number of copies of the target template must be equivalent to the known number of competitive standard molecules spiked into this particular reaction, enabling quantification of target molecules.

In order to perform competitive PCR, a control standard as described in the previous paragraph must be constructed for each target mRNA to be quantified. Several methods have been devised for this purpose, and indeed the construction of standards is also facilitated by PCR. In Chapter 3, for example, a series of overlapping PCRs is designed so that a small deletion of a few base-pairs (bp) is created in the target cDNA sequence. The target is PCR-amplified in two separate fragments, leaving an intervening region of a few bp between them. The two fragments are annealed together via overlapping complementary regions tagged onto the PCR primers. Thus, when the two fragments are annealed together, the intervening region is deleted. The annealed fragments are then PCR-amplified as a single product for use as the competitive standard. The advantage of making such minor alterations to the target cDNA is that the standard will be almost identical to the target, so that there is not likely to be a difference in the amplification efficiency between both. This is crucial to the validity of competitive PCR, which depends on equal competition between both templates. However, because of the close similarity between the PCR products obtained from target and competitor, the high resolution of a DNA-sequencing gel or column is required to separate the products for determination of the equivalence point.

A common approach to generating a competitive standard is to make a larger deletion, usually of about 30%, in the target. This method permits differentiation between amplification products of target and competitor on a standard agarose gel *(6)*. The advantage of this strategy is that once the standard has been constructed, no deviation from the standard RT-PCR protocol is required; the equivalence point is simply detected on a standard agarose gel. The competitor is usually of sufficiently similar size and sequence composition to the target to result in identical efficiencies of amplification, but this should always be checked. Chapter 7 demonstrates a rapid protocol for constructing a DNA standard. The standard is derived from the target template by PCR using a composite sense primer; the composite primer binds to an internal site in the target, but has the "regular" sense primer sequence tagged onto its 5' end. PCR with this composite primer and the regular anti-sense primer and will generate a truncated product—yet it is one with the regular sense and anti-sense primer sites at its ends *(7)*. In Chapter 10, composite primers are used to create a standard by the MIMIC approach; a piece of DNA unrelated to the target sequence (the "MIMIC") is amplified by a pair of composite primers containing sequences specific for the MIMIC DNA, but with the target primer sequences tagged on *(8)*. Once again, this results in a MIMIC fragment with the primer sites of the target incorporated into its ends. The same MIMIC DNA can be used to construct a standard for any chosen target. Although methods that derive the standard from the target sequence offer the advantage of a competitor and standard with a similar sequence composition, and therefore may amplify with similar efficiencies, the MIMIC-based approach avoids the formation of heteroduplexes during competitive PCR.

Heteroduplexes can arise when the competitive standard is generated by making a large deletion in the center of the target. When target and standard are co-amplified, a hybrid can occur because of annealing of one strand of the target with one strand of the standard. In this heteroduplex, the portion of the target that is absent from the standard remains unannealed, and loops out to form a bulky secondary structure *(6,9)*. This bulky heteroduplex has a slower electrophoretic mobility than either the target or standard, and forms a third, higher band on the gel. Because the heteroduplex consists of one strand each of the target and the standard, its formation does not appear to bias the ratio of target:standard, and therefore should not affect quantification of the target.

In order to generate a suitable deletion, another approach is to clone the target template into a plasmid, and then use unique restriction sites to excise an appropriate fragment. Cloning also permits an RNA copy of the competitive standard to be transcribed via the T7, T3, or SP6 RNA polymerase promoter on the plasmid vector *(6)*. Indeed, promoters for RNA polymerases can also be incorporated into a competitive standard generated by PCR by linking the promoter sequence onto the anti-sense primer. The advantage of an RNA standard is that it can be spiked into the RT-PCR at the cDNA synthesis stage, so that the efficiency of the RT step, as well as the PCR, is controlled. However, there is no amplification in the RT step, and the efficiency does not vary substantially between similar samples. Thus, DNA standards which are easier to construct are commonly used *(7)*.

An alternative approach to generating a deletion is to introduce a small sequence change in the target. This enables differential detection of PCR products from target and competitor by use of two separate hybridization probes that are specific for the area of sequence difference. This approach generates a standard of identical size and sequence composition to the target, so that amplification efficiencies for both should be identical. Although this technique introduces an additional step to the process, because a hybridization is required for analysis, this step nevertheless increases the specificity of the detection. Use of hybridization detection eliminates the need for analysis of PCR products by gel electrophoresis, and makes the technique amenable to enzyme-linked immunosorbent assays (ELISA) format. PCR-ELISA involves trapping the PCR products through immobilized "capture probes" in the wells of a microtiter plate. The captured PCR products are then detected and quantified using a specific hybridization probe, which is labeled to permit colorimetric measurement by ELISA. PCR-ELISA is the basis of many commercially available quantitative PCR assays, such as the Roche assays used to quantify the RNA viruses HIV and HCV ***(10)*** in patient sera (reviewed in Chapter 11; *see* Chapter 12). This is a clinically useful application of RT-PCR; in addition to providing a highly sensitive diagnostic test, quantitative RT-PCR tests enable the viremia level to be monitored in response to therapy. RT-PCR also provides material for genotype analysis of the virus present, and allows the presence and sequence diversity of variant viral "quasispecies" to be analyzed ***(11)*** (*see* Chapter 13). Also useful in virology research, RT-PCR can easily be adapted to the detection of negative-strand (anti-sense) RNA produced as a replicative intermediate by certain RNA viruses, such as picornaviruses. RT-PCR detection of anti-sense RNA is also useful in experimental situations to check for expression of anti-sense RNA in cell lines transfected with anti-sense constructs ***(12)*** (*see* Chapter 25).

The particular usefulness of competitive PCR is that it essentially allows quantification regardless of the plateau in PCR kinetics; the internal competitive standard will compete equally with the target throughout the PCR, thus controlling for changes in the PCR kinetics. This enables standard agarose-gel electrophoresis—which is normally only sufficiently sensitive to detect products at a relatively late stage in PCR, often beyond the linear phase—to be used for detection and quantification. However, if a detection technique is used which is sufficiently sensitive to detect PCR products early on, during the linear phase of the PCR, then direct measurement of PCR product can allow quantification of the target template without the need for a competitive control. In Chapter 8, HPLC is used to quantify PCR products directly in the linear phase with sensitivity and remarkable reproducibility, enabling direct quantification of template mRNA *(13)*. This approach is suited to high-throughput situations, and automated detection allows analysis of about 100 samples per 12 h of HPLC run. Chapter 9 describes another remarkably sensitive technique for quantification of PCR products, using fluorescent-labeled detection probes. Time-resolved fluorometry is used to measure the bound probe, making the assay amenable to microtiter-plate format *(14,15)*. PCR-ELISA also allows measurement of PCR products during the linear phase of PCR, enabling direct quantification without the need for using a competitive standard *(16)* (*see* Chapter 24). Even these sensitive techniques can include a competitive standard for added refinement of quantification.

A method that is rapidly growing as a technique for quantitative PCR is real-time PCR. A fluorescent DNA-binding dye is included in the PCR, and using a specially designed instrument, the accumulation of PCR product can be monitored in real-time during the PCR. For example the Roche Lightcycler involves performing PCR in a thin-walled, light-transparent cuvet, in an air-heated and -cooled thermal-cycling chamber. Product is continuously monitored within each sample cuvet by a fluorescence detector, and is usually detectable at early cycle numbers. By real-time monitoring, the kinetics of the reaction are followed so that product yield can be measured in the linear phase. By including a dilution series of a known concentration of a target template, a standard curve is obtained, from which the template concentrations in the test samples are quantified. Chapter 12 demonstrates the use of real-time RT-PCR in an important clinical application: quantifying HCV viremia levels in patient sera.

Even without the construction of competitive standards, or the use of advanced instrumentation, good, semi-quantitative data can be obtained from standard RT-PCR by performing limited PCR cycles (*see* Chapter 19). Careful optimization of the cycle number for each specific target can allow detection of PCR products before entering into the late stages of the plateau. The cycle number can be tailored depending on the relative abundance of the target mRNA. The validity of results from semi-quantitative RT-PCR can be confirmed by using multiple techniques; these may include detection of the corresponding protein by Western blot or immunofluorescence flow cytometry analysis, or detection of the mRNA and protein by *in situ* hybridization and immunohistochemistry, respectively. Analysis of results from the various techniques can provide a comprehensive picture of the level of expression in different samples. In Chapter 19, for example, this approach was employed to demonstrate that Fas ligand (CD95L)—a mediator of immune downregulation—was expressed more frequently in liver metastases compared to matched primary tumors in human colon cancer *(17)*.

Quantitative or semi-quantitative RT-PCR facilitates differential expression of an individual mRNA to be measured in different samples. Microarray technology enables differential expression of hundreds or thousands of different mRNAs to be analyzed simultaneously. Chapter 18 demonstrates how this approach can be used to investigate alterations in gene expression that occur during the transformation of normal cells to cancerous cells. RNA is isolated from both types of tissue, amplified by a reverse transcriptase (RT)-RNA-polymerase strategy, and fluorescently labeled. The labeled RNA is then hybridized to a microarray containing immobilized probes for hundreds of mRNAs. Analysis of the hybridization results helps to identify genes that are upregulated, downregulated, or unaltered in the transformation process. RT-PCR is frequently used to confirm differences in expression levels of individual genes identified by microarray analysis.

In addition to measuring alterations in gene expression at various stages of the transformation process, RT-PCR has many other applications in cancer research and diagnosis. RT-PCR for mRNAs encoding various tumor markers, including carcinoembryonic antigen (CEA), has frequently been used to detect the presence of tumor micrometastases in patient bone marrow and in the circulation *(18)*. RT-PCR also facilitates the detection of mutations in oncogenes or tumor-suppressor genes, such as APC and BRCA-1, using the protein truncation test *(19, 20)*. The mRNA is amplified by RT-PCR using a sense primer that has sequences for a T7 promoter and a ribosome-binding site tagged onto its 5' end. This facilitates T7 RNA polymerase-mediated transcription of RNA from the amplified products, which is then translated in vitro using a labeled amino acid such as ^{35}S-methionine. The size of the protein is analyzed by polyacrylamide gel electrophoresis (PAGE) and autoradiography. Since most mutations result in the premature introduction of a stop codon, synthesis of a truncated protein indicates the presence of a mutation. The use of PCR with primers that incorporate promoter sequences for RNA polymerases also facilitates the generation of RNA template for other genetic analyses, such as the RNase cleavage assay. This technique, which is related to SSCP, involves annealing RNA from the test sample with a complementary RNA strand from a normal control. A sequence difference, such as a mutation or polymorphism, results in a heteroduplex when the strands are annealed. The looped-out, single-stranded portion of the heteroduplex is amenable to cleavage by RNase, so that detection of the presence of cleavage products by gel electrophoresis indicates the presence of sequence changes relative to the control. In Chapter 20, this technique is used to analyze polymorphisms in the gene for tumor necrosis factor-α (TNF-α), an inflammatory cytokine, in rheumatoid arthritis patients *(21)*.

4. *In Situ* Localization and Quantification of mRNA Expression

Although quantitative or semi-quantitative RT-PCR provide useful data on the level of expression of specific mRNAs, this information has limited value when RNA from complex tissue is analyzed. No information is obtained regarding which cells within the tissue are expressing the mRNA. A number of strategies are now available for determining the cellular source of mRNA expression *in situ* within tissue sections. However, *in situ* techniques are generally not quantitative, so that a type of biological equivalent of the Heisenberg uncertainty principle exists, i.e.—it is difficult to simultaneously measure magnitude and position of mRNA. However, a reasonable approach is to use RT-PCR to quantify the global mRNA level within the tissue, and to use *in situ* techniques to determine which cells are expressing the mRNA. Signals from *in situ* assays also provide qualitative or semi-quantitative data regarding levels of expression within different cells. This combined approach can allow determination of alterations in specific mRNA expression, both in level and localization, in biopsy samples of diseased versus normal tissues *(9)*.

A new methodology has recently emerged, which facilitates the analysis of mRNA expression by RT-PCR in specific cell subsets within a complex tissue. This technique—called laser capture microdissection, or LCM—uses a laser to melt cellular material from the targeted cell, or groups of cells, within the tissue onto a plastic matrix, from which nucleic acid (or proteins) can then be extracted for analysis *(2)*. Under a microscope, the laser can be accurately aimed at the desired cells, and the resolution of the lasers currently in use (with a laser beam of approx 7 µm in diameter) is such that even a single individual cell can be microdissected from a human tissue section. If RNA is isolated from the LCM-captured cellular material, RT-PCR can be used to analyze specific mRNA expression. Although LCM offers single-cell resolution, the amount of RNA recovered often does not permit accurate mRNA detection by even the most sensitive RT-PCR assays. Usually, several cells are microdissected (1000s are often required to obtain reproducible results), so that LCM enables mRNA expression to be assessed in groups of cells, or specific cellular structures within the tissue. In Chapter 15, for example, LCM is used to analyze mRNA expression specifically within microdissected nests of melanoma tumor cells. Such an approach provides a more accurate view of mRNA expression by the tumor cells, since if whole tumor tissue had been used, this would also contain a high proportion of normal cells, such as stromal cells and lymphocytes surrounding the tumor nests.

As with other *in situ* techniques for mRNA detection, LCM-RT-PCR is prone to difficulties because of the inherent instability of mRNA within clinical tissue specimens, as a result of the presence of RNases. In contrast, analysis of genomic DNA from microdissected samples is more readily reproducible, because of the relative stability of DNA. In Chapter 16, for example, LCM and PCR are combined to detect the presence of the pathogenic microbe *Mycobacterium paratuberculosis* specifically within microdissected granulomata in colonic tissue. LCM can also provide protein for Western blot analysis or ELISA, and LCM has even allowed microarray analysis of mRNA expression within specific microdissected cell types (neurons) *(22)*. Since LCM can provide material for quantitative techniques such as quantitative RT-PCR, it may appear to provide a solution to the location-and-magnitude problem in mRNA analysis. However, it must be cautioned that mRNA stability may vary dramatically between different tissue samples during processing for LCM. Also, factors such as the thickness of the tissue sections used, the area of cells captured, and the efficiency of laser-transfer of material from each sample may all impact on the mRNA quantification.

Although there are difficulties inherent in measuring mRNA levels within specific cell types in complex solid tissues, there is less difficulty when the mixed cell populations are in suspension, as is the case with blood cells, for example. Techniques are available to separate and isolate different cell subsets within populations of blood cells, based on cell-surface markers differentially expressed by the various cell subclasses. Antibodies to these markers are available in column or magnetic bead formats to enable subset enrichment or purification. Fluorescence-activated cell sorting (FACS) is a more sophisticated approach, enabling the isolation of cell subsets to be achieved with very high purity. In chapter 24, RT-PCR performed on FACS-purified cells enabled expression of specific mRNAs—encoding cannabinoid receptors in this example—to be quantified in specific lymphocyte subsets *(16)*.

An ingenious strategy has emerged which harnesses the remarkable sensitivity of PCR to the detection of non-nucleic acid macromolecules, such as proteins. The technique, termed immuno-PCR, involves the covalent attachment of a DNA fragment to an antibody specific for the target macromolecule. After binding of the chimeric antibody to its target, PCR is performed using primers specific for the DNA fragment attached to the antibody. The amplified DNA can then be detected by a hybridization probe specific for the DNA fragment. Thus, PCR introduces a massive amplification step in the antibody detection. Chapter 14 describes how this approach can be adapted for the *in situ* detection of macromolecules within tissue sections, with extraordinary sensitivity. This approach allows the detection of a low-abundance hepatitis B antigen within clinical liver biopsy specimens *(23)*. In addition to its use in LCM and *in situ* immuno-PCR, more conventional approaches for the *in situ* detection of mRNA are also facilitated by PCR. Chapter 17 describes two PCR-based techniques to enable the rapid generation of long probes that are useful for standard *in situ* hybridization *(24)*.

5. RT-PCR in Immunology

One of the benefits of RT-PCR is that it provides ample material for the study of diversity within RNA populations. As already mentioned, Chapter 13 shows how RT-PCR facilitates the study of diversity of viral RNA sequences in sera from individual HCV patients. Sequence diversity is of particular importance in immunology, where sequence diversity among T-cell-receptor (TCR) and immunoglobulin (Ig) mRNAs provides valuable information about immune responses. Without RT-PCR, this molecular diversity would be difficult to analyze. Chapter 21 demonstrates how RT-PCR amplification of mRNA encoding a variable region within the T-cell receptor can be combined with single-strand conformation polymorphism (SSCP) and nucleotide-sequence analysis to assess the clonal expansion of T cells during an immune response *(25)*. A similar approach can be applied to the study of clonal expansion of B cells by analyzing sequence diversity among Ig variable-gene mRNAs.

The development of the ability to generate monoclonal antibodies (mAbs) represented a milestone in molecular biology. mAbs are useful as highly specific probes for individual proteins, applied to various immunodetection techniques. mAbs can also be developed that exert functional activity against target proteins; for example, antibodies have been developed that can block specific receptor-ligand interaction, or that can neutralize the activity of a target protein. Some function-altering mAbs—such as those that neutralize the activity of the pro-inflammatory cytokine TNF-α—are finding clinical utility in the treatment of inflammatory diseases such as rheumatoid arthritis and Crohn's disease *(26)*. Some mAbs can mimic the function of specific protein ligands by binding to the ligand's receptor in a way that triggers receptor signaling (e.g., the various mAbs that trigger the death receptor Fas/CD95 *(27)*). mAbs can also be developed that can enhance the activity of specific proteins—for example, by altering or stabilizing the protein's structure—as in the case of mAbs that can restore the activity of certain mutant forms of the tumor-suppressor protein *p53* *(28)*).

The ability to fully exploit mAbs in biomedical research has recently been enhanced by RT-PCR and recombinant DNA technology. Making bacteriophage libraries of cDNAs encoding antibody fragments facilitates the development of recombinant mAbs, without the need for immunization of animals or the generation of hybridomas. The cloned antibodies are exposed on the bacteriophage coat ("phage display"), so that clones expressing the required antibody are selected by binding to immobilized antigen *(29)*. Chapter 22 provides a detailed protocol for the construction of a phage display mini-library of antibodies derived from tumor-infiltrating B cells. Various sets of specific, degenerate oligonucleotide primers are used to PCR-amplify the various antibody variable genes (V_H, V_κ, and V_λ), using tumor-derived cDNA as a template. An assembly strategy is used to randomly combine the amplified VH sequences with either the V_κ or V_λ sequences, with a spacer encoding a flexible linker in between. This combination of one heavy-chain variable region (V_H) with one light-chain variable region (V_κ or V_λ) generates single-chain antibody fragments (scFvs), which represent the repertoire of antibodies expressed by the intratumoral B cells. Thus, a mini-library is obtained that is enriched for anti-tumor antibodies. The library can be further enriched by performing DNA sequence analysis of the different variable-region cDNAs, and selecting the most abundantly represented sequences for assembly of the scFvs. The sequence data allows degenerate primers to be used that specifically amplify the most abundant sequences. The recombinant scFv DNA sequences are cloned into a bacteriophage vector, so that the scFvs are "displayed" on the bacteriophage coat. Selection is then performed, to isolate phage-antibody clones that recognize a specific target antigen(s).

A useful consequence of recombinant antibody technology is the ability to clone and express recombinant antibody fragments from plasmid vectors within mammalian cells. For example, this allows intracellular expression of an antibody ("intrabody") that neutralizes a specific intracellular target protein ***(30)***. Antibodies can also be used that can restore the function of mutant proteins. Certain short, amino-acid consensus sequences which can be fused to the end of the recombinant scFv, can actually target the scFv to specific intracellular compartments. Thus, the scFv can be directed to the endoplasmic reticulum, or to the nucleus. Other peptide sequences can be added to the scFv, such as a leader sequence to facilitate secretion from bacterial clones, or a "tag" sequence to facilitate affinity purification of the scFvs from bacterial cultures. Usually, the scFv is derived from a well-characterized hybridoma that has already been selected for its production of the specific antibody to be expressed in the cell. Chapter 23 describes the assembly of an scFv by RT-PCR amplification of the variable heavy and light chains expressed in the specific hybridoma. Techniques are provided for the cloning, expression, and purification of the engineered scFv.

6. RT-PCR in cDNA Cloning

The protein-encoding version of thousands of important human genes have been isolated from cDNA libraries. A cDNA library is constructed by isolating mRNA from the cell or tissue that expresses the target protein, converting all the mRNA into cDNA copies via the enzyme RT, and cloning all the cDNAs into bacterial cells through a plasmid or bacteriophage vector. The library is then screened with a probe to detect a clone encoding the protein of interest. The probe is usually either a mixed oligonucleotide probe based on partial amino-acid sequence data for the protein, or alternatively an antibody specific for the protein of interest. For the latter approach, an expression vector must be used to enable the cloned cDNA to be expressed at the protein level in *Escherichia coli*. Because they are derived from mRNAs, cDNAs do not contain introns, and this facilitates expression of the encoded polypeptides in prokaryotic cells.

Although the human genome project will facilitate the cloning, expression, and functional characterization of previously unknown human genes, cDNA cloning still represents a useful method of gene discovery or identification led by prior knowledge of the protein. It is especially useful in cases where amino-acid sequence data is lacking or difficult to obtain, but where an antibody can be raised against the protein of interest. In fact, cDNA libraries facilitate the identification of protein antigens in certain diseases. cDNA cloning has led to the identification of numerous tumor-specific antigens, for example, by screening tumor-derived cDNA libraries with antibodies from autologous patient sera (the SEREX approach) ***(31)***. Such information may facilitate the development of tumor vaccines. cDNA cloning also facilitates the discovery of interesting and useful enzymes from other higher and lower eukaryotes ***(32)***. As already mentioned, "phage display" libraries of cDNAs encoding antibody-variable regions facilitates the development and engineering of recombinant antibodies. This approach can lead to the development of new antibodies, can assist in the study of humoral responses (e.g., anti-tumor responses), and can facilitate intracellular expression of antibodies with functional properties.

The construction of a cDNA library is a multistep procedure, with inefficiencies and loss of material inherent in each step. Frequently, the yield of recombinant clones obtained at the end of the cDNA cloning procedure is poor, often more than an order of magnitude below the number of clones required to ensure a representative cDNA library ($> 10^6$ clones are believed to be required to ensure complete representation of all the mRNAs in a typical human tissue). PCR can be used to amplify the cDNA prior to cloning, thus permitting the construction of large cDNA libraries, even from minute amounts of starting tissue ***(33)***. Chapters 26 and 27 provide some techniques that have been optimized for the PCR-amplification and subsequent cloning of cDNA to facilitate the construction of representative cDNA libraries. Restriction sites can be incorporated in the primers used to amplify the cDNA, in order to facilitate cloning. One risk involved in generating PCR-based cDNA libraries is that the PCR amplification may introduce bias in the representation of clones. This could arise if some of the starting cDNA templates amplify with greater efficiency than others, so that poorly amplifying sequences can be outcompeted and either underrepresented or even completely lost in the resultant cDNA library. The inclusion of a DNA denaturant—such as dimethyl sulfoxide

(DMSO) or formamide—in the PCR appears to minimize such bias, presumably by "smoothing out" secondary structures and denaturing (GC)-rich regions that give rise to poor amplification efficiency in some cDNAs. RT-PCR is particularly useful in the construction of cDNA libraries when the source of cell or tissue is limited; RT-PCR-based cDNA libraries have even been constructed from single cells *(34)*.

7. Conclusion

This book demonstrates very specific and diverse ways in which RT-PCR can be applied to biomedical research. Although no single volume can comprehensively cover the diversity of ways in which this remarkable technique can be applied, the goal of this volume is to encourage the reader to expand the use of RT-PCR in their research, in ways they may not have previously considered.

References

1. Kumazaki, T., Mitsui, Y., Hamada, K., Sumida, H., and Nishiyama, M. (1999) Detection of alternative splicing of fibronectin mRNA in a single cell. *J. Cell. Sci.* **112,** 1449–1453.
2. Sirivatanauksorn, Y., Drury, R., Crnogorac-Jurcevic, T., Sirivatanauksorn, V., and Lemoine, N. R. (1999) Laser-assisted microdissection: applications in molecular pathology. *J. Pathol.* **189,** 150–154.
3. Greber-Platzer, S., Schatzmann-Turhani, D., Cairns, N., Balcz, B., and Lubec, G. (1999) Expression of the transcription factor ETS2 in brain of patients with Down syndrome—evidence against the overexpression-gene dosage hypothesis. *J. Neural. Transm. Suppl.* **57,** 269–281.
4. Greber-Platzer, S., Balcz, B., Cairns, N., and Lubec, G. (1999) c-fos expression in brains of patients with Down syndrome. *J. Neural. Transm. Suppl,* **57,** 75–85.
5. Goode, T., O'Connell, J., Ho, W. Z., O'Sullivan, G. C., Collins, J. K., Douglas, S. D., et al. (2000) Differential expression of neurokinin-1 receptor by human mucosal and peripheral lymphoid cells. *Clin. Diagn. Lab. Immunol.* **7,** 371–376.
6. O'Connell, J., Goode, T., and Shanahan, F. (1998) Quantitative measurement of mRNA expression by competitive RT-PCR. *Methods Mol. Biol.* **92,** 183–193.
7. O'Connell, J., Bennett, M. W., Nally, K., O'Sullivan, G. C., Collins, J. K., and Shanahan, F. (2000) Interferon-gamma sensitizes colonic epithelial cell lines to physiological and therapeutic inducers of colonocyte apoptosis. *J. Cell. Physiol.* **185,** 331–338.
8. Lai, J. P., Douglas, S. D., Zhao, M., and Ho, W. Z. (1999) Quantification of substance P mRNA in human mononuclear phagocytes and lymphocytes using a mimic-based RT-PCR. *J. Immunol. Methods.* **230,** 149–157.
9. Goode, T., O'Connell, J., Anton, P., Wong, H., Reeve, J., O'Sullivan, G. C., et al. (2000) Neurokinin-1 receptor expression in inflammatory bowel disease: molecular quantitation and localisation. *Gut* **47,** 387–396.

10. Fanning, L., Kenny, E., Sheehan, M., Cannon, B., Whelton, M., O'Connell, J., et al. (1999) Viral load and clinicopathological features of chronic hepatitis C (1b) in a homogeneous patient population. *Hepatology* **29,** 904–907.
11. Mullan, B., Kenny-Walsh, E., Collins, J. K., Shanahan, F., and Fanning, L. J. (2001) Inferred hepatitis C virus quasispecies diversity is influenced by choice of DNA polymerase in reverse transcriptase-polymerase chain reactions. *Anal. Biochem.* **289,** 137–146.
12. Yeung, M. C., Liu, J., and Lau, A. S. (1996) An essential role for the interferon-inducible, double-stranded RNA- activated protein kinase PKR in the tumor necrosis factor-induced apoptosis in U937 cells. *Proc. Natl. Acad. Sci. USA.* **93,** 12451–12455.
13. Bachmair, F., Huber, C., and Daxenbichler, G. (1999) Quanititation of gene expression by means of HPLC analysis of RT-PCR products. *Clin. Chim. Acta.* **279,** 25–34.
14. Halminen, M., Sjoroos, M., Makela, M. J., Waris, M., Terho, E., Lovgren, T., et al. (1999) Simultaneous detection of IFN-gamma and IL-4 mRNAs using RT-PCR and time-resolved fluorometry. *Cytokine* **11,** 87–93.
15. Sjoroos, M., Ilonen, J., and Lovgren, T. (2001) Solid-phase PCR with hybridization and time-resolved fluorometry for detection of HLA-B27. *Clin. Chem.* **47,** 498–504.
16. Marchand, J., Bord, A., Penarier, G., Laure, F., Carayon, P., and Casellas, P. (1999) Quantitative method to determine mRNA levels by reverse transcriptase-polymerase chain reaction from leukocyte subsets purified by fluorescence-activated cell sorting: application to peripheral cannabinoid receptors. *Cytometry* **35,** 227–234.
17. Mann, B., Gratchev, A., Bohm, C., Hanski, M. L., Foss, H. D., Demel, G., et al. (1999) FasL is more frequently expressed in liver metastases of colorectal cancer than in matched primary carcinomas. *Br. J. Cancer* **79,** 1262–1269.
18. Miyamoto, A., Fujiwara, Y., Sakon, M., Nagano, H., Sugita, Y., Kondo, M., et al. (2000) Development of a multiple-marker RT-PCR assay for detection of micrometastases of hepatocellular carcinoma. *Dig. Dis. Sci.* **45,** 1376–1382.
19. van der Luijt, R., Khan, P. M., Vasen, H., van Leeuwen, C., Tops, C., Roest, P., et al. (1994) Rapid detection of translation-terminating mutations at the adenomatous polyposis coli (APC) gene by direct protein truncation test. *Genomics* **20,** 1–4.
20. Kraus, C., Gunther, K., Vogler, A., Hohenberger, W., Pfeiffer, R. A., and Ballhausen, W. G. (1998) Rapid RT-PCR-based protein truncation test in the screening for 5' located mutations of the APC gene. *Mol. Cell Probes* **12,** 143–147.
21. Waldron-Lynch, F., Adams, C., Shanahan, F., Molloy, M. G., and O'Gara, F. (1999) Genetic analysis of the 3' untranslated region of the tumour necrosis factor shows a highly conserved region in rheumatoid arthritis affected and unaffected subjects. *J. Med. Genet.* **36,** 214–216.

22. Luo, L., Salunga, R. C., Guo, H., Bittner, A., Joy, K. C., Galindo, J. E., et al. (1999) Gene expression profiles of laser-captured adjacent neuronal subtypes. *Nature Med.* **5,** 117–122.
23. Cao, Y., Kopplow, K., and Liu, G. Y. (2000) In-situ immuno-PCR to detect antigens. *Lancet* **356,** 1002–1003.
24. Bennett, M. W., O'Connell, J., O'Sullivan G, C., Roche, D., Brady, C., Kelly, J., et al. (1999) Expression of Fas ligand by human gastric adenocarcinomas: a potential mechanism of immune escape in stomach cancer. *Gut* **44,** 156–162.
25. Kurokawa, M., Furukawa, H., Yabe, T., Matsui, T., Toda, M., Hamada, C., et al. (1999) Frequency of clonally expanded T cells evaluated by PCR from a single cell. *J. Immunol. Methods* **224,** 203–208.
26. Baert, F. J., D'Haens, G. R., Peeters, M., Hiele, M. I., Schaible, T. F., Shealy, D., et al. (1999) Tumor necrosis factor alpha antibody (infliximab) therapy profoundly downregulates the inflammation in Crohn's ileocolitis. *Gastroenterology* **116,** 22–28.
27. Trauth, B. C., Klas, C., Peters, A. M., Matzku, S., Moller, P., Falk, W., et al. (1989) Monoclonal antibody-mediated tumor regression by induction of apoptosis. *Science* **245,** 301–305.
28. Caron de Fromentel, C., Gruel, N., Venot, C., Debussche, L., Conseiller, E., Dureuil, C., et al. (1999) Restoration of transcriptional activity of p53 mutants in human tumour cells by intracellular expression of anti-p53 single chain Fv fragments. *Oncogene* **18,** 551–557.
29. Desai, S. A., Wang, X., Noronha, E. J., Kageshita, T., and Ferrone, S. (1998) Characterization of human anti-high molecular weight melanoma-associated antigen single-chain Fv fragments isolated from a phage display antibody library. *Cancer Res.* **58,** 2417–2425.
30. Cochet, O., Gruel, N., Fridman, W. H., and Teillaud, J. L. (1999) Ras and p53 intracellular targeting with recombinant single-chain Fv (scFv) fragments: a novel approach for cancer therapy? *Cancer Detect. Prev.* **23,** 506–510.
31. Tureci, O., Sahin, U., and Pfreundschuh, M. (1997) Serological analysis of human tumor antigens: molecular definition and implications. *Mol. Med. Today* **3,** 342–349.
32. Healy, V., Doonan, S., and McCarthy, T. V. (1999) Purification, characterization and cDNA cloning of an endo-exonuclease from the basidiomycete fungus Armillaria mellea. *Biochem. J.* **339,** 713–720.
33. Lambert, K. N., and Williamson, V. M. (1993) cDNA library construction from small amounts of RNA using paramagnetic beads and PCR. *Nucleic Acids Res.* **21,** 775–776.
34. Belyavsky, A., Vinogradova, T., and Rajewsky, K. (1989) PCR-based cDNA library construction: general cDNA libraries at the level of a few cells. *Nucleic Acids Res.* **17,** 2919–2932.

2

The Basics of RT-PCR

Some Practical Considerations

Joe O'Connell

1. Introduction

The basic reverse-transcriptase-polymerase chain reaction (RT-PCR) technique is used routinely and widely in most biomedical research laboratories. The fundamental considerations for such basics as primer design, good laboratory set up and practice, and techniques for performing RT-PCR have been fully examined elsewhere *(1,2)*. Thus, I will include only observations from my own experience in this chapter. Although the purpose of this chapter is to help the reader in setting up their own PCR assays, and to provide tips on effective PCR laboratory set up, it should be noted that each individual chapter in this volume contains complete experimental detail on the protocols described.

2. RNA ISOLATION: "Home-Made" Preps vs Commercial Kits

In-house techniques for RNA isolation have been widely employed. These involve homogenizing tissue specimens in a guanidine thiocyanate lysis buffer, followed by phenol extraction and alcohol precipitation (for example, *see* Chapter 7) *(3)*. A number of kits for RNA extraction are now commercially available; these usually obviate the need for phenol extraction, instead involving spun-column chromatography or magnetic bead separation to purify the RNA. As such, these methods are considerably more rapid and convenient than the in-house methods. However, it should be noted that most techniques for quick RNA isolation, whether in-house or using commercially available kits, result in some contamination with genomic DNA. The best way to purify RNA with-

From: *Methods in Molecular Biology, vol. 193: RT-PCR Protocols*
Edited by:J. O'Connell © Humana Press Inc., Totowa, NJ

out DNA contamination is probably by ultracentrifugation of RNA lysates through cesium chloride density gradients *(4)*, a technique that is obviously not suited to routine requirements for RT-PCR. The simplest solution to this problem is the use of intron-spanning primers for the RT-PCR, so that PCR products derived from the mRNA are easily distinguishable from the larger, intron-containing products derived from contaminating genomic DNA. If the target region of mRNA to be amplified does not contain introns, DNA should be eliminated from the RNA, using RNase-free DNase prior to performing RT-PCR. It is usually advisable to repeat the RNA extraction to remove the DNase, which could otherwise hinder cDNA synthesis.

3. Primers for cDNA Synthesis: Oligo-dT, Random Hexamers, or Gene-Specific

An additional factor for consideration in RT-PCR is the choice of primer to synthesize the cDNA. Although oligo (dT) primers may seem to be the best choice, since they will specifically prime from the poly A tail of mRNA, it should be noted that most mammalian mRNAs contain extensive non translated regions (usually from one to several kb) downstream from their coding regions. If PCR primers are located in the coding region, particularly near the 5' end of the mRNA, then the use of oligo (dT) primers necessitates that full-length, or nearly full-length cDNA copies are synthesized from the 3' terminus of the mRNA. Because of secondary structures in mRNA, synthesis of such long cDNAs is not always effective. However, random hexanucleotide primers prime randomly from sites throughout the mRNA, making it more likely that 5' sequences will be represented in the resultant cDNA. Alternatively, a gene-specific primer can be synthesized which will prime only from the target mRNA, so that the cDNA is enriched for the cDNA of interest prior to PCR. This may be particularly useful when the target mRNA is at a low copy-number. Although the anti-sense PCR primer can also be used for cDNA priming, a better approach is to synthesize a separate primer immediately downstream from the anti-sense PCR primer site. The advantage of this method is that the PCR primers will be "nested" relative to the cDNA primer; thus, any nonspecifically primed cDNA should not be amplified during PCR. Because cDNA synthesis is usually performed at relatively low temperatures (e.g., 37°C or 42°C), the specific primer sequence should be shorter than a typical PCR primer, so that its annealing temperature is not considerably higher than the cDNA synthesis temperature; this minimizes non specific priming. For most applications, random hexamers work well, and offer the particular advantage that a single cDNA prep. can be used for PCR detection of many different cDNAs.

4. Primer Design

As previously mentioned, the best strategy for RT-PCR is to select primers that span intron(s), so that mRNA-specific PCR products are obtained. For this, either a complete or partial genomic sequence is required, which indicates the position of at least one intron in the gene. Performing an alignment between the genomic and cDNA sequences may be helpful for this purpose. Even if the genomic sequence is not available, any reasonably sized target region within the mRNA (300–400 bp) has a reasonable probability of spanning an intron. This can be tested empirically by performing PCR with the primers using genomic DNA as a template, to determine whether the product is larger than that obtained from cDNA. (Note: since introns can be large, frequently > 1 kb, intron-spanning primers frequently do not amplify the larger genomic sequence efficiently, and no product is obtained from genomic DNA).

In the early days of PCR, primers were often selected on a fairly random basis (the "let's stick two forks in the sequence" approach). Primers selected in this way often worked very well. More recently, computer programs have become available that are designed to select the most efficient primers within a given sequence. These programs incorporate a number of theoretical considerations that determine the optimal primer pairs, including:

1. Minimal self-complementarity between both primers in the pair, or within each individual primer; self-complementarity (leading to dimers or hairpin loops) can yield a high level of primer artifacts during PCR.
2. The primers should have a matched annealing temperature. If the optimal annealing temperature for primer 1 is substantially lower than that of primer 2, PCR may need to be performed at an annealing temperature well below the optimal annealing temperature for primer 2; this could allow mis-priming by primer 2.
3. The primers should have a good stability of binding to their target sequences; also, in theory, there should be a gradient of increasing binding strength from the 3' to the 5' end. Even transitory binding of the 3' end of a primer to a non identical target sequence can lead to non-specific priming. By keeping the strength of binding at the 3' end low (i.e., a relatively [AT]-rich sequence), it is believed to bind only to a perfectly matched target sequence, and the stronger affinity of the 5' end (i.e., a relatively[GC]-rich sequence) will clamp the primer in place.
4. There should be a good difference between the calculated annealing temperature of the primers and the annealing temperature of the PCR product; this favors binding of the primers to their target sequences over re-annealing of the PCR product itself.

The program will usually select and rank multiple potential primer pairs. Select the best option that spans an intron in the genomic sequence.

A useful analysis to perform at this point is a homology search for the selected primer sequences against a sequence database (e.g., perform a BLAST search of the Entrez Nucleotide database, which incorporates sequences from the GenBank database; this service is available on the Entrez-PubMed website at: http://www.ncbi.nlm.nih.gov/PubMed). This will identify any homologous sequence that the primers may also amplify (such as a closely related member of a gene family). Some homology with other genes is acceptable, particularly if the region of homology does not involve the 3' end of the primer; a perfect match at this end can lead to nonspecific priming, whereas even a single nucleotide difference at the 3' end will render it unlikely to prime the nonidentical target. Unless both primers show strong homology to the related target, nonspecific priming may not be a problem. Another advantage of the homology search is that it cross-checks the primer sequences against multiple versions of the target sequence. For example, databases frequently contain several cDNA and/or genomic sequences for a particular gene, which are independently submitted by different laboratories. Since occasional single-nucleotide discrepancies have been known to occur between different published versions of the same gene sequence, the homology search will establish whether the primer matches all known versions of the target sequence.

One potential problem to consider when selecting primers is the occurrence of processed pseudogenes for the gene of interest. Certain genes have been shown to have additional, redundant copies in the genome that are similar to the mRNA sequence (i.e., minus intron sequences) (β-actin is a notable example, which is often used as a control for equivalence of RT-PCR efficiency between samples). These sequences may represent evolutionary relics of mRNA sequences incorporated into the genome at some point via viral RT activity. If the target is known to have a processed pseudogene, even if intron-spanning primers are designed, they will yield a product from the pseudogene in contaminating genomic DNA identical in size to the mRNA-specific product. If the pseudogene sequence is known, it may be possible to design primers that do not amplify the pseudogene. Alternatively, genomic DNA must be eliminated from the RNA preps.

5. PCR Reaction Conditions

Once the optimal primers have been rationally selected, usually very little optimization of the PCR itself is necessary. In my experience, most primer pairs that are selected using software such as the Lasergene Primerselect program (DNASTAR Inc., Madison, WI) work well with a standard set of PCR conditions. We routinely use PCR primers at a final concentration of 0.1 μ*M* each, dNTPs at 50 μ*M*, $MgCl_2$ at 1.5 m*M*, and 1.0 U of Taq DNA polymerase per 50 µl reaction. These are all somewhat minimal concentrations, at the lower ends of the potential concentration ranges. This is preferable because excessive concentrations of primers, dNTPs or the DNA polymerase are more likely to promote mis-priming, leading to nonspecific PCR products. Better specificity is obtained with moderate reagent concentrations. We commonly use the following program of thermal cycling: denaturation at 96°C for 15 s; annealing at 55°C for 30 s, and extension at 72°C for 3 min. Although the calculated annealing temperature of the selected primers may be somewhat greater or less than 55°C, in our studies this annealing temperature has worked well for a long list of different primer pairs. We always perform "hot start," by heating the reaction at 80°C for about 1 min before adding the polymerase. This prevents nonspecific priming from occurring during reaction setup, which can occur if all the reaction components were mixed at room temperature—i.e., well below the optimal annealing temperature of the primers. Although heat-stable DNA polymerases usually only reach optimal activity at 72°C, any nonspecific products that are generated at the lower temperature would be efficiently amplified during the subsequent PCR, leading to substantial background. An alternative method for performing hot-start allows the complete PCR setup to be performed at the bench, at room temperature. Wax beads are available that can be inserted in the reaction tube before pipeting in the polymerase, so that the polymerase is physically separated from the rest of the reaction components. When the tube reaches high temperature in the thermocycler, the wax melts, allowing mixing of the polymerase with the rest of the reaction components. Other methods for performing hot-start involve the use of anti-Taq neutralizing antibodies, which prevent activity of the polymerase until antibody binding is lost at higher temperatures. Alternatively, a chemically modified Taq can be used, which only becomes active at high temperatures. We typically perform 30–40 cycles, depending on the relative abundance of the mRNA transcript being detected. PCR product specificity should ideally be confirmed, either by restriction mapping or by DNA sequence analysis. If the PCR product is to be cloned, a proofreading thermostable DNA polymerase (e.g., Pwo, Pfu, Vent, or UlTma) should be used in the PCR to minimize misincorporation errors.

6. Contamination: Detection, Precautions, and Remedies

The source of contamination in PCR is almost invariably amplicons from a previous run. Because amplicons are present in enormous quantity at the end of the PCR, it is easy for the DNA to become aerosolized and spread as contamination throughout the laboratory environment. Although the contamination of reagent stocks leads to false-positives in all reactions, including negative controls, environmental contamination is usually random, affecting only some tubes in the same run. This type of contamination is thus insidious, because the negative control tubes are often "clean;" therefore, unexpected positive results can be caused simply by contamination. If a contamination problem is suspected, it is sometimes helpful to perform multiple negative-control PCRs, so that at least some tubes should pick up the sporadic contamination. Strictly, every tube in a PCR run should have an individual negative control, to increase the ability to detect random contamination.

Proper handling of the PCR product is the key to avoiding contamination problems. Separate laboratory areas (preferably separate rooms) should be allocated for setup of the PCR and analysis of the PCR products. Traffic between these areas should be minimized, particularly from the analysis area back to the setup area. Separate pipettors, pipet tips, and reaction tubes should be kept in both areas, and materials such as notebooks and markers should not be moved between both areas. A separate lab coat should be worn in each area. It is essential to allocate a separate pipettor for handling of the finished PCR product only (such as loading gels); also, a separate pipetor should be kept next to the thermal cycler to be used for adding the DNA polymerase to the PCR during hot-start, and for no other purpose. As already mentioned, sporadic contamination is usually the result of distribution of aerosolized amplicons in the environment. Therefore, care must be taken after the PCR tube is opened at the end of the run and the products are pipetted out for analysis. Pipet tips and tubes used for handling the PCR products should be carefully disposed, in a closed container; tubes that contained PCR products should be re-capped before disposing. Also, electrophoresis buffer and gel-staining solutions should be carefully disposed as soon as possible after they have been used, since they represent large volumes containing PCR products leached from the gel.

Despite the best efforts at prevention, most researchers encounter PCR contamination at some point, particularly when multiple runs are performed over a long period of time using the same primers. When contamination occurs, a number of steps can remedy the situation. Since contamination is usually environmental, one simple solution that can work is to temporarily move the PCR setup area to a different area of the laboratory—ideally a different room—where there is less likelihood of contamination with PCR amplicons. The move should involve the use of fresh batches of reagents and different pipettors and tips. The use of filtered, "aerosol-resistant" pipet tips during PCR setup may reduce the risk of contamination. In time, the original PCR setup area should be rid of the contamination; scrubbing the bench and surfaces (e.g., using 10% household bleach) may speed up the decontamination process. If contamination is widespread, and other measures fail to clear the contamination, a quick solution is to make a different set of primers for the target, which will not amplify the contaminant amplicon. However, the cause of the contamination should be investigated so that the problem doesn't arise again. When repetitive PCRs using the same primer pair are envisaged, it may be worth considering additional precautions to prevent contamination. Although it adds to the expense of the PCR, the use of uracil-N-glycosylase (UNG) is a reliable way to minimize the risk of contamination *(5)*. If dUTP is substituted for dTTP in the PCR, the dU-containing amplicons are destroyed by treatment with UNG. Thus, UNG treatment of the PCR reaction prior to beginning thermal cycling will eliminate any contaminant, dU-containing amplicons from previous runs, without affecting the template DNA/cDNA, or the primers. The UNG enzyme is denatured during the initial denaturation step in the PCR, so that newly synthesized PCR products are not degraded. A cheaper alternative to the use of UNG is to ultraviolet-(UV)-irradiate the PCR reaction tube prior to addition of the template DNA. For example, treatment of the reaction tube with UV light of 254 nm wavelength and intensity of 30,000 $\mu J/cm^2$ for 2×2 min can destroy any contaminating, double-stranded DNA. However, unlike UNG, which destroys contaminant amplicons in the reaction tube immediately prior to beginning the PCR, there is a pipetting step after the UV treatment (adding the DNA template) which allows a window for the introduction of contamination. Of course, if hot-start is performed, the risk of contamination during the addition of the DNA polymerase is still present.

7. Conclusion

A wide range of variations on the basic RT-PCR method have been employed in various situations. As such, RT-PCR is a fairly robust technique, tolerating a wide range in the different reaction conditions. Hopefully, the parameters discussed in this chapter will help in the design and setup of RT-PCR reactions, and will inform good PCR laboratory practice.

References

1. Lo, Y. M. D. (1998) Introduction to the polymerase chain reaction. *Methods Mol. Biol.* **16,** 3–10.
2. Lo, Y. M. D. (1998) Setting up a PCR laboratory. *Methods Mol. Biol.* **16,** 11–17.
3. Chomczynski, P., and Sacchi, N. (1987) Single-step method of RNA isolation by acid guanidinium thiocyanate-phenol-chloroform extraction. *Anal. Biochem.* **162,** 156–159.
4. Chirgwin, J. M., Przybyla, A. E., MacDonald, R. J., and Rutter, W. J. (1979) Isolation of biologically active ribonucleic acid from sources enriched in ribonuclease. *Biochemistry* **18,** 5294–5299.
5. Longo, M. C., Berninger, M. S., and Hartley, J. L. (1990) Use of uracil DNA glycosylase to control carry-over contamination in polymerase chain reactions. *Gene* **93,** 125–128.

II

Highly Sensitive Detection and Analysis of mRNA

3

Using the Quantitative Competitive RT-PCR Technique to Analyze Minute Amounts of Different mRNAs in Small Tissue Samples

Susanne Greber-Platzer, Brigitte Balcz, Christine Fleischmann, and Gert Lubec

1. Introduction

The reverse-transcriptase-polymerase chain reaction (RT-PCR) can be used to determine minute amounts of mRNAs in tissue samples by co-amplification of a quantified amount of a competitive sequence (internal standard), so-called quantitative competitive (qc) RT-PCR. The first description of qcRT-PCR was provided by Wang et al. *(1)*. As an internal standard, an RNA template with the same primer sites as the target mRNA was reverse-transcribed. The amplified PCR products differed in size and were separated by agarose gel electrophoresis. The presence of ^{32}P-labeled 5' primer allowed quantification by scintillation counting.

Determination of an unknown amount of any cDNA can also be performed by the use of an external standard, either based on a housekeeping gene or by serial dilutions of known amounts of an RNA standard quantified separately from the target RNA. Such approaches are subject to inter-sample variation in amplification efficiency. To overcome these problems, a known amount of a standard is included in the sample, and competitive co-amplification of target RNA and standard is followed. The inclusion of an internal standard as DNA can control the PCR efficiency *(2,3)*, but the inclusion of an internal standard as RNA can also control the reverse-transcription efficiency, and therefore provides more reliable quantitation *(4)*.

Construction of the competitor, the so-called internal standard, is the main part of the qcRT-PCR technique. To act as competitor, it is important that the standard shows equivalence with the target mRNA with regard to primary and

From: *Methods in Molecular Biology, vol. 193: RT-PCR Protocols*
Edited by: J. O'Connell © Humana Press Inc., Totowa, NJ

secondary structure, as well as primer affinity. This means that the length and sequence of the internal standard must be similar to the target, and contains only a slight modification to distinguish between the internal standard and target mRNA.

To guarantee an optimal internal standard, it is advisable to produce a synthetic RNA (sRNA) standard by slight modification of the wild-type target cDNA sequence. Modification is achieved by either insertion or deletion of a few basepairs (bp), or manipulation of a restriction enzyme site. The change of a restriction site makes post-PCR processing necessary, but a simple change of few bp (deletion or insertion) allows direct quantification of the PCR products.

As an alternative to the modification procedure, some groups have used as an internal standard the homologous gene from another species based on interspecies sequence differences. Because of the high degree of homology between many interspecies sequences, it will be difficult to distinguish between the target mRNA and the internal standard after co-amplification ***(5,6)***. A simple detection based on fragment size differentiation is impossible, and therefore secondary structure differences are used to distinguish between the internal standard and the target mRNA sequence.

Since highly sensitive capillary electrophoresis was introduced in the detection of DNA and RNA, quantification of very low concentrations are no longer a problem. Moreover, quantification is performed with high reproducibility and specificity by co-amplification of a competitive standard ***(7,8)***. Using this approach, it is now possible to detect target mRNA concentrations in the attogram level in minute sample vols.

For qc RT-PCR, a few criteria must be considered. It is necessary:

1. To use a competitor as *synthetic RNA* to control reverse transcription and PCR steps.
2. To use an *internal standard* to avoid inter-sample variation in amplification efficiency.
3. To achieve *close homology* between competitor and target mRNA.
4. To *modify* the sequence as *little* as possible (insertion, deletion, or change of a restriction enzyme site).
5. To localize the modification *far away from the primer sites.*

In our protocol, the internal standards show a deletion of a few bp in standards constructed for the housekeeping gene human β-Actin, the immediate early gene cFOS, the transcription factors ETS-2, and the synaptosomal receptor gene SNAP-25. The deleted region never exceeded 20% of the finally amplified and analyzed target DNA product size, which varied from 160–245 bp. Detection was performed by a capillary electrophoresis (Genetic Analyzer, ABI Prism 310 Applied Biosystems, PE, Foster City, CA) using 5' FAM-labeled primers.

2. Materials

2.1. Reagents

1. RNAzol B (Molecular Research Center, Inc., Cincinnati, OH).
2. Total human brain RNA (Clontech).
3. pSP64 poly(A) vector (Promega Corporation, Madison, WI).
4. TA Cloning Kit INFαF (Invitrogen Corporation, San Diego, CA).
5. Restriction enzymes and rapid DNA ligation kit (Boehringer Mannheim GmbH, Mannheim, Germany).
6. Oligo(dT) cellulose columns; DNA and RNA mol-wt standards; yeast tRNA and cRNA; SP6 transcription kit; restriction enzymes (Life Technologies, Gibco-BRL, Gaithersburg, MD).
7. cDNA synthesis reagents: RNase inhibitor; oligo $d(T)_{16}$; deoxynucleotide 5' triphosphates (dNTPs); Moloney Murine Leukemia Virus (MMLV) reverse transcriptase (RT); PCR reagents: dNTPs; primers; $MgCl_2$; AmpliTaq DNA polymerase (PE Applied Biosystems, and Roche Molecular System Inc., Branchburg, NJ).
8. PCR product purification kit (Qiagen Inc., Santa Clarita, CA).
9. Reagents for the Genetic Analyzer: deionized formamide; POP4; 10x buffer with EDTA and Genescan 500-TAMRA (PE Applied Biosystems)
10. cDNA templates and sequences. A pBluescript vector containing the ETS-2 cDNA was obtained as a gift from Prof. Watson. All other cDNA sequences were produced by reverse transcription of total RNA isolated from human brain tissue (Clontech). The accession numbers of the cDNA sequences used are: human ETS-2 cDNA: J04102 ***(9)***; human β-Actin cDNA: NM 001101 ***(10)***; human cFOS cDNA: V01512 ***(11)***; human SNAP-25 cDNA: NM003081 ***(12)***.
11. PCR primers. The primer sequences for ETS-2 were selected by the OLIGO program of Rychlik and Rhoads ***(13)***. The primer sequences for the other three cDNAs were selected by the Primer Express program (Perkin-Elmer). Primers were synthesized commercially (Applied Biosystems GmbH, Weiterstadt, Germany).

2.2. Instruments

1. PCR was performed on a Perkin-Elmer GeneAmp PCR System 9600, Perkin-Elmer GeneAmp PCR System 2400, or a BioRad PCR system.
2. The PCR products labeled with 5-carboxy-fluorescein (FAM) were analyzed on the ABI-310 Genetic Analyzer (PE Applied Biosystems) containing a 47 cm × 50 μm long electrophoresis capillary using performance-optimized polymer 4 (POP4) and as matrix standard Genescan 500-TAMRA.

3. Methods

3.1. Synthesis of the Internal Standards for Human β-actin, Human cFOS, Human ETS-2, and Human SNAP-25

We constructed four different internal standards by slight modification of the wild-type sequences from the human β-actin, cFOS, ETS-2, and SNAP-25 cDNAs. The reverse transcription followed standard procedures using Oligo d(T)16 primer and MMLV RT (Perkin-Elmer). Construction of each standard generally involves a PCR assembly approach: first, the target is amplified in two fragments, with a short intervening region between the fragments. This intervening region will be deleted in the internal standard. The use of mutually complementary, composite primers at the 3' end of the first fragment and the 5' end of the second fragment allows PCR products from both fragments to be annealed together, minus the intervening region. The annealed fragments are then PCR-amplified to generate a single fragment representing the original target minus the short deletion between the two fragments. This is then used to generate the internal standard RNA (*see* **Note 1**). This basic approach is illustrated in **Fig. 1A**.

Construction of the four standards here illustrates variations that can be made in the basic protocol.

1. For the ETS-2 standard, the target ETS-2 template is derived from a plasmid clone. Restriction sites are introduced into the standard to facilitate cloning into a plasmid vector containing RNA polymerase promoter sites. This enables the generation of RNA copies of the standard by in vitro transcription from the plasmid.
2. Construction of the β-actin standard involves a similar approach to ETS-2, except the target template is derived from cDNA from human tissue. Instead of cloning the standard DNA, an SP6-promoter site and a polyA tail are incorporated into the standard DNA during PCR-amplification; a poly-adenylated RNA copy of the standard can then be generated by direct in vitro transcription of the PCR-constructed DNA standard. The cFOS standard is also constructed by this approach.
3. Construction of the SNAP-25 standard involves a similar approach to the β-actin and cFOS standards. In this case, use is made of an internal *Eco*N1 restriction site to assemble the DNA standard from one fragment which incorporates an *Eco*N1 site at its 3' end during PCR-amplification, and a second fragment which is cleaved at the natural *Eco*N1 site in the target. Both fragments are ligated together via the *Eco*N1 sites, with omission of an intervening sequence to generate the standard DNA.

3.2. cDNA Production

3.2.1. ETS-2 cDNA from the Bluescript Vector

1. Linearize the cloning vector containing the cDNA of interest (for ETS-2, we used a pBluescript vector containing the ETS-2 cDNA) by restriction digestion with an appropriate enzyme: *Eco*RI for the pBluescript-ETS-2 vector. Perform the digestion in a final vol of 50 µL containing restriction buffer, 20 U of *Eco*RI (Gibco-BRL), and pBluescript-ETS-2 DNA purified from an overnight culture. Digest at 37°C for 1 h.

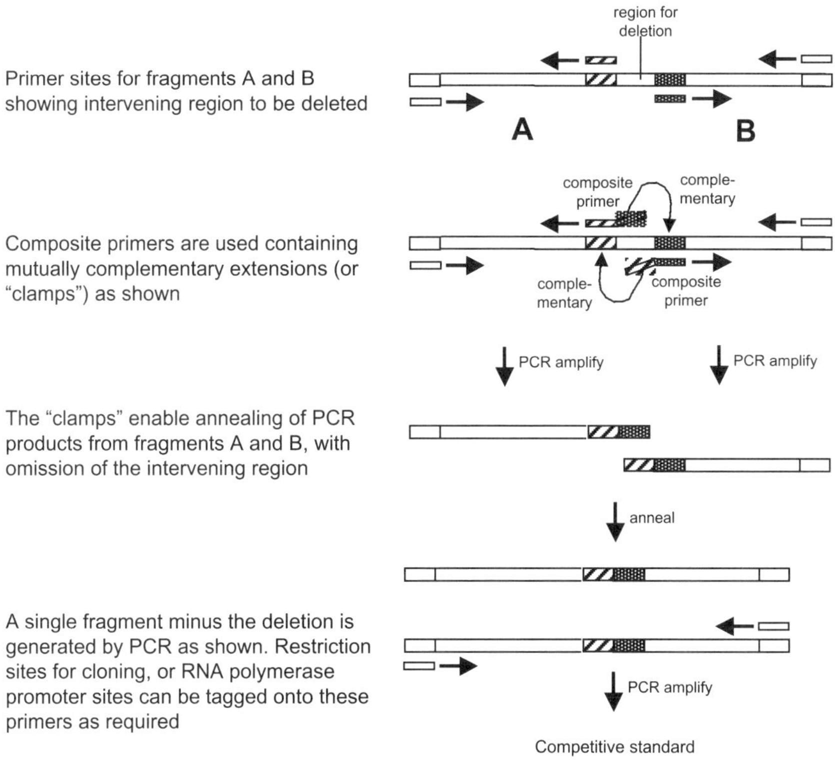

Fig. 1A. The basic strategy for generating deletions via PCR assembly using clamps.

2. Purify the digested DNA by triple phenol-chloroform extraction followed by precipitation by the addition of 1/10 vol of 3.0 *M* sodium acetate, pH 5.3, and 2.5 vol of ethanol. Allow the precipitation to occur at –20°C overnight. Collect the precipitate by centrifugation for 30 min at 8000*g* at 4°C. After washing in 75% ethanol, dry the pellet at 55°C for 10 min. Dissolve the pellet by adding 50 µL of autoclaved water and heating at 65°C for 10 min.
3. Determine the amount of linearized pBluescript-ETS-2 DNA by UV absorbance spectrophotometry (optical density $[OD]_{260/280}$).

3.2.2. Human β-Actin, cFOS, and SNAP-25 cDNAs are Amplified from Human Brain Total RNA

1. In the absence of a recombinant plasmid clone, templates for the other three genes of interest, β-Actin, cFOS, and SNAP25, can be obtained by reverse transcription from human brain total RNA (Clontech). Perform reverse transcription using the standard procedure (GeneAmp™ RNA-PCR kit), in a final vol of 20 µL containing:

a. buffer PE II.
b. 5 mM $MgCl_2$.
c. 1 mM each dNTP.
d. 1 U RNase inhibitor.
e. 2.5 U MMLV RT.
f. 2.5 μM Oligo d(T)16
g. 1 μg total RNA.

Incubate the reaction for 10 min at room temperature, 15 min at 42°C, 5 min at 99°C, and 5 min at 5°C.

2. Use an aliquot of the cDNA (2–5 μL) for PCR-amplification to produce DNA fragments of human β-Actin, cFOS, and SNAP-25 (*see* **Subheadings 3.3.2. to 3.3.4.**).

3.3. Production of Two Overlapping cDNA Fragments of the Target (Fragment A and Fragment B)

3.3.1. Production of Two zHuman ETS-2 cDNA Fragments with Overlapping Ends

1. For ETS-2, the linearized pBluescript vector containing human ETS-2 cDNA is used as template to amplify two DNA fragments (fragment A and fragment B, *see* **Fig. 1B**). Fragment A and B are neighboring sequences on the ETS-2 cDNA separated by only few bp (18 bp). Using composite PCR primers, the 3'end of fragment A and the 5'end of fragment B contain a mutually-complementary overlap (or bp "clamp"), so that the two fragments can be annealed together following PCR. The composite primers are such that the 3'end of fragment A is linked with the reverse-complement sequence of the forward primer of fragment B, and the 5'end of fragment B is linked with the reverse-complement sequence of the reverse primer of fragment A (*see* **Table 1**, bp clamp in **bold letters**). Following PCR-amplification, both DNA fragments contain an overlapping sequence of 43 bp. Thus, fragments A and B can be annealed together, minus the intervening 18-bp region.
Perform PCR in a final vol of 50 μL containing:

a. PCR buffer with 1.5 mM $MgCl_2$ (HT Biotechnology).
b. 150 μM each dNTP.
c. 400 μM each primer.
d. 0.2 U Taq polymerase (HT Biotechnology).
e. 100 ng linearized pBluescript vector containing human ETS-2 cDNA.

Use the following thermal cycles: initial denaturation for 3 min at 95°C followed by 30 cycles of: 15 s at 95°C, 15 s at 60°C and 30 s at 72°C. Perform a single final elongation step for 7 min at 72°C.

2. Analyze the amplified fragments A and B by agarose gel electrophoresis (*see* **Fig. 2**).

B

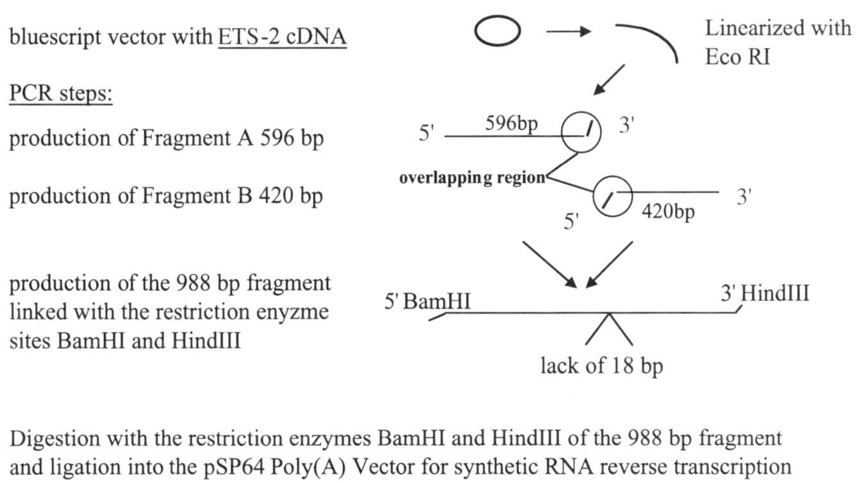

Fig. 1B. The different steps in the construction of the internal standard of human ETS-2 are shown. After linearization of the bluescript vector containing human ETS-2 cDNA, fragment A and fragment B are produced in the first PCR step. The reverse sequence of the forward primer of fragment B is linked to fragment A, and the reverse sequence of the reverse primer of fragment A is linked to fragment B; this produces an overlapping region (clamp). In the second PCR step, fragments A and B are annealed via the clamp, and amplified to produce a single 1006-bp fragment. The 1006-bp fragment has a deletion of the 18-bp intervening region between fragments A and B. In the second PCR step, the primers contain linked restriction sites (*Bam*HI and *Hind*III) to facilitate cloning of the PCR product into the pSP64 poly(A) vector.

3.3.2. Production of Two Human β-Actin cDNA Fragments with Overlapping Ends

1. As a first step, a 1601-bp fragment of human β-actin is amplified from brain cDNA by PCR using primer-pair 1 (*see* **Table 2**). The 1601-bp fragment is then used for construction of the internal standard.
   Sequences for primer-pair 1 are described in **Table 2**. Perform PCR in a final vol of 50 μL containing:
   a. PCR buffer II without $MgCl_2$ (Perkin-Elmer).
   b. 2 m*M* $MgCl_2$.
   c. 150 μ*M* each dNTP.
   d. 600 μ*M* each primer.
   e. 2 U Taq polymerase (Perkin-Elmer).
   f. 5 μL cDNA.
   Use the following thermal cycles: initial denaturation for 3 min at 94°C followed

Table 1
Human ETS-2 Primer-Pairs

Primer-Pair	Primer sequence	cDNA Position of the primer	Length of the amplified fragment (bp)
Primer-pair A forward	5'-GCAGCGGCAGGATGAATGAT-3'	281 bp	
Reverse	5'-**CTCTGTGCCAAAACCTAATGT** AGGAACGGAGGTGAGGTGTGAA-3'	855 bp	596 bp
Primer-pair B forward	5'-**TTCACACCTCACCTCCGTTCCT** ACATTAGGTTTTGGCACAGAG-3'	874 bp	420 bp
Reverse	'-GGCTTATTGAGGCAGAGAGAC-3'	1271 bp	
Primer-pair 1 forward 5'*Bam*HI	5'-**GGAAGCTT**GCAGCGGCAGGATGAATGAT-3'	281 bp	Wildtype 1006 bp/
Reverse 5'*Hind*III	5'-GGGATCCGGCTTATTGAGGCAGAGAGAC-3'	1271 bp	internal standard 988 bp
Primer-pair 2 forward	5'- TGGAGTGAGCAACAGGTATG-3'	607 bp	Wild-type 665 bp/
Reverse	5'-GGCTTATTGAGGCAGAGAGAC-3'	1271 bp	internal standard 647 bp
Primer-pair 3 forward 5'Fam-labeled	5'-CTGGAGCTGGCACCTGACTT-3'	733 bp	Wildtype 284 bp/
Reverse	5'-GACTTGGGGAACATCTGAAACT-3'	1016 bp	internal standard 266 bp

Linked sequences are given in bold letters (overlapping region, restriction enzyme sites *Bam*HI and *Hind*III).

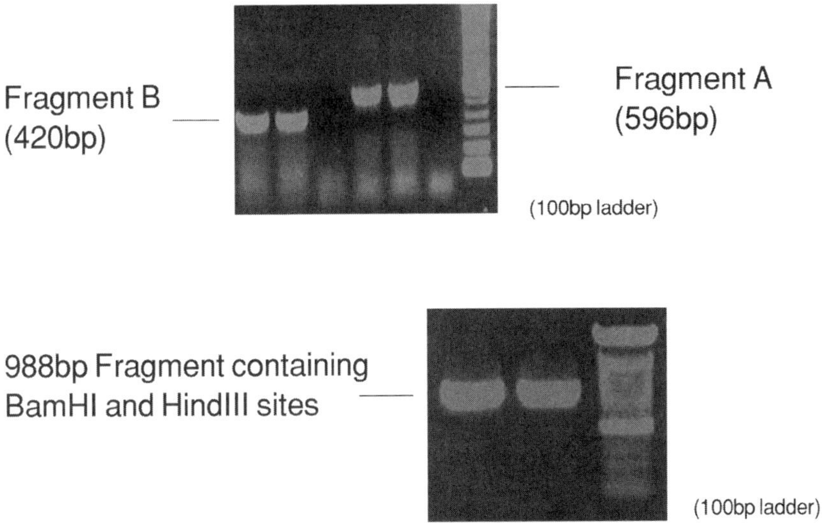

Fig. 2. Agarose gels showing the ETS-2 fragments A and B (first gel), and the 1006-bp single ETS-2 fragment (second gel). As a size marker, the 100-bp ladder from Gibco-BRL was used.

by 30 cycles of: 30 s at 94°C, 30 s at 60°C and 60 s at 72°C. Perform a single final elongation step for 7 min at 72°C.

2. The 1601-bp fragment of the human β-actin cDNA must be purified to avoid any production of unspecific DNA fragments.
   Purification is performed by column extraction (e.g., Qiagen®: PCR purification kit), and the concentration is measured spectrophotometrically.
3. The purified 1601-bp fragment of the human β-actin cDNA is used as template to PCR-amplify fragment A (974 bp) and fragment B (625 bp). Fragments A and B are separated by a 41-bp sequence. The primer-pair sequences include composite primers to allow for a bp clamp between fragments A and B as previously described (*see* **Table 2**, bp clamp in **bold letters**). As a result, fragments A and B contain an overlapping region of 39 bp that allows annealing.
   Perform PCR in a final vol of 50 μL containing:
   a. buffer II (Perkin-Elmer®).
   b. 2 m*M* $MgCl_2$.
   c. 150 μ*M* each dNTP.
   d. 600 μ*M* each primer.
   e. 1 U Taq polymerase.
   f. 100 ng purified 1601-bp fragment.
   Use the following thermal cycles for both fragments: initial denaturation for 3 min at 94°C followed by 30 cycles of: 20 s at 94°C, 20 s at 60°C and 40 s at 72°C. Perform a single final elongation step for 7 min at 72°C.

Table 2
Human β-actin Primer-Pairs

Primer-Pair	Primer sequence	cDNA Position of the primer	Length of the amplified fragment (bp)
Primer-pair 1 forward	5'-GCCAGCTCACCATGGATGAT-3'	31 bp	1601 bp
Reverse	5'-GCACGAAGGCTCATCATTCA-3'	1631 bp	
Primer-pair A forward	5'-GCCAGCTCACCATGGATGAT-3'	31 bp	
Reverse	5'-**CGCTCAGGAGGAGCAATGAT** TCTGCATCCTGTCGGCAAT-3'	984 bp	974 bp
Primer-pair B forward	5'-**ATTGCCGACAGGATGCAGA** ATCATTGCTCCTCCTGAGCG-3'	1026 bp	625 bp
Reverse	5'-GCACGAAGGCTCATCATTCA-3'	1631 bp	
Primer-pair 2 forward 5SP6-promoter	5'-**GTATCATACACATACGATTTAGGTGACAC** TATAGAACCATGTACGTTGCTATCCAGGC-3'	433 bp	Wild-type 1277 bp/ internal standard 1236 bp
Reverse 5'poly(T)	5'-**AATTCGGTTTTTTTTTTTTTTTTTT TTTTTTGGGAG**GCACGAAGGCTCATCATTC-3'	1631 bp	
Primer-pair 3 forward	5'-CCATCATGAAGTGTGACGTG-3'	883 bp	Wild-type 225 bp/ internal standard 184 bp
Reverse 5'Fam labeled	5'-ACATCTGCTGGAAGGTGGAC-3'	1106 bp	

Linked sequences are given in bold letters (overlapping region, SP6-promoter region, poly(T)tail).

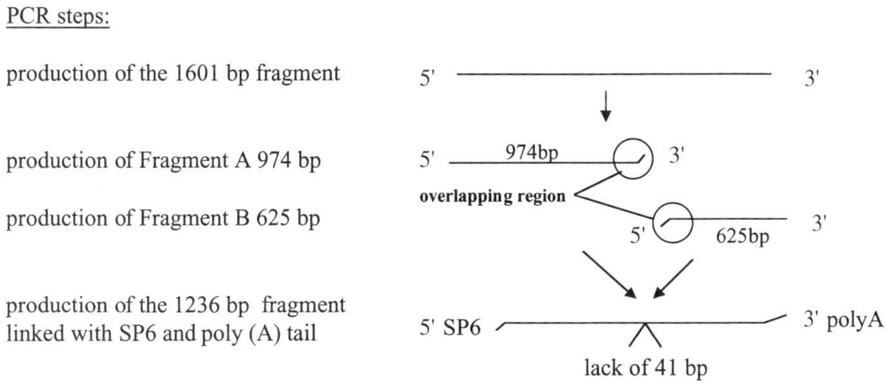

Purification of the 1236 bp fragment for synthetic RNA reverse transcription with SP6 Polymerase

Fig. 3. The different steps in the construction of the internal standard of human β-actin are shown. In the first PCR step a 1601-bp fragment of human β-actin is amplified from human brain cDNA. The fragment is used as template to produce fragments A and B in the second PCR step. The reverse sequence of the forward primer of fragment B is linked to fragment A, and the reverse sequence of the reverse primer of fragment A is linked to fragment B, producing an overlapping clamp region. In the third PCR step, fragments A and B are annealed via the clamp, and amplified to produce a single 1236-bp fragment. The 1236-bp fragment has a deletion of the 41-bp intervening region between fragments A and B. In the third PCR step, the primers contain linked SP6-promoter and poly(A) tail sequences to facilitate direct in vitro transcription of RNA from the PCR product.

Figure 3 shows the construction of fragments 1, A and B; **Fig. 4** shows the PCR products analyzed by agarose gel electrophoresis.

3.3.3. Production of Two Human cFOS Fragments with Overlapping Ends

1. Two DNA fragments (fragment A, 530 bp and fragment B, 620 bp) are PCR-amplified from cFOS cDNA in separate reactions. Fragments A and B are separated by 13 bp. The primer-pair sequences include composite primers to allow for a bp clamp between fragments A and B as previously described (*see* **Table 3**, bp clamp in **bold letters**). As a result, fragments A and B contain an overlapping region of 60 bp, which allows annealing.
   Perform PCR in a final vol of 50 μL containing:
   a. PCR buffer II without $MgCl_2$ (Perkin-Elmer®).
   b. 2 m*M* $MgCl_2$.

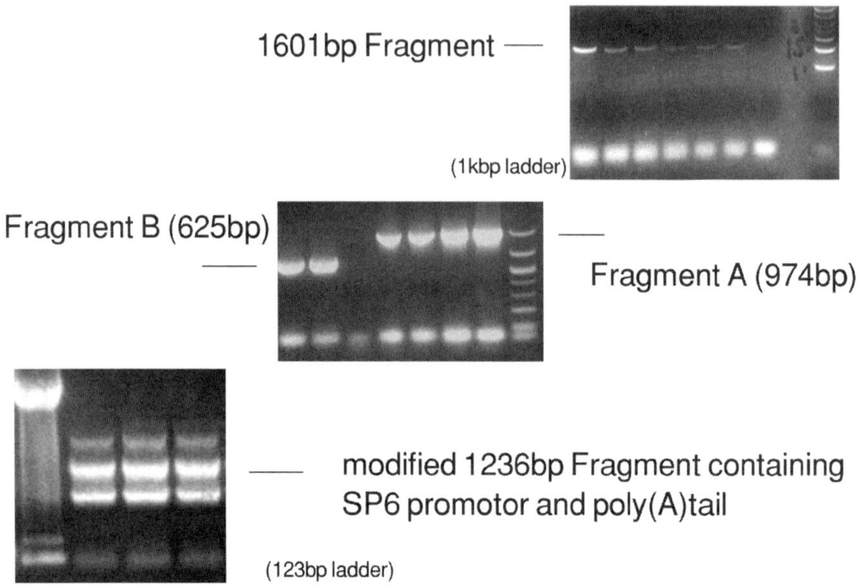

Fig. 4. Agarose gels showing the 1601 bp β-actin fragment (first gel), fragments A and B (second gel) and the 1236-bp single β-actin fragment containing an SP6-promoter sequence and poly(A) tail (third gel).

c. 200 m*M* each dNTP.
d. 800 µ*M* each primer.
e. 3 U Taq polymerase.
f. 5 µL cDNA.

Use the following thermal cycles for both fragments: initial denaturation for 3 min at 94°C followed by 40 cycles of: 30 s at 94°C, 20 s at 56°C and 30 s at 72°C. Perform a single final elongation step for 7 min at 72°C.

Figure 5 shows the construction of fragments A and B; **Fig. 6** shows the PCR products analyzed by agarose gel electrophoresis.

3.3.4. Production of Two Human SNAP-25 Fragments with an EcoNI Restriction Enzyme Site

1. Two SNAP-25 DNA fragments are PCR-amplified from SNAP-25 cDNA in separate reactions using primer-pairs A and B (*see* **Table 4**).

   The reverse primer for construction of fragment A has an *Eco*NI restriction site linked to its 5'end; this incorporates an *Eco*NI site onto the 3'end of fragment A (438 bp).

   Fragment B consists of a 732-bp sequence of the SNAP-25 cDNA ,including the entire sequence of fragment A, and an additional sequence of 313 bp, which contains an *Eco*NI restriction enzyme site at position 656 bp.

Table 3
Human cFOS Primer-Pairs

Primer-Pair	Primer-sequence	cDNA position of the Primer	Length of the amplified fragment (bp)
Primer-pair A forward	5'-GGCTTCAACGCAGACTACGA-3'	166 bp	
Reverse	5'-**CGCTTGGAGTGTATCAGTCAT**TGGCAATCTC GGTCTGCAAAGCAGACTTCTCATCTTCTC-3'	655 bp	530 bp
Primer-pair B forward	5'-**GAGAAGATGAGAAGTCTGCTTTGCAGACCGA GATTGCCAA** TGACTGATACACTCCAAGCG -3'	689 bp	620 bp
Reverse	5'-AGCGAGTCAGAGGAAGGCTCAT-3'	1268 bp	
Primer-pair 1 forward 5'SP6-promoter	5'-**GTATCATACACATACGATTTAGGTGACACTA TAGAA**GGCTTCAACGCAGACTACGA-3'	166 bp	
Reverse 5'poly(T)	5'-**GAATTCGGTTTTTTTTTTTTTTTTTTTTT TTTTTGGGAG**AGCGAGTCAGAGGAAGGCTCAT-3'	1268 bp	Wild-type 1182 bp/ internal standard 1169 bp
Primer-pair 2 forward	5'-CTGGCGTTGTGAAGACCATGA-3'	479 bp	Wild-type 234 bp/ internal standard 221 bp
Reverse—Fam labeled	5'-AGCAGGTTGGCAATCTCGGT-3'	713 bp	

Linked sequences are given in bold letters (overlapping region, SP6-promoter region, poly(T)tail).

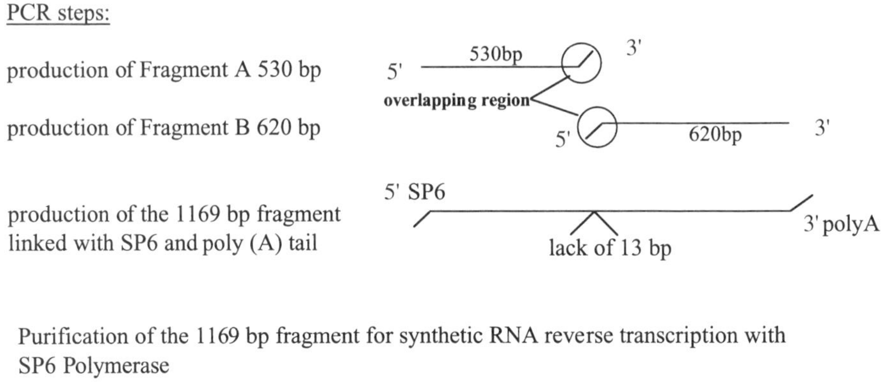

Fig. 5. The different steps in the construction of the internal standard of human cFOS are shown. In the first PCR step, fragments A and B are amplified from human brain cDNA. The reverse sequence of the forward primer of fragment B is linked to fragment A, and the reverse sequence of the reverse primer of fragment A is linked to fragment B, producing an overlapping clamp region. In the second PCR step, fragments A and B are annealed via the clamp, and amplified to produce a single 1169-bp fragment. The 1169-bp fragment has a deletion of the 13-bp intervening region between fragments A and B. In the second PCR step, the primers contain linked SP6-promoter and poly(A) tail sequences to facilitate direct in vitro transcription of RNA from the PCR product.

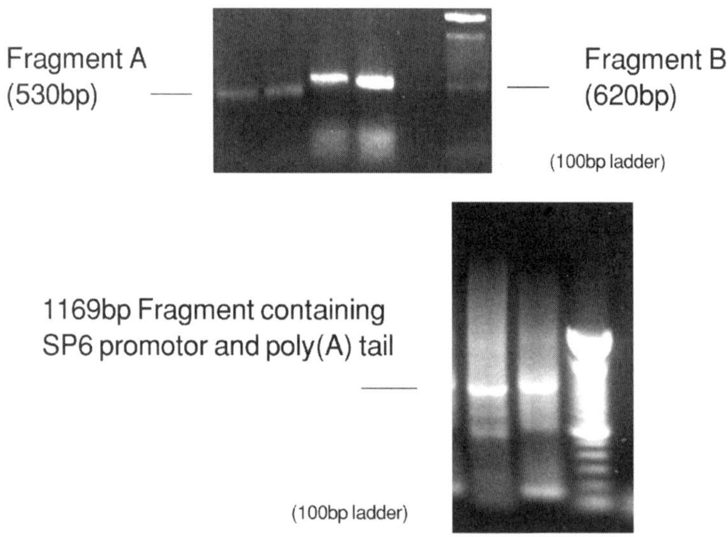

Fig. 6. Agarose gels showing the cFOS fragments A and B (first gel), and the 1169-bp single cFOS fragment containing an SP6-promoter sequence and poly(A) tail (second gel).

Table 4
Human SNAP-25 Primer-Pairs

Primer-Pair	Primer-sequence	cDNA Position of the primer	Length of the amplified segment (bp)
Primer-pair A forward	5'-GCAATGAGCTGGAGGAGATG-3'	228 bp	438 bp
Reverse 5'linked *Eco*NI	5'-**CTTAAGTCCTGCTCTAGG**CATCCAT TTCATTTTCTCGGG-3'	647 bp	
Primer-pair B forward	5'- GCAATGAGCTGGAGGAGATG-3'	228 bp	732 bp
Reverse	5'- CACAGGACAACATTCACAATGG-3'	959 bp	
Primer-pair 1 forward 5' SP6-promoter	5'-**GTATCATACACATACGATTTAGGTGACA CTATAGAA**ACCAGTTGGCTGATGAGTCG-3'	261 bp	Wild-type 752 bp/ internal standard 748 bp
Reverse 5' poly (T)	5'-**GAATTCGGTTTTTTTTTTTTTTTTTTTTTTTT TTTGGGAG**GTGTTATGTGTGCAGACCTACTAGAT-3'	923 bp	
Primer-pair 2 forward	5'-CTGGAACGCATTGAGGAAGG-3'	374 bp	Wild-type 426 bp/ internal standard 422 bp
Reverse	5'-ACGTTGGTTGGCCTCATCA-3'	799 bp	
Primer-pair 3 forward	5'-CTGCTCGTGTAGTGGACGAA-3'	555 bp	Wild-type 174 bp/ internal standard 170 bp
Reverse	5'-TCTGTGTATCGATCTCATTGCC-3'	728 bp	

Linked sequences are given in bold letters (*Eco*NI, SP6-promoter region, poly(T)tail).

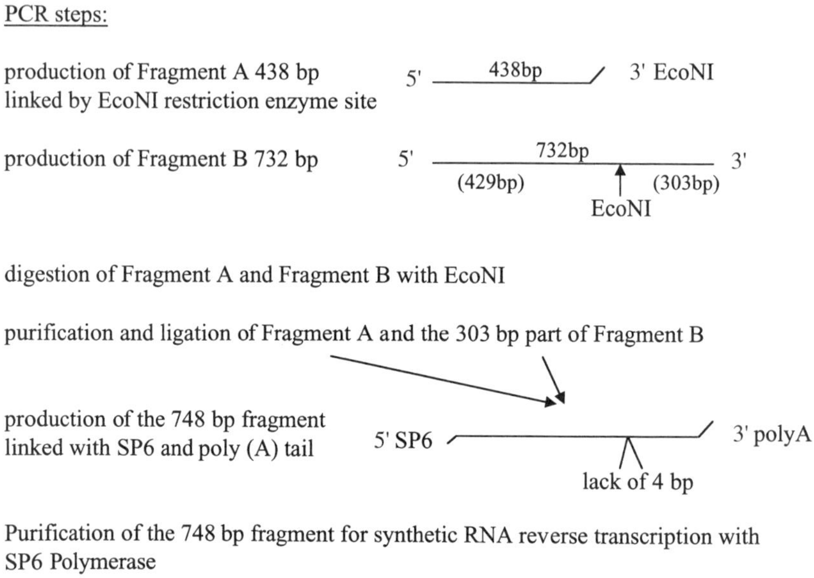

Fig. 7. The different steps in the construction of the internal standard of human SNAP-25 are shown. In the first PCR step, fragments A and B are amplified from human brain cDNA. The reverse primer used to amplify fragment A has a linked *Eco*N1 restriction site, so that fragment A contains an EcoN1 site at its 3' end. Fragment B contains a natural *Eco*NI restriction enzyme site. After digestion with the restriction enzyme *Eco*NI, fragment A and the short part of the digested fragment B (303 bp) are purified and ligated together. The ligation product has a deletion of the 4-bp intervening region between fragments A and the short part of fragment B. In the second PCR step, the ligated product is amplified using primers containing linked SP6-promoter and poly(A) tail sequences to facilitate direct in vitro transcription of RNA from the PCR product.

Perform PCR in a final vol of 50 µL containing:

Fragment A:

a. PCR buffer II without $MgCl_2$ (Perkin-Elmer®).
b. 2 m*M* $MgCl_2$.
c. 200 µ*M* each dNTP.
d. 600 µ*M* each primer.
e. 2 U Taq polymerase.
f. 5 µL cDNA.

Use the following thermal cycles for both fragments: initial denaturation for 3 min at 95°C followed by 40 cycles of: 30 s at 95°C, 30 s at 60°C and 30 s at 72°C. Perform a single final elongation step for 7 min at 72°C.

Fragment B:

a. PCR buffer II without $MgCl_2$ (Perkin-Elmer®).
b. 2 m*M* $MgCl_2$.

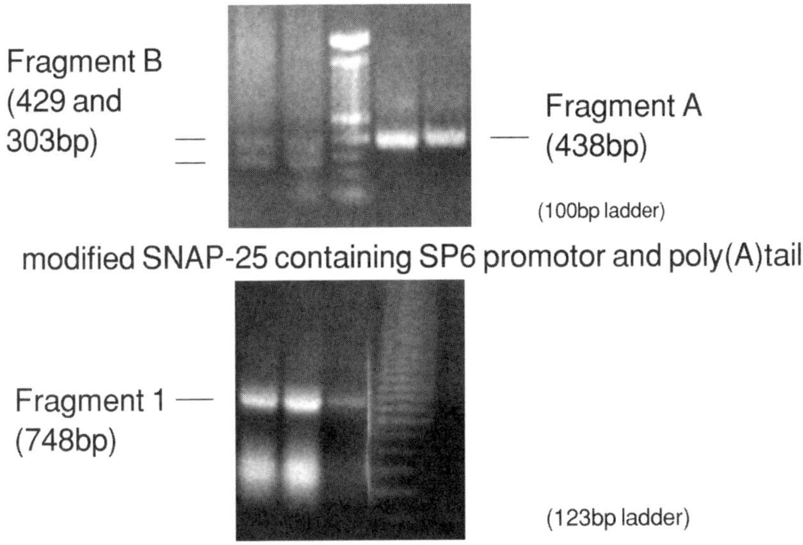

Fig. 8. Agarose gel showing the SNAP-25 fragments A and B (429-bp and 303-bp parts) after digestion with the restriction enzyme *Eco*NI (first gel). The second gel shows the 748-bp single SNAP-25 fragment containing an SP6-promoter sequence and poly(A) tail.

c. 250 μ*M* each dNTP.
d. 600 μ*M* each primer.
e. 2 U Taq polymerase.
f. 5 μL cDNA.

Use the following thermal cycles for both fragments: initial denaturation for 3 min at 95°C followed by 40 cycles of: 30 s at 95°C, 30 s at 63°C and 45 s at 72°C. Perform a single final elongation step for 7 min at 72°C.

Figure 7 shows the construction of fragments A and B; **Fig. 8** shows the PCR products analyzed by agarose gel electrophoresis.

3.3.5. Purification of Fragments A and B from ETS-2, β-Actin, cFOS, and SNAP-25

1. Purify the PCR-amplified DNA fragments A and B for each of the four target cDNAs on 1% low-melting agarose gels (*see* **Note 2**). Use a standard purification procedure as recommended by the manufacturers of the low melting agarose. Briefly:
   a. Excise the bands.
   b. Weigh the slices.
   c. Dissolve the slices in 5× TE buffer [10 m*M* Tris-HCl, 1 m*M* ethylenediamine-tetraacetic acid (EDTA), pH 8] at 95°C for 5 min.
   d. Cool the solution to room temperature.

e. Triple-extract the DNA with phenol-chloroform.
f. Precipitate the DNA with ethanol by adding 0.2 vol of 10 *M* ammonium acetate and 2 vol of ethanol at 4°C.

After 10 min at room temperature, centrifuge at 8000*g* for 35 min at 4°C. Wash the pellet in 70% ethanol, dry at 55°C for 10 min, and redissolve the purified DNA fragments in 50 µL of autoclaved, bi-distilled H_2O.

2. To facilitate construction of the modified SNAP 25 fragment, digest both fragments A and B with the restriction enzyme *Eco*NI, and ligate fragment A to the smaller sequence of fragment B (303 bp). Digest with the restriction enzyme *Eco*NI by mixing the following:
   a. 10× buffer.
   b. 1.5 U *Eco*NI.
   c. 1 µg of DNA for each fragment.

   Incubate the digest for 1 h at 37°C. After the digestion, separate the cleavage fragments on a 1% low melting agarose gel. Excise and purify fragment A and the smaller part of fragment B (303 bp) from the gel.
3. Quantify the purified products by ultraviolet (UV) absorbance spectrophotometry. In our example, the purified PCR-amplified fragments were in the following concentration ranges:
   a. 15–30 ng/µL for the two ETS-2 fragments.
   b. 70 ng/µL for the two β-actin fragments.
   c. 30–80 ng/µL for the two cFOS fragments.
   d. 35–51 ng/µL for the two SNAP 25 fragments.

3.4. Protocols for the Assembly of the DNA Standards for ETS-2, β-Actin, cFOS, and SNAP 25 and Linkage of SP6 Promoter and poly(A) Tail

This section describes the different protocols to assemble the internal standards; the in vitro transcription to synthetic RNA using SP6 RNA polymerase; and purification of the RNA standards via oligo d(T) affinity chromatography.

3.4.1. PCR Amplification of a Modified DNA Fragment of **ETS-2** and its ligation to the pSP64 poly(A) vector for in vitro transcription of synthetic RNA.

3.4.2. PCR amplification of a modified DNA fragment of **β-actin** using primer-pairs with 5'linkage of the SP6-promoter and poly(T) tail.

3.4.3. PCR amplification of a modified DNA fragment of **cFOS** using primer-pairs with 5'linkage of the SP6-promoter and poly(T) tail.

3.4.4. Ligation of two DNA fragments of **SNAP-25** following *Eco*NI digestion, and PCR amplification using primer-pairs with 5'linkage of the SP6-promoter and poly(T) tail (*see* **Notes 3** and **4**).

3.4.1. Assembly of the Modified ETS-2 Product (Deletion of 18 bp)

1. Mix fragment A (596 bp) and fragment B (420 bp) and PCR-amplify a single product using a single primer-pair. The fragments will anneal via the overlapping

bp clamp during the PCR to facilitate amplification of a single product. To generate the single PCR product, use the forward primer that was used to generate fragment A and the reverse primer used to amplify fragment B (*see* **Fig. 1A**). To facilitate subsequent cloning, include a *Bam*HI site linked to the forward primers and a *Hind* III site linked to the reverse primer (**Table 1**, primer-pair 1).

Use equal concentrations (50–80ng) of both DNA fragments in the following PCR amplification reaction producing a DNA fragment of 1006 bp in length. The difference between the 1006-bp DNA fragment and the wild-type ETS-2 cDNA is the deletion of 18 bp between nucleotide position 856 and 874 (*see* **Table 1**, primer-pair 1 and **Fig. 1B**).

Perform PCR in a final vol of 50 μL containing:

a. 50–80 ng of each template (fragments A and B)10× buffer containing 1.5 m*M* $MgCl_2$.
b. 200 μ*M* dNTPs.
c. 600 μ*M* each primer (*see* **Table 1**, primer pair 1).
d. 0.2 U Taq polymerase.

Use the following thermal cycles: initial denaturation for 3 min at 95°C followed by 30 cycles of: 20 s at 95°C, 15 s at 60°C and 40 se at 72°C. Perform a single final elongation step for 7 min at 72°C.

2. Purify the PCR product using a PCR purification kit (Qiagen® PCR purification kit).
3. Digest the 1006-bp fragment (modified ETS-2 product) and the pSP64/Poly(A) vector with the restriction enzymes *Hind*III and *Bam*HI as follows:

a.	volume: 40 μL	volume: 50 μL
b.	20 μL modified ETS-2 product	25 μL pSP64(polyA) vector isolated from an overnight culture
c.	1 μL *Hind*III	1 μL *Hind*III
d.	1 μL BamHI	1 μL *Bam*HI
e.	10× buffer	10× buffer

Incubate at 37°C for 1 h.

4. Purify the modified ETS-2 product by phenol-chloroform extraction and ethanol precipitation. The purified, modified ETS-2 product should be resuspended in the concentration range 50–100 ng/μL.
5. Purify the linearized pSP64(polyA) vector from a 1% low-melting agarose gel as described in **Subheading 3.3.5.** Resuspend the purified vector in a concentration range of 50–150 ng/μL.
6. Next, the modified ETS-2 product and the pSP64 Poly(A) vector are ligated together. Prepare a ligation mix as follows:
   a. 150 ng digested pSP64(polyA) vector.
   b. 500 ng digested modified ETS-2 product (insert).
   c. 10× ligase buffer.
   d. 5 U T4 DNA ligase.

   Incubate at 15°C for 15 h; inactivate the ligase at 65°C for 15 min, and store the ligation at –20°C.

7. Transform the ligated DNA into competent *Escheria coli* cells (Gene Hunting System) using the protocol recommended by the supplier of the competent cells. The vector contains a β-lactamase (Amp^r) coding region, and therefore transformed cell colonies are selected on ampicillin agar plates overnight.
8. Pick colonies and culture overnight in 5 mL of TB medium at 37°C with shaking. Purify plasmids from the selected bacterial clones using standard procedures.
9. Linearize the plasmids by digestion with the restriction enzyme *Eco*RI. A number of colonies are picked to ensure that an insert-containing, recombinant plasmid is obtained. The presence of an insert can be determined by agarose gel electrophoresis after linearization of the plasmid.
   Prepare a restriction digest in a final vol of 50 µL as follows:
   a. 25 µl extracted plasmid.
   b. 10× restriction buffer.
   c. 1.5 µL *Eco*RI.
   Incubate at 37°C for 1 h.
10. Purify the linearized, recombinant plasmid by triple phenol-chloroform extraction followed by ethanol precipitation. Add 1/10 vol of 3 *M* sodium acetate, pH 5.3, and 2.5 vol of ethanol; incubate at –20°C overnight. Centrifuge at 13,000*g* for 35 min at 4°C. Wash the DNA pellet in 70% ethanol, dry, and redissolve the plasmid in 50 µL of H_2O at 65°C for 10 min.
    The concentration range of the redissolved plasmid should be in the range of 0.2–1 µg/µL.
11. Confirm the deletion of the 18-bp region (nucleotides 856–873) by direct DNA-sequence analysis.
    The plasmid is now ready for in vitro transcription to generate an RNA copy of the standard.

3.4.2. Assembly of the Modified β-Actin Product (Deletion of 41 bp)

1. PCR is used to assemble a single product from the annealed β-actin fragments A (974 bp) and B (625 bp), as described in **Subheading 3.4.1.1.** for ETS-2. A single product of 1236 bp is obtained. In this instance, an SP6-promoter sequence and a poly (A) tail are linked to the PCR primers (*see* **Table 2**, primer-pair 2, and **Fig. 3**). This allows direct in vitro transcription of RNA from the PCR product, without the need for cloning into a plasmid.
   Perform PCR in a final vol of 50 µL containing:
   a. Buffer II (Perkin-Elmer®).
   b. 2 m*M* $MgCl_2$.
   c. 200 µ*M* each dNTP.
   d. 600 µ*M* each primer.
   e. 1 U Taq DNA polymerase.
   f. 30–50 ng of each template fragment A and B.
   Use the following thermal cycles: initial denaturation for 3 min at 95°C followed by 30 cycles of: 30 s at 95°C, 30 s at 60°C and 60 s at 72°C. Perform a single final elongation step for 7 min at 72°C.

2. Purify the assembled PCR product (1236 bp) from a 1% low-melting agarose gel as described in **Subheading 3.3.5.**
3. Confirm the deletion of the 41-bp region (nucleotides 985–1025) by direct DNA-sequence analysis.

3.4.3. Assembly of the Modified cFOS Product (Deletion of 13 bp)

1. The assembly strategy for the cFOS fragments A (530 bp) and B (620 bp) is the same as that used for β-actin as described in **Subheading 3.4.2.** A single 1169-bp fragment is obtained with a linked SP6-promoter sequence and a poly (A) tail. For primer sequences, *see* **Table 3**, primer-pair 1; also *see* **Fig. 5**.
   Perform PCR in a final vol of 50 µL containing:
   a. Buffer II (Perkin-Elmer®).
   b. 2 m*M* $MgCl_2$.
   c. 250 µ*M* each dNTP.
   d. 600 µ*M* each primer.
   e. 2 U Taq DNA polymerase.
   f. 30-50 ng of each fragment A and B.
   Use the following thermal cycles: initial denaturation for 3 min at 95°C followed by 40 cycles of: 30 s at 95°C, 20 s at 55°C and 45 s at 72°C. Perform a single final elongation step for 7 min at 72°C.
2. Purify the assembled PCR product (1169 bp) from a 1% low-melting agarose gel as described in **Subheading 3.3.5.**
3. Confirm the deletion of the 13-bp region (nucleotides 676–688) by direct DNA-sequence analysis.

3.4.4. Assembly of the SNAP-25 Product (Deletion of 4 bp)

1. Equal concentrations of both the purified, *Eco*NI-digested products, fragment A (425 bp), and the smaller product of fragment B (303 bp) are ligated together to yield a 728-bp fragment (*see* **Fig. 7**). The ligated product lacks an intervening 4-bp sequence (nucleotides 648–651) between the two fragments.
   Prepare a ligation mix in a final vol of 20 µL containing:
   a. 10× ligase buffer.
   b. 60 U DNA ligase.
   c. 500 ng of each fragment.
   Incubate at 15°C for 24 h; inactivate the ligase at 65°C for 15 min.
2. Purify the ligated DNA by double phenol-chloroform extraction followed by ethanol precipitation.
3. The ligation product is used as template to PCR-amplify a 748-bp fragment; the PCR allows incorporation of an SP6-promoter and a poly A tail onto the product. For primer sequences, *see* **Table 4**, primer-pair 1; *also see* **Fig. 7**.
   Perform PCR in a final vol of 50 µL containing:
   a. Buffer II (Perkin-Elmer®).
   b. 2 m*M* $MgCl_2$.
   c. 200 µ*M* each dNTP

   d. 500 μ*M* each primer.
   e. 2 U Taq DNA polymerase.
   f. 30–50 ng DNA ligation product.

   Use the following thermal cycles: initial denaturation for 3 min at 95°C followed by 40 cycles of: 30 s at 95°C, 30 s. at 58°C and 30 s at 72°C. Perform a single final elongation step for 7 min at 72°C.
4. Purify the assembled PCR product (748 bp) from a 1% low-melting agarose gel as described in **Subheading 3.3.5.**
5. Confirm the deletion of the 4 bp region (nucleotides 648–651) by direct DNA-sequence analysis.

3.5. Synthesis and Purification of the RNA Internal Standards

1. Synthesize the internal RNA standard for ETS-2 by transcribing in vitro from the pSP64 poly(A) vector, using SP6 RNA polymerase. For all the other modified fragments of β-actin, cFOS, and SNAP-25, transcribe in vitro directly from the purified PCR products using SP6 RNA polymerase (*see* **Note 5**). Perform in vitro transcription by using either 3 μg of the linearized plasmid containing the modified ETS-2 product, or 300–500 ng of each modified product containing an SP6-promoter sequence and a poly(A) tail. Mix a reaction in a final vol of 25 μL containing:
   a. Transcription buffer (200 m*M* Tris-HCl, 30 m*M* MgC12, 10m*M* spermidine hydrochloride)
   b. 10 m*M* DTT.
   c. 500 μ*M* each rNTP.
   d. RNase inhibitor, 10 U.
   e. SP6 polymerase, 35 U.
   f. 3 μg plasmid or 300–500 ng modified PCR fragment.

   Incubate at 40°C for 1 h. Eliminate the DNA by digestion with 20–30 U of DNase I at 37°C for 15 min.
2. Purify the in vitro transcribed RNA by triple phenol-chloroform extraction for RNA (i.e., using water-saturated phenol at pH 4.3). Ethanol-precipitate the RNA by adding 1/10 vol of 3 *M* sodium acetate, pH 4.3, and 2.5 vol of absolute ethanol; leave at –20°C overnight. Collect the precipitated RNA by centrifugation at 8000*g* for 40 min at 4°C. Wash the pellet carefully twice in 80% ethanol, dry at room temperature for approx 30 min or at 55°C for 10 min. Redissolve the pellet and store at –20°C.
3. Purify the RNA standards containing poly(A) tails by oligo(dT) cellulose column chromatography (Gibco-BRL) following the manufacturer's protocol. Quantify the purified RNA standard by UV absorbance spectrophotometry. In our example, the RNA standards were in the following concentration ranges: ETS-2 sRNA: 1.2 ng/μL; β-actin, cFOS, and SNAP-25 sRNAs: approx 200–300 pg/μL (*see* **Note 6**).
4. Store the sRNA at –80°C in silanized tubes (silanization solution®, Serva, Germany) to prevent RNA adhesion to plasticware; in our experience, sRNA should not be stored for longer than 3 mo. Also, divide the sRNA into several aliquots

and add yeast tRNA (1mg/mL) as a carrier. Carrier tRNA does not cause background amplification; RT-PCR reactions performed with yeast tRNA as template were negative (*see* **Notes 7** and **8**).

3.6. Isolation of Total RNA from Small Samples of Human Tissues

1. Isolate RNA from the tissue of interest using a kit such as the RNAzol™ B-RNA isolation method (Molecular Research Centre, Inc.). In our example, this led to the isolation of pure, undegraded RNA from human myocardial biopsies and brain tissue. We used myocardial tissue in the range of 2–6 mg. We used brain tissue from five different regions in the range of 50 mg dissolve the RNA pellets in DEPC-treated, bi-distilled, autoclaved water. We found that the concentration range of total RNA obtained was 15–100 ng/µL from the myocardial tissue samples, and about 1 µg/µL from the brain tissue. For storage at –80°C, divide the total RNA into aliquots with 10 ng total RNA/µL containing yeast tRNA (1 mg/ml) as carrier in silanized tubes (*see* **Notes 7** and **8**).

3.7. Competitive Quantitative RT-PCR

β-Actin and cFOS quantification is performed with a single round of PCR using a Fam-labeled primer. ETS-2 and SNAP-25 quantification requires nested PCR, because of the lower concentration range in small tissue samples.

3.7.1. Competitive Reverse Transcription (RT) Step

1. Make serial dilutions of the synthetic RNA internal standards for ETS-2, β-actin, cFOS, and SNAP-25, and add an aliquot of each dilution to a fixed amount of total tissue RNA (10 ng). In our example, the concentration range of synthetic RNA amounts used for quantification in 10 ng total RNA of brain tissue were as follows:
   a. Human ETS-2 sRNA: 0.01–100 fg/µL.
   b. Human β-actin sRNA: 2–2000 pg/µL.
   c. Human cFOS sRNA: 0.2–2000 fg/µL.
   d. Human SNAP-25 sRNA: 0.01–1000 fg/µL.
2. Perform reverse transcription (RT step) in a 20 µL final vol containing:
   a. Buffer II (Perkin-Elmer®).
   b. 4 m*M* $MgCl_2$.
   c. 1 m*M* each dNTP.
   d. 1U RNase inhibitor.
   e. 2.5 µ*M* Oligo d(T)16.
   f. 2.5 U/µL MMLV RT.
   g. 1 µL (10 ng) total RNA.
   h. known amount of sRNA internal standards.

   The RT reaction was incubated for 10 min at room temperature; 15 min at 42°C; 5 min at 99°C and 5 min at 5°C. Store the cDNA samples at –80°C.

3.7.2. Competitive PCR Step

An aliquot of 2–5 µL of the cDNA sample is amplified by PCR for ETS-2, β-actin, cFOS and SNAP-25. Because of varying PCR protocols and annealing temperatures, qcRT-PCR was performed separately for each target mRNA. Otherwise, it is possible to perform qcRT and PCR for more than one target in the same reaction tube (*see* **Note 9** and **10**).

1. Perform PCR using labelled primers on 2 or 5 µL of the cDNA template for amplification of β-actin and cFOS, respectively, and 5 µL for the nested PCR amplification of ETS-2 and SNAP-25. Two µL of the primary PCR is used as template for the secondary (nested) PCRs. A 5'-FAM-labeled primer is used in the final PCR, whether primary or secondary, in each case. In addition to regular RT-PCR negative controls, some additional controls are also necessary. To control for cross-contamination, perform RT-PCR on each internal standard and total RNA sample separately. Also, perform controls on RNase-treated samples to check for the presence of contaminating DNA templates.
2. Perform all PCRs in a final vol of 50 µL containing:
   a. Buffer II (Perkin-Elmer®).
   b. 2 m*M* $MgCl_2$.
   c. 150 µ*M* each dNTP.
   d. 400 µ*M* each primer.
   e. 2 U Taq polymerase.
   f. cDNA.

   The thermal cycles are as follows: initial denaturing step for 3 min at 94°C; 25 cycles of: 30 s denaturation at 94°C, 30 s annealing at 60°C, 60 s elongation at 72°C; a final single elongation step for 7 min at 72°C. Differences are: 30 and 35 cycles are performed for the β-actin and c-FOS PCRs, respectively, and 40 cycles of primary and 35 cycles of nested PCR are performed for SNAP-25. Also, for the c-FOS PCR, an annealing temperature of 54°C is used. Primer sequences and product sizes for the target and internal standard amplicons are shown in **Tables 1–4**. For ETS-2, primer pair 2 is used for the primary PCR and primer pair 3 for the nested PCR (**Table 1**). For β-actin, primer pair 3 is used for the PCR (**Table 2**). For c-FOS, primer-pair 3 is used for the PCR (**Table 3**). For SNAP-25, primer-pair 2 is used for the primary PCR and primer-pair 3 is used for the nested PCR (**Table 4**).

3.8. Quantification by the ABI Prism 310 Genetic Analyzer (PE Applied Biosystems) Based on Capillary Electrophoresis Detecting Fluorescence-Labeled Products

1. Separate the amplified PCR products by capillary electrophoresis using the ABI Prism 310 Genetic Analyzer. Use Genescan 500-TAMRA (Perkin-Elmer) as internal size standards (basepair ladder); mix 0.5 µL of the Genescan 500–TAMRA with 19.5 µL of deionized formamide, and add 0.5 to 1 µL of the FAM-labeled PCR products. Prior to running the samples as per the manufacturer's instruc-

tions, first denature the DNA for 2 min at 94°C followed by snap-cooling on ice for 5 min.

2. Use the Genetic Analyzer program to determine the peak height and area, and the Genotyper program to calculate the concentration of the unknown elements. Peak height or area of the wild-type PCR product is compared to the peak height or area the internal standard, and the ratio of the two is obtained. The unknown mRNA concentration is then calculated by multiplying this ratio by the known amount of internal standard (*see* **Note 11**). Quantification is most accurate when the ratio of internal standard to wild-type PCR products is close to 1.

3.8.1. Analysis of Results

1. ETS-2 in human brain tissue. **Figure 9** demonstrates an example of an ETS-2 analysis showing the 266-bp peak of the internal standard and the 284-bp peak of the wild-type PCR products. We analyzed ETS-2 mRNA concentrations in five different regions of the human brain. The mean values are 38.5–65.1 (S.D.) attogram/10ng total RNA in the cerebellum; 10.95–21.1 (S.D.) attogram / 10 ng total RNA in the occipital lobe; 5.4–8.1 (S.D.) attogram / 10 ng total RNA in the parietal lobe; 16.9–26.7 (S.D.) attogram/10 ng total RNA in the frontal lobe, and 2.4–4.1 (S.D.) attogram/10 ng total RNA in the temporal lobe.
2. β-Actin in human brain tissue. **Figure 10** demonstrates an example of a β-actin analysis showing the 184-bp peak of the internal standard and the 225-bp peak of the wild-type PCR products. We analyzed β-actin mRNA concentration in five different regions of the human brain. Mean values are 81.4–82 (S.D.) picogram/ 10 ng total RNA in the cerebellum; 26.4–42.7 (S.D.) picogram/10 ng total RNA in the occipital lobe; 18.7–15.5 (S.D.) picogram/10 ng total RNA in the parietal lobe; 37–52 (S.D.) picogram/10 ng total RNA in the frontal lobe, and 30.1–31.2 (S.D.) picogram/10 ng total RNA in the temporal lobe.
3. cFOS in human brain tissue. **Figure 11** demonstrates an example of a cFOS analysis showing the 221-bp peak of the internal standard and the 234-bp peak of the wild-type PCR products. We analyzed cFOS mRNA concentration in five different regions of the human brain. Mean values are 74.1–79 (S.D.) femtogram/10 ng total RNA in the cerebellum; 40.8–24.5 (S.D.) femtogram/10 ng total RNA in the occipital lobe; 23.8 –28.7 (S.D.) femtogram/10 ng total RNA in the parietal lobe; 16.9–7.9 (S.D.) femtogram/10 ng total RNA in the frontal lobe, and 4.1–4.7 (S.D.) femtogram/10 ng total RNA in the temporal lobe.
4. SNAP-25 in human brain tissue. **Figure 12** demonstrates an example of a SNAP-25 analysis showing the 170-bp peak of the internal standard and the 174-bp peak of the wild-type PCR products. We are currently performing the analysis of SNAP-25 in two different regions of the human brain. Preliminary results show a concentration range of 50–500 attogram SNAP-25 in 10 ng total RNA of human brain samples.

   To rule out general transcriptional changes, the analyzed concentrations of ETS-2, cFOS, and SNAP-25 can be normalized vs the housekeeping gene β-actin.

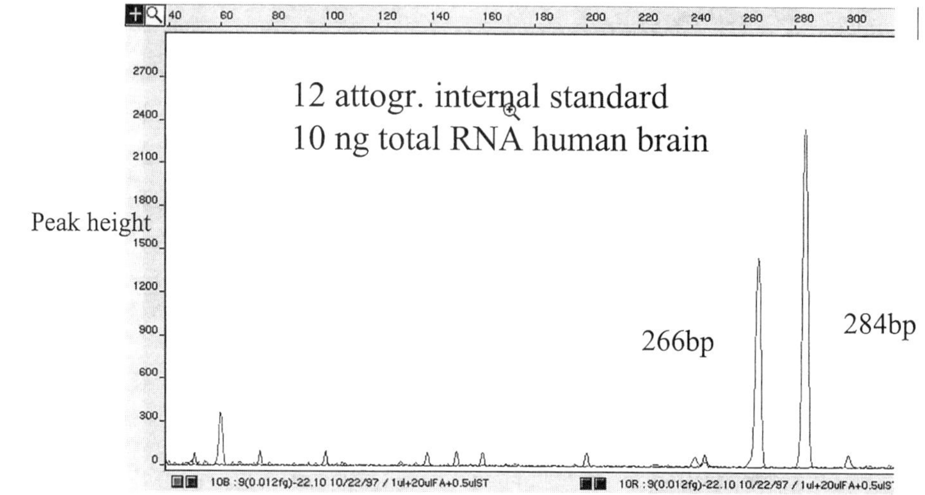

Fig. 9. Quantitation of ETS-2 mRNA in human brain tissue by competitive RT-PCR. RT-PCR products for the ETS-2 target and internal standard were detected and quantified using the ABI Prism 310 Genetic Analyzer (Perkin-Elmer Applied Biosystems®). The ratio of RT-PCR products for target to internal standard (known template concentration) enabled calculation of the unknown target ETS-2 mRNA template concentration. The peaks of the Genescan 500-TAMRA size markers are also shown.

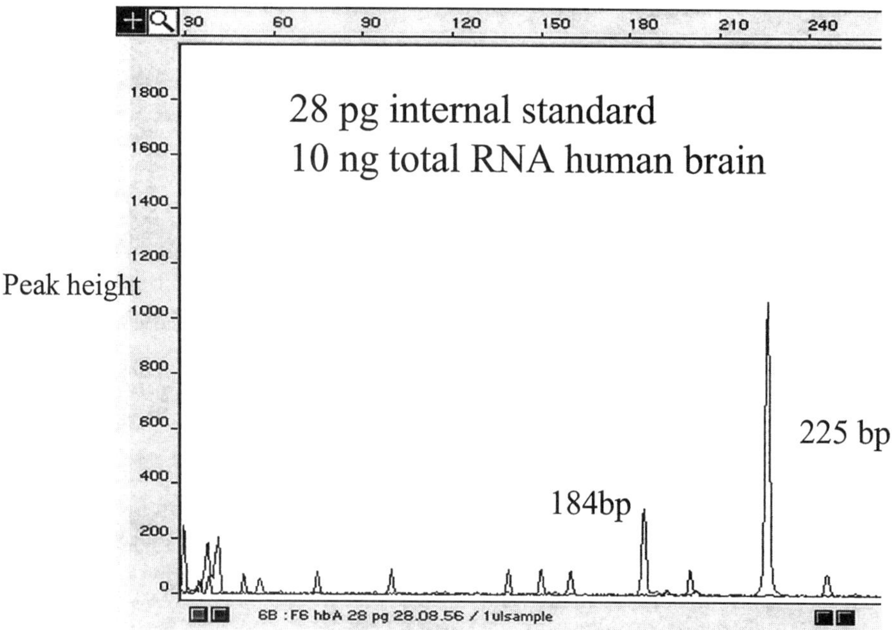

Fig. 10. Quantitation of β-actin mRNA in human brain tissue by competitive RT-PCR. RT-PCR products for the β-actin target and internal standard were detected and quantified using the ABI Prism 310 Genetic Analyzer (Perkin-Elmer Applied

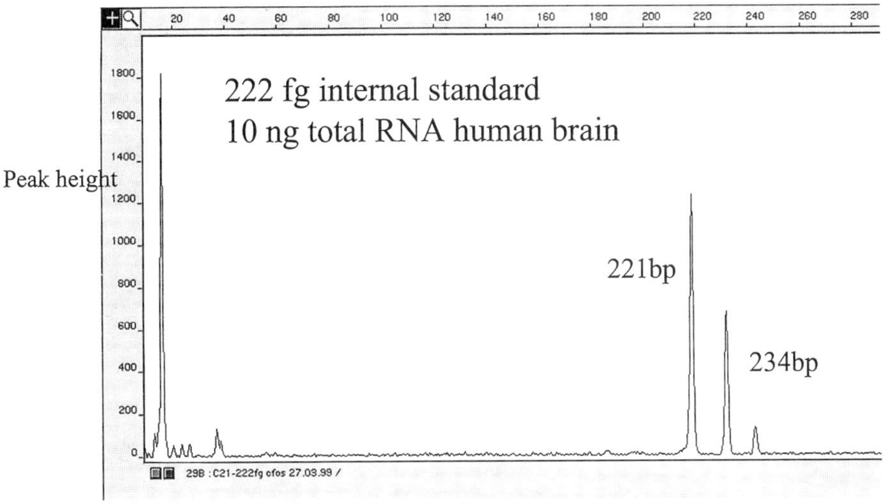

Fig. 11. Quantitation of cFOS mRNA in human brain tissue by competitive RT-PCR. RT-PCR products for the cFOS target and internal standard were detected and quantified using the ABI Prism 310 Genetic Analyser (Perkin-Elmer® Applied Biosystems®). The ratio of RT-PCR products for target to internal standard (known template concentration) enabled calculation of the unknown target cFOS mRNA template concentration.

4. Notes

1. Synthetic RNA as competitor should be closely homologous to the target mRNA; the modified region should be as small as possible, and should never exceed 20% of the target template. The modification should not affect the secondary structure, or the primer sites, thus ensuring equal annealing ability of the primer pair to the internal standard and the target sequence. Primer pairs for the competitive PCR should be localized as far as possible away from the modified region. It is best to localize the modified region near the center of the target template.
2. Separation on low-melting agarose gel offers the ability to isolate only the desired fragment and to avoid any contamination.
3. Ligation of two DNA fragments via a restriction site (as opposed to annealing via an overlapping "clamp" region) offers the advantage that covalently joined copies of the modified fragment are created. However, PCR is still necessary to amplify the quantity of the modified fragment, and also to incorporate the SP6-promoter and poly A tail sequences. The PCR amplification provides a sufficient

[continued] Biosystems®). The ratio of RT-PCR products for target to internal standard (known template concentration) enabled calculation of the unknown target β-actin mRNA template concentration. The peaks of the Genescan 500-TAMRA size markers are also shown.

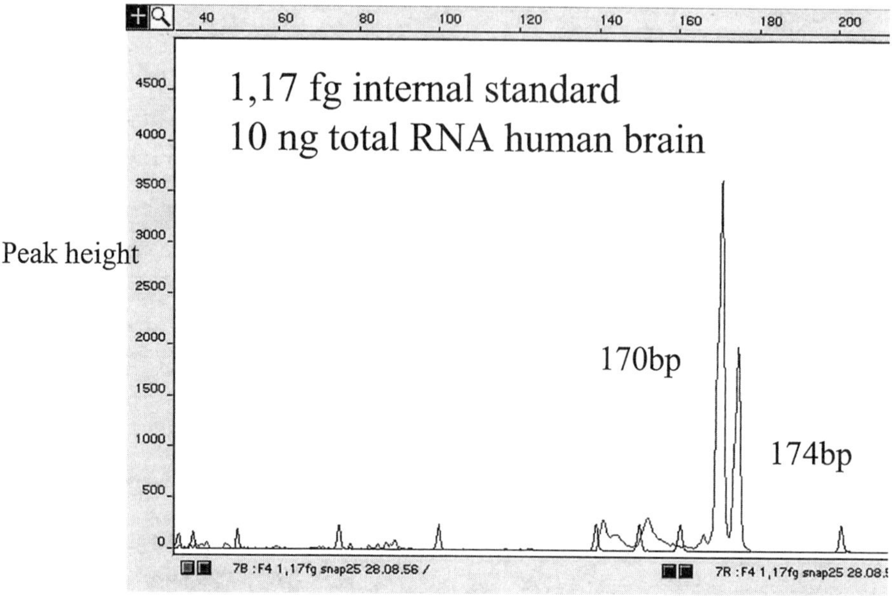

Fig. 12. Quantitation of SNAP-25 mRNA in human brain tissue by competitive RT-PCR. RT-PCR products for the SNAP-25 target and internal standard were detected and quantified using the ABI Prism 310 Genetic Analyzer (Perkin-Elmer Applied Biosystems®). The ratio of RT-PCR products for target to internal standard (known template concentration) enabled calculation of the unknown target SNAP-25 mRNA template concentration. The peaks of the Genescan 500-TAMRA size markers are also shown.

template for several in vitro transcription reactions. It is advisable to synthesize RNA every 3 mo because of RNA degradation during storage.

4. We observed poor amplification in PCRs, where the primers are located directly at the ends of the target internal control template. Such problems do not arise if the primers are positioned a few bp away from the ends.
5. Synthetic RNA production from a purified PCR product containing the SP6-promoter region and a poly(A) tail is rapid. Synthetic RNA production from a vector containing the SP6-promoter region and a poly(A) tail is more involved, since multiple steps are required, such as cell culture of plasmid-containing bacteria, plasmid purification, and linearization.
6. A better yield of synthetic RNA (sRNA) is obtained from inserts cloned into the pSP64 poly(A) vector than from the SP6-promoter-linked PCR products. However, cloning adds a good deal of additional labor to the construction of the standard. The PCR assembly strategy is easier to perform, and the lower yield of sRNA is generally not an obstacle, since only small amounts are required per competitive RT-PCR reaction.

7. To avoid absorption of RNA to plasticware it is necessary to dilute small concentrations in yeast tRNA (1 mg/mL) and to use silanized, autoclaved, thin-walled PCR tubes (0.2 mL) for storage of the synthetic RNA.
8. Dissolved synthetic RNA must be stored at –80°C, for a maximum of 3 mo. After this period, degradation of RNA occurs, and we can no longer observe linearity and reproducibility of the analysis (weak and varying peaks detected by capillary electrophoresis).
9. Quantification of different mRNAs can be performed from 10 ng of total RNA by direct or nested PCR amplification reactions, using a mixture of multiple synthetic RNAs in varying concentrations of a known amount.
10. To perform multiplex PCR for the quantification of two or more fragments of interest, the annealing temperatures of the primer pairs and the PCR protocols should be similar. The resolution of fragments of equal size can be achieved by use of different fluorescence-labeled primers for each PCR product, including the target and internal standard fragment.
11. A minimum of three reactions containing different internal standard concentrations should be analyzed for each sample to confirm linearity and reproducibility of the analyzed concentration of the target mRNA.

References

1. Wang, A.M., Doyle, M.V., and Mark D.F. (1989) Quantitation of mRNA by the polymerase chain reaction. *Proc. Natl. Acad. Sci. USA* **86,** 9717–9721.
2. Diviacco, S., Norio, P., Zentilin, L., Menzo, S., Clementi, M., Biamont, G., et al. (1992) A novel procedure for quantitative polymerase chain reaction coamplification of competitive templates. *Gene* **122,** 313–320.
3. Gilliland, G., Perrin, S., Blanchard, K., and Bunn, H.F. (1990) Analysis of cytokine mRNA and DNA: detection and quantitation by competitive polymerase chain reaction. *Proc. Natl. Acad. Sci. USA* **87,** 2725–2729.
4. Zhou, N.-M., Matthys, P., Polacek, C., Fiten, P., Sato, A., Billiau, A., et al. (1997) A competitive RT-PCR Method for the Quantitative Analysis of Cytokine mRNAs in Mouse Tissues. *Cytokine* **9,** 212–218.
5. Hess, J.F., Borkowski, J.A., Young, G.S., Strader, C.D., and Ransom, R.W. (1992) Cloning and pharmacological characterization of a human bradykinin (BK-2) receptor. *Biochem. Biophys. Res. Commun.* **184,** 260–268.
6. de Breuil, R.M., Patel, J.M., and Mendelow, B.V. (1993) Quantification of beta-actin-specific mRNA transcripts using xeno-competitive PCR. *PCR Methods Appl.* **3,** 57–59.
7. Ulfelder, K.J., Schwartz, H.E., Hall, J.M., and Sunzeri, F.J. (1992) Restriction fragment length polymorphism analysis of ERB B2 oncogene by capillary electrophoresis. *Anal. Biochem.* **200,** 260–267.
8. Murata, T., Takizawa, T., Funaba, M., Fujimura, H., Murata, A., and Torii, K. (1997) Quantitation of mouse and rat β-actin mRNA by competitive polymerase chain reaction using capillary electrophoresis. *Anal. Biochem.* **244,** 172–174.
9. Watson, D.K., McWilliams, M.J., Lapis, P., Lautenberger, J.A., Schweinfest,

C.W., and Papas, T.S. (1988) Mammalian ets-1 and ets-2 genes encode highly conserved proteins. *Proc. Natl. Acad. Sci. USA* **85,** 7862–7866.

10. Vandekerckhove, J., and Weber, K. (1978) Mammalian cytoplasmic actins are the products of at least two genes and differ in primary structure in at least 25 identified positions from skeletal muscle actins. *Proc. Natl. Acad. Sci. USA* **75,** 1106–1110.
11. VanStraaten, F., Müller, R., Curran, T., BanBeveren, C., and Verma, I.M. (1983) Complete nucleotide sequence of a human c-onc gene: deduced amino acid sequence of the human c-fos protein. *Proc. Natl. Acad. Sci. USA* **80,** 3183–3187.
12. Zhao, N., Hashida, H., Takahashi, N., and Sakaki, Y. (1994) Cloning and sequence analysis of the human snap 25 cDNA. *Gene* **145,** 313–314.
13. Rychlik, W., and Rhoads, R.E. (1989) A computer program for choosing optimal oligonucleotide primers for filter hypbridization, sequencing and in vitro amplification. *Nucleic Acids Res.* **17,** 8543–8551.

4

Detection of mRNA Expression and Alternative Splicing in a Single Cell

Tsutomu Kumazaki

1. Introduction

We have developed a method to detect mRNA expression in a single cell without isolating RNA from the cell *(1)*. The method includes the following manipulations: detach cells, pick a single cell, destroy the cell by heating, reverse-transcribe RNA, and PCR-amplify. Using this method, we succeeded in detecting the expression of an abundant mRNA for glyceraldehyde 3-phosphate dehydrogenase (GAPDH). This is a simple and easy method to detect the expression of abundant mRNAs. However, the method failed to detect fibronectin (FN) mRNA in fibroblasts—which express significant amounts of FN—probably because the expression level of FN mRNA was not adequate for the single-step method. Next, we developed another method *(2)*. This new method is composed of two steps, a variation of nested PCR. Like the old method, the first step includes the following manipulations: detach cells, pick a single cell, destroy the cell by freezing and thawing, reverse-transcribe RNA, and PCR-amplify. The second step is composed of PCR amplification with another nested PCR-primer set. Using this nested PCR technique, we could detect the expression of FN. We also succeeded in analyzing alternative splicing of FN mRNA in a single cell.

These methods can be applied to many fields. In developmental biology, for example, the expression of a critical gene for development of an embryo may be detectable if only a single cell or a few cell(s) are selected from the embryo.

2. Materials

All materials were stored at –20°C, except for PCR beads which were stored at 4°C.

From: *Methods in Molecular Biology, vol. 193: RT-PCR Protocols*
Edited by: J. O'Connell © Humana Press Inc., Totowa, NJ

1. Dithiothreitol (DTT) and human placenta-derived RNase inhibitor (HPRI) were purchased from Sigma-Aldrich (USA) and Takara Shuzou (Japan), respectively.
2. Oligo(dT) primer and random hexamer primer were purchased from Amersham-Pharmacia-Biotech (USA) and Takara Shuzou, respectively.
3. Sequence-specific primers were synthesized by Amersham-Pharmacia-Biotech. Primer sequences for all primers used are shown in **Note 1**.
4. rTth DNA polymerase with reverse transcription buffer (10 m*M* Tris-HCl, pH 8.3, 90 m*M* KCl, and 1 m*M* $MnCl_2$) and chelating buffer (5% glycerol, 10 m*M* Tris-HCl, pH 8.3, 100 m*M* KCl, 0.75 m*M* EGTA, 0.05% Tween 20, and 2.5 m*M* $MgCl_2$) was purchased from PE Biosystems (USA).
5. The mixture of Moloney Murine leukemia virus (MMLV) reverse transcriptase (RT) and Taq DNA polymerase with reaction buffer was purchased from Life Technologies (USA).
6. PCR beads were purchased from Amersham-Pharmacia-Biotech.

3. Methods

3.1. Method 1: Detection of mRNA Expression in a Single Cell by Direct RT-PCR (1)

1. Detach cells cultured at a sparse density from the culture dish by scraping them with a rubber policeman.
2. Isolate single cells by drawing a vol of 0.5 µL by micropipet under an inverted light-microscope.
3. Add isolated cells to 9.5 µL of RT premixture containing reverse transcription buffer (PE Biosystems, 1×), 0.2 m*M* deoxynucleotide triphosphates (dNTPs), 1 m*M* oligo(dT) primer, 5 m*M* random hexamer primers, and 1.25 units of the heat-stable rTth DNA polymerase (*see* **Notes 2** and **3**).
4. Heat the cells at 100°C for 1 min (to destroy the cell and release its RNA).
5. Incubate the mixture at 70°C for 15 min and then at 42°C for 5 min to perform reverse transcription reaction.
6. Add 40 µL of chelating buffer containing 35 pmol each of primers GAPDH(S1) and GAPDH(AS1) (*see* **Note 1** for the primer sequences and **Note 3**).
7. Conduct a 60-cycle PCR amplification with parameters of 94°C for 1 min, 60°C for 2 min and 72°C for 3 min, which is followed by a final extension at 72°C for 15 min (*see* **Note 4**).
8. Analyze the PCR products by electrophoresis on an agarose gel (1.2–1.8% agarose gel is used). **Figure 1** shows a representative result obtained by this method (*see* **Note 5**).

3.2. Method 2: Detection of Alternative Splicing in a Single Cell by Two step RT-PCR (2)

This method is composed of two phases. The first (**steps 1–9**) is RT-PCR, which is similar to method 1 (**Subheading 3.1.**) with some changes. The most important change is that the cDNA is synthesized using the sequence-specific

Fig. 1. Detection of GAPDH mRNA by Method 1 (**Subheading 3.1.**) (modified with permission from **ref. *1***). Lane 1, 123-bp DNA ladder marker; lane 2 and 3, PCR products from different single cells.

anti-sense PCR primer. The second phase (**steps 10–13**) is PCR amplification with a nested primer pair, which is set in the first PCR product.

1. Detach cells cultured at a sparse density from the culture dish by scraping them with a rubber policeman.
2. Isolate single cells by drawing a vol of 1 µL by micropipet under an inverted light-microscope.
3. Add isolated cells to 9 µL of water containing 5 m*M* DTT and 2 units/µL of HPRI. Both DTT and HPRI prevent RNase digestion in RNA *(3)*.
4. Quickly freeze the cells on dry ice (the cells can be stored at –70°C until used).
5. Quickly thaw the single cells (10 µL) in a water bath. This freezing and thawing destroys the cells. Stand the tubes containing the cells in an ice bath soon after thawing.
6. Add 15 µL of RT-PCR mixture containing RT-PCR buffer (Life Technologies), final concentration=1×, 0.5 µL of MMLV RT and Taq DNA polymerase mixture (Life Technologies), and 0.5 pmol each of primers EDB(S1) and EDB(AS1), or GAPDH(S2) and GAPDH(AS2) (*see* **Note 1** for the primer sequences and **Note 3**). EDB is a region of the FN gene that contains introns that can be differentially spliced in the FN mRNA.
7. Perform RT at 55°C for 30 min.
8. Inactivate the RT by incubating at 94°C for 2 min. MMLV RT, but not Taq DNA polymerase, can be heat-inactivated, differing from the rTth polymerase outlines in **Subheading 3.1.**

9. Conduct a 40-cycle PCR amplification with parameters of 94°C for 30 s, 60°C for 30 s and 72°C for 2 min, followed by a final extension at 72°C for 5 min. This first-round PCR product can be stored at –20°C.
10. Take two aliquots (3 µL) of the RT-PCR product (*see* **Note 6**).
11. Add each aliquot to a PCR bead tube (Amersham-Pharmacia-Biotech) with 22 µL water containing 2 pmol of either primer pair EDB(S2) and EDB(AS2), or GAPDH(S3) and GAPDH(AS3). The PCR bead reconstitutes the reaction mixture to 10 m*M* Tris-HCl, pH 9.0, 50 m*M* KCl, 1.5 m*M* $MgCl_2$, 0.2 m*M* each of dATP, dCTP, dGTP, and dTTP, and 1.5 U of Taq DNA polymerase when a total of 25 µL of water is added.
12. Conduct the second step (nested) PCR amplification over 40 cycles (for GAPDH) or 55 cycles (for FN) with parameters of 94°C for 30 s, 60°C for 30 s and 72°C for 1 min, followed by a final extension at 72°C for 5 min.
13. Analyze the PCR products by electrophoresis on an agarose gel. **Figure 2** shows a representative result obtained by this method.

4. Notes

1. The primer sequences (S=sense primer and AS=anti-sense primer) used are as follows:

   GAPDH(S1) TTCATTGACCTCAACTACAT,
   GAPDH(AS1) GTGGCAGTGATGGCATGGAC,
   GAPDH(S2) TCTTCTTTTGCGTCGCCAGC,
   GAPDH(AS2) GAGGCAGGGATGATGTTCTG,
   GAPDH(S3) TGAAGGTCGGTGTCAACGGA,
   GAPDH(AS3) GATGGCATGGACTGTGGTCAT,
   EDB(S1) TGGTTATAGAATTACCACAACC,
   EDB(AS1) CAGAAAAGTCAATGCCAGTTG,
   EDB(S2) GGTCCATGCTGATCAGAGCTC,
   EDB(AS2) CAGGTGACACGCATGGTGTCTG. Sizes of PCR products from the primer pairs are as follows: GAPDH(S1)-GAPDH(AS1): 443 bp, GAPDH(S2)-GAPDH(AS2): 671 bp, GAPDH(S3)-GAPDH(AS3): 533 bp, EDB(S1)-EDB(AS1): 489 bp and 762 bp, EDB(S2)-EDB(AS2): 189 bp and 462 bp. GAPDH(S3) and GAPDH(AS3) primers are set inside GAPDH(S2)-GAPDH(AS2) product, and similarly, EDB(S2) and EDB(AS2) primers are designed inside EDB(S1)-EDB(AS1) product. In Method 1 (**Subheading 3.1.**), the primer pair for GAPDH (S1-AS1) can be changed to the primer pair used in Method 2 (**Subheading 3.2.**)(S3-AS3).
2. Because Mn^{2+} concentration is the critical factor for RT reaction and its optimal concentration is in a narrow range, it is very important to maintain its optimal concentration of 1 m*M*. The reaction time can be shortened as in Method 2 (**Subheading 3.2.**).
3. When performing Method 2, we found that the PCR primer concentration had to be kept low, because higher concentrations tended to give a smeared gel pattern after PCR. Also, the use of random primers in the cDNA synthesis yielded a

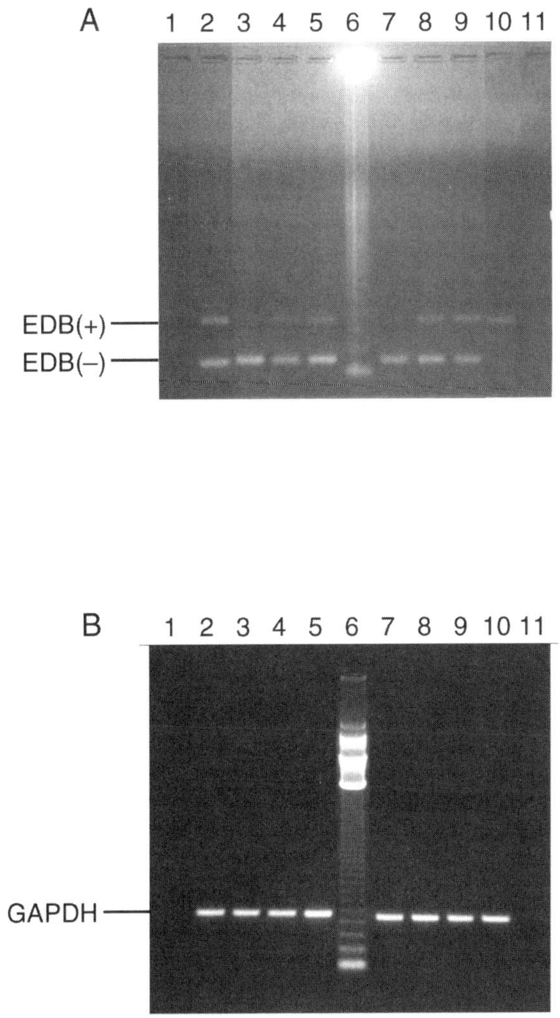

Fig. 2. Agarose gel analysis of PCR products obtained by Method 2 (**Subheading 3.2.**) (modified with permission from **ref.** *2*) **(A)** Analysis of the EDB site in FN. EDB(+) represents the 462-bp band and EDB(–) represents the 189-bp band, corresponding to two alternatively spliced variants of the FN mRNA. Lanes 1 and 11, negative controls incubated with RNase A and without primers, respectively; lanes 2–5 and 7–10, single cell analysis; lane 6, 123-bp DNA ladder marker. Lanes 3 and 7 show detection of only the 189-bp band, demonstrating the presence of only the EDB(–) splice variant in each cell; lane 10 shows only the 462-bp band, corresponding to the EDB(+) splice variant, in that particular cell. Other individual cells show the presence of both bands, indicating the simultaneous expression of both the EDB(–) and EDB(+) splice variants in these single cells (lanes 2, 4, 5, 8, and 9).**(B)** Analysis of GAPDH. GAPDH corresponds to the 533-bp band. Lanes 1 and 11, negative controls incubated without cell and without reverse transcriptase, respectively; lanes 2–10, same as in **(A)**.

smeared pattern of PCR products. Thus, when performing Method 1, one can probably obtain a better result if the PCR primers are used at a lower concentration, or if the oligo(dT) (or indeed, random primer) in the cDNA synthesis is replaced by a sequence-specific primer.

4. The yield of PCR product is highly dependent on the combination of available machine and tube; the cycle number can be varied accordingly.
5. In both methods, the reaction contains a whole cell, which means a contamination of genomic DNA. However, we could not detect any band of expected size which could be derived from genomic DNA. Thus, the contamination is negligible.
6. Using four aliquots of the first step RT-PCR product of Method 2, we simultaneously analyzed alternative splicing at three sites in FN and GAPDH control *(2)*.

References

1. Kumazaki, T., Hamada, K., and Mitsui, Y. (1994) Detection of mRNA expression in a single cell by direct RT-PCR. *Biotechniques* **16,** 1017–1019.
2. Kumazaki, T., Mitsui, Y., Hamada, K., Sumida, H., and Nishiyama, M. (1999) Detection of alternative splicing of fibronectin mRNA in a single cell. *J. Cell Sci.* **112,** 1449–1453.
3. Klebe, R. J., Grant, G. M., Grant, A. M., Garcia, M. A., Giambernardi, T. A., and Taylor, G. P. (1996) RT-PCR without RNA isolation. *Biotechniques* **21,** 1094–1100.

5

Nested RT-PCR

Sensitivity Controls are Essential to Determine the Biological Significance of Detected mRNA

Triona Goode, Wen-Zhe Ho, Terry O'Connor, Sandra Busteed, Steven D. Douglas, Fergus Shanahan, and Joe O'Connell

1. Introduction

Although standard reverse transcriptase-polymerase chain reaction (RT-PCR) is a remarkably sensitive technique, its sensitivity can be further increased by performing "nested" RT-PCR. This involves taking an aliquot of the product from the primary RT-PCR, and using it as a template for a secondary round of PCR amplification. To avoid further amplification of primer-dimer artifacts or nonspecific products generated in the primary PCR, a different set of primers is employed in the secondary PCR. These "nested" primers are internal to the primers used in the primary PCR, yielding a somewhat shorter amplicon (**Fig. 1**). Because their sequences are different from the primary set of primers, they will not amplify artifacts or nonspecific products generated in the primary PCR, enabling product specificity to be maintained over the high number of amplification cycles combined in the primary and secondary PCRs.

Nested RT-PCR is an extremely sensitive technique, and has been used in several applications for which maximum sensitivity is crucial. For example, nested RT-PCR is often used to detect RNA viruses, such as Human Immunodeficiency Virus (HIV) and Hepatitis C Virus (HCV), in serum samples in which the viral titer is frequently less than 100 viral genomes per mL of serum. Nested RT-PCR for tumor-associated mRNAs has been used to detect tumor cells in the circulation, often when there is less than one tumor cell per 10^6 white blood cells (WBC), or to detect micrometastases in bone marrow. Nested RT-PCR is therefore a powerful diagnostic tool to determine the presence or

From: *Methods in Molecular Biology, vol. 193: RT-PCR Protocols*
Edited by: J. O'Connell © Humana Press Inc., Totowa, NJ

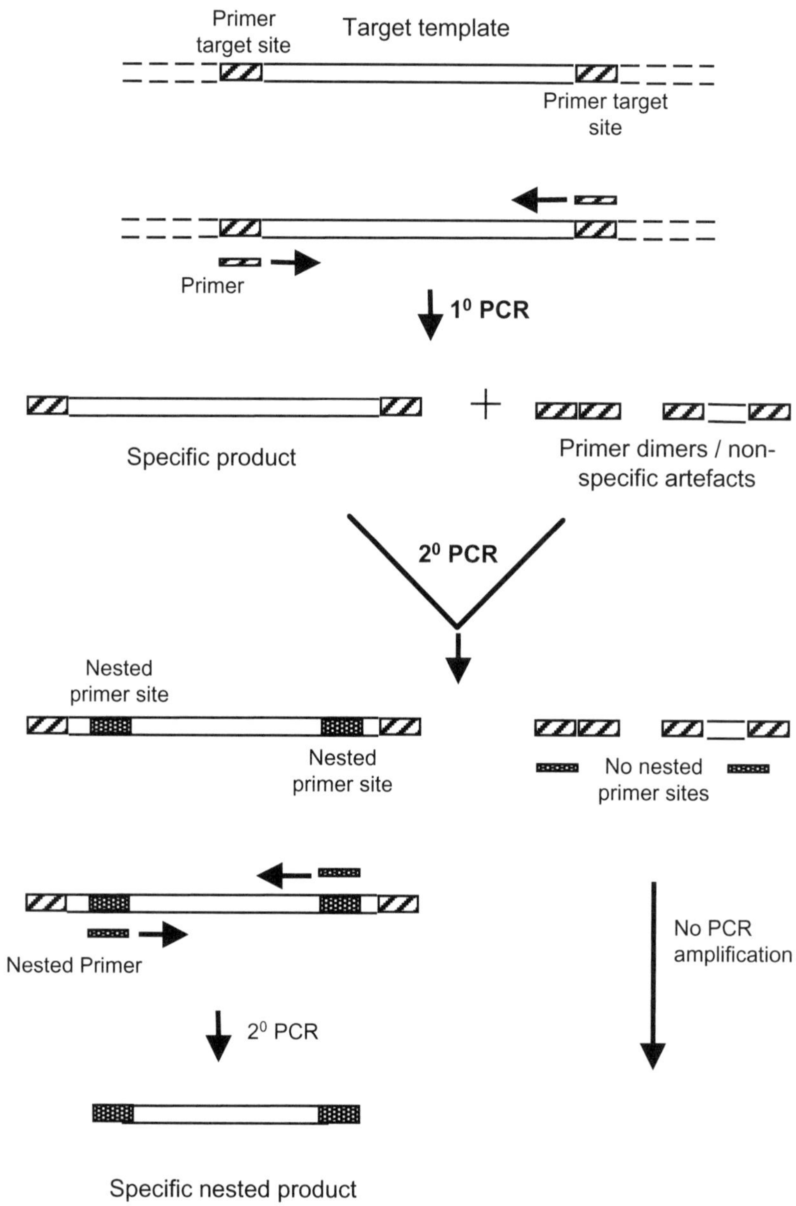

Fig. 1. Nested RT-PCR. The sensitivity of standard RT-PCR can be increased by performing a secondary, or "nested" PCR on an aliquot (usually 1%) of the products from the primary PCR. The secondary PCR uses a different set of primers, "nested" or internal to those used in the primary PCR. These primers will not find priming sites on any primer dimers or nonspecific artifacts generated in the primary PCR. Thus, the nested primers will only prime any specific product generated in the primary PCR, thus helping to maintain PCR specificity through the large number of cycles of the combined primary and secondary PCRs. Reprinted with permission from **ref. *(16)***.

absence of viruses or tumor cells with extreme sensitivity. However, in other research applications, the extreme sensitivity of nested RT-PCR can lead to overestimation of the biological significance of detected mRNA. Care is needed in the interpretation of results, and it is essential to define the level of sensitivity that is being achieved. In the present study, we demonstrate a specific application of nested RT-PCR, and demonstrate how the results were placed in the context of data from other techniques in order to properly interpret the nested RT-PCR results.

Substance P (SP) has been implicated in peripheral and mucosal neuroimmunoregulation. However, until recently, there was confusion regarding immunocyte expression of the receptor for SP, neurokinin-1 receptor (NK-1R), and specifically whether there was differential NK-1R expression in the mucosal vs the peripheral immune system. Previous studies of NK-1R expression by peripheral lymphoid cells have used different techniques, yielding conflicting results. Using relatively crude radioligand-binding assays, SP-binding sites have been identified on a subset of human T lymphocytes *(1)*, and on human monocytes/macrophages *(2)*. Recent nested RT-PCR studies have demonstrated NK-1R mRNA expression by human peripheral blood lymphocytes (PBL), monocytes, and monocyte-derived macrophages (MDM) *(3,4)*. Lipopolysaccharide (LPS) -activated murine macrophages have also been shown to express NK-1R mRNA *(5)*. However, others have reported the absence of SP-binding sites on human PBL *(6)*. A non-neurokinin receptor for SP has been identified on human monocytes *(7)*. SP has also been shown to activate human T cells receptor-independently *(8)*.

We directly examined the expression of NK-1R in human lamina propria mononuclear cells (LPMC) isolated from the colonic mucosa, peripheral-blood mononuclear cells (PBMC), PBL, monocytes, and MDM. When we analyzed peripheral lymphoid cells for NK-1R mRNA expression by RT-PCR, an apparent disparity of results also arose. Using standard RT-PCR, NK-1R mRNA expression was only evident in LPMC, but not in PBMC, PBL, monocytes, or MDM. This would suggest that NK-1R is expressed in mucosal and not peripheral lymphoid cells. Using sensitive nested RT-PCR, NK-1R mRNA expression was detected in all cell populations: LPMC, PBMC, PBL, monocytes, and MDM. This suggested that NK-1R expression in lymphoid cells was not restricted to mucosal sites, but also occurred in the periphery. In order to resolve this paradox, sensitivity controls, and additional analyses were initiated.

Here, we present techniques for the isolation of mucosal and peripheral lymphoid subsets for RT-PCR analysis, the basic nested RT-PCR technique for detection of NK-1R mRNA, an assessment of the sensitivities of standard vs nested NK-1R RT-PCR, and an example of the type of additional data needed to interpret nested RT-PCR results.

2. Materials

1. Cell lines and human cells: the NK-1R-positive IM-9 B lymphoblastoid cell line is available from the European Collection of Cell Cultures (ECACC), Salisbury, UK. PBMC are obtained from a sample of heparinized blood taken from a volunteer, and purified over Ficoll-Hypaque density gradients (Sigma Chemical Co., St. Louis, MO). LPMC are isolated from the mucosa of a surgically resected specimen of the human colon. It is important to ensure that ethical approval from the appropriate ethics committee covers such use of human tissue for research purposes.
2. Cell-culture media and conditions: IM-9 cells were cultured in Dulbecco's modified Eagle's medium (DMEM; Gibco-BRL, Grand Island, NY) supplemented with 10% fetal calf serum (FCS; Gibco-BRL), 2 m*M* L-glutamine (Gibco-BRL), and antibiotics: gentamicin, penicillin, streptomycin, and fungizone (Gibco-BRL). Cells were maintained at 37°C in a humidified atmosphere containing 5% CO_2. PBL were cultured in RPMI-1640 medium (Gibco-BRL) containing 10% FCS, with or without 1 µg/mL phytohemaglutinin (PHA) (Sigma) for 72 h. Two different activation procedures were employed to activate PBMC. PBMC (1×10^6/mL) were cultured with either PHA (3 µg/mL; Sigma) or PMA (10 ng/mL; Sigma) plus ionomycin (500 ng/mL; Sigma), in DMEM with 10% FCS at 37°C in a humidified 5% CO_2 atmosphere for 48 and 24 h, respectively. Freshly isolated monocytes were plated in 48-well culture plates at a density of 2.5×10^5 cells/well in DMEM containing 20% FCS. The total length of time in culture for fresh monocytes is no more than 48 h, and MDM refer to 7- to 10-day cultured monocytes in vitro. Monocyte and MDM viability was monitored by trypan blue exclusion and maintenance of cell adherence.
3. Medium for dissociation of epithelial cells: Chelating medium: calcium-magnesium-free Hanks balanced salt solution (HBSS) (Gibco-BRL) containing 1 m*M* ethylenediaminetetraacetic acid (EDTA), 50 µg/mL gentamicin, 100 U/mL penicillin, 100 µg/mL streptomycin, and 0.25 µg/mL fungizone (Gibco-BRL); enzymatic medium: 0.5 mg/mL collagenase and 1 mg/mL hyaluronidase (Sigma Chemical Co.) in RPMI-1640 containing 10% FCS, 2 m*M* L-glutamine, 100 U/mL penicillin, 100 µg/mL streptomycin, 0.25 µg/mL fungizone, and 50 µg/mL gentamicin (Gibco-BRL).
4. Primer design: PCR primers are designed using a program for selecting primer pairs, such as the DNASTAR Lasergene Primerselect program (DNASTAR Inc., Madison, WI). Primers are chosen that show no significant homology to any other genes in the European Molecular Biology Laboratories (EMBL) DNA sequence database (available on CD-ROM from DNASTAR Inc.). Primer pairs are chosen to span introns in their genomic sequences, thus ensuring mRNA-specific amplification.

   Standard, or primary, NK-1R PCR was performed using the following sense and anti-sense primers, respectively: TGACCGCTACCACGAGCAAGTCTC and ATAGTCGCCGGCGCTGATGAAG, which correspond to nucleotides 699–722 and 993–972 of human NK-1R cDNA respectively, and yield an amplification product size of 295 bp. Nested PCR for NK-1R utilized the following primers: TCTCTGCCAAGCGCAAGGTGGTCA (sense) and GAAGGCATGCTTG-AAGCCCAGACG (anti-sense) which correspond to nucleotides 719–742 and 960–937 of human NK-1R cDNA respectively, and yield an amplification product size of 242 bp.

Control β-actin PCR utilized the following primers: CCTTCCTGGGCATGGAG TCCTG (sense) and GGAGCAATGATCTTGA TCTTC (anti-sense) corresponding to nucleotides 794–815 and 995–975 of human β-actin cDNA, respectively.

5. Access to a recent version of the EMBL DNA sequence database (e.g., on CD-ROM from DNASTAR) to perform a homology search of the selected primers.
6. Access to automated DNA synthesis for primer synthesis.
7. RNA isolation: Prepare guanidine thiocyanate cell lysis buffer: 4.0 M guanidine thiocyanate (Sigma Chemical Co.), 25 m*M* sodium citrate, pH 7, 0.5% sarcosine; add 7.2 μL of β-mercaptoethanol per mL. For extraction, use molecular-biology-grade, water-saturated phenol (Sigma); ethanol (Sigma) is used for precipitation of the RNA, and 70% ethanol for washing the dried pellet. RNA pellets are redissolved in water pretreated with diethyl pyrocarbonate (DEPC; Sigma) to inactivate any contaminating ribonucleases. DEPC, which is highly toxic, is added at 0.1% to a bottle of distilled water in a fume-hood, and the lid is replaced. The water is left to incubate with the DEPC for several hours with occasional vigorous shaking to disperse the DEPC throughout the water. Finally, the DEPC-treated water is autoclaved. Heat breaks down DEPC to ethanol and carbon dioxide, so autoclaved DEPC-water is non toxic. (Several kits are also commercially available for quick RNA isolation (e.g., RNeasy® kit from Qiagen, Crawley, UK). Nucleic acid quantification strips can be used to quantify RNA (e.g. nucleic acid quantitation kit, Invitrogen BV, the Netherlands).
8. cDNA synthesis: Avian myeloblastoma virus (AMV) RT and its buffer, and the ribonuclease inhibitor RNasin, are obtained from Promega (Madison, WI); random hexanucleotide primers (Boehringer Mannheim, GmbH, Mannheim, Germany) are used at 125 n*M* in the cDNA synthesis.
9. PCR: 10 m*M* dNTP stocks, Taq DNA polymerase, and its buffer are obtained from Promega; a proofreading, heat-stable DNA polymerase such as UlTma DNA polymerase is obtained from Perkin-Elmer (Norwalk, CT).
10. Electrophoresis: molecular-biology-grade agarose (Promega); ethidium bromide solution (Sigma); *Hae*III-digested ϕX174 DNA size markers (Promega).

3. Methods

3.1. Cell Isolation and Culture

1. PBMC: Isolate PBMC from heparinized blood by centrifugation over Ficoll-Hypaque gradients (Sigma Chemical Co.). Recover cells and wash twice in DMEM (Gibco-BRL).
2. Monocytes: Monocytes can be isolated according to previously described techniques ***(10,11)***. Briefly, incubate mononuclear cells with DMEM in a 2% gelatin-coated flask for 45 min at 37°C, followed by removal of nonadherent

cells with DMEM. Detach adhered monocytes with EDTA. Following the initial purification, > 97% of the cells should be monocytes. Monocyte yield can be determined by immunofluorescence flow cytometry for nonspecific esterase using monoclonal antibody (mAb) against CD14 (Leu-M3), and low-density lipoprotein specific for monocytes and macrophages.

3. PBL: Collect nonadherent cells (i.e., PBL) from the gelatin-coated flasks, and wash 3× with phosphate-buffered saline (PBS).
4. LPMC: LPMC can be isolated from the colonic mucosa by the technique originally described by Bull and Bookman ***(12)***. The entire procedure is performed in 6 h without interruption. Wash the surgically resected human colonic specimen in saline, and remove any adherent debris. Using a forceps, strip the mucosa free from the underlying musculature and cut into small pieces (1 × 3 cm). To remove the epithelium, incubate the tissue in a shaking water bath at 37°C for 30 min in calcium-magnesium-free HBSS containing 1 m*M* ethylenediaminetetraacetic acid (EDTA) and 50 µg/mL gentamicin. Also add 100 U/mL penicillin, 100 µg/mL streptomycin, and 0.25 µg/mL fungizone as a 1% fungi-bact solution. Repeat the incubations every 30 min with fresh medium until the supernatant is free of epithelial cells. Mince the remaining tissue into smaller pieces and digest in a shaking water bath at 37°C for 1 h with 0.5 mg/mL collagenase and 1 mg/mL hyaluronidase (Sigma Chemical Co.) in RPMI-1640 containing 10% FCS, 2 m*M* L-glutamine, 1% fungi-bact solution, and 50 µg/mL gentamicin. Collect the supernatant and pellet the cells by centrifugation (500*g* for 5 min), resuspend in the same medium without the enzyme solution, and filter through a 60 µm nylon mesh to remove particulate material. Centrifuge the filtrate over a Ficoll-Hypaque density gradient. Recover LPMC, wash twice in DMEM, and resuspend in DMEM supplemented with 10% FCS and antibiotics as detailed above. LPMC isolations can be checked by immunofluorescence flow cytometry, and should contain >95% CD45 positivity.
5. Cell lines: As a positive control for NK-1R, culture IM-9B lymphoblastoid cells in DMEM supplemented with 10% FCS, 2 m*M* glutamine, and antibiotics. Maintain cells at 37°C in a humidified atmosphere containing 5% CO_2.
6. Activation of PBMC: Culture PBMC (1×10^6/mL) with PHA (3 µg/mL) in DMEM with 10% FCS at 37°C in a humidified 5% CO_2 atmosphere for 48 h.
7. Generation of MDM: Plate freshly isolated monocytes in 48-well culture plates at a density of 2.5×10^5 cells/well in DMEM containing 20% FCS. The total length of time in culture for fresh monocytes is no more than 48 h, and MDM refer to 7- to 10-d cultured monocytes in vitro. Monitor monocyte and MDM viability by trypan blue exclusion and assessment of cell adherence.

3.2. Detection of NK-1R mRNA Expression by Standard and Nested RT-PCR

1. Isolate total RNA from the various cell preps by first lysing in 0.5 mL of guanidine thiocyanate lysis buffer ***(13)***. Add 50 µL of 2 *M* sodium acetate, pH 4.0, 0.5 mL of water-saturated phenol, and 0.1 mL of a mixture of chloroform and isoamyl

alcohol (ratio of 49:1). Mix the tubes after addition of each reagent, and mix the final suspension vigorously. Leave the tube on ice for 15 min, then centrifuge at 10,000*g* for 20 min in a microfuge. Remove the RNA-containing aqueous phase to a new tube, and add 2–2.5 vol of absolute ethanol. Leave on ice for 15 min to precipitate the RNA. Collect the precipitate by centrifugation at 10,000 *g* for 30 min in a microfuge. Wash the pellet in 70% ethanol, dry, and redissolve the RNA in 10–20 µL of DEPC-treated water. Several kits are also commercially available for quick RNA isolation (*see* **Note 1**).

2. Synthesize cDNA using the AMV RT (Promega) and random hexanucleotide primers (Boehringer Mannheim, GmbH). Perform the cDNA synthesis in a final reaction vol of 30 µL containing: 1 U of AMV RT, 40 U of RNasin, 0.5 m*M* each of the four dNTPs, either 125 nM random hexanucleotide primers or 500 nM NK-1R-specific anti-sense primer GGATTTCATTTCCAGCCCCT, and approx 1 µg of total RNA. Incubate for 90 min at 42°C. Dilute the cDNA by adding 70 µL of T.E. (10 m*M* Tris-HCl, pH 8.0, 1 m*M* EDTA) and store at –20°C (*see* **Note 2**).
3. Standard or primary PCR. Perform NK-1R-specific PCR using 1% or 10% of the cDNA as template. The reaction should contain 1.5 m*M* $MgCl_2$, 50 µ*M* dNTPs, 0.1 µ*M* each primer and 1 U of Taq DNA polymerase (Promega Corp.) per 50 µL reaction. Perform negative controls by either omitting RT from cDNA synthesis or by omitting cDNA from the PCR amplifications. As a positive control, use RNA from cells that are known to abundantly express NK-1R mRNA: the IM-9B lymphoblastoid cell line. Perform "hot start" to increase the specificity of the PCR amplification. Use the following thermal cycling program: denaturation at 96°C for 15 s, annealing at 60°C for 30 s and extension at 72°C for 1.5 min for 45 cycles.
4. Nested PCR. Perform secondary or "nested" PCR using 1% of the amplification products obtained from the primary PCR above as template. Perform the secondary PCR amplification using identical reagent concentrations and thermal cycles as detailed above for the primary PCR, except use the nested primer pair and perform 35 cycles (*see* **Note 3**).
5. Control β-actin PCR: use identical reagent concentrations as described for the NK-1R PCR above, except use β-actin-specific primers and the following thermal cycles: denaturation at 96°C for 15 s, annealing at 55°C for 30 s, and extension at 72°C for 3 min for 35 cycles (*see* **Note 4**).
6. Analyze all PCR products by electrophoresis through 2% agarose gels and viewing under ultraviolet (UV) light after ethidium bromide staining (**Fig. 2**). Product specificities can be confirmed by restriction mapping or by DNA sequence analysis, e.g., using an ABI Prism 310 Genetic Analyzer (Perkin-Elmer).
7. Sensitivity controls. Set up a dilution series of a purified target template to determine the sensitivity limit of the nested RT-PCR assay. Ideally, the sensitivity of the primary and nested RT-PCR assays should be determined using a dilution series of a known quantity of an RNA transcript of the target. This controls for the efficiency of cDNA synthesis as well as PCR amplification (*see* **Note 5**). For the NK-1R PCRs, we used an NK-1R RNA transcript in vitro transcribed from a

plasmid clone that we had previously generated (*see* **Note 6**). Using this dilution series, we found that the primary RT-PCR assay was capable of detecting less than 50 NK-1R transcripts, and the nested RT-PCR assay was capable of detecting less than 10 NK-1R transcripts in the starting RNA template (*see* **Note 7**).

8. Quantitation of NK-1R mRNA expression by quantitative competitive RT-PCR (qcRT-PCR). This was performed using a qcRT-PCR assay that we had previously developed. In brief, development of the assay involved construction of a competitive internal RNA standard for NK-1R mRNA *(14)*. This standard was identical to the target NK-1R sequence, except for an internal deletion of 71 bp. Construction of the competitive standard involved cloning a NK-1R PCR product (324 bp) into pBluescript (Stratagene, La Jolla, CA) at the *Eco*RV site. This NK-1R PCR product was obtained using the following sense and anti-sense primers, respectively: GACTCCTCTGACCGCTACCA and GGATTTCATTTCCA GCCCCT corresponding to nucleotides 691–710 and 1014–995 of the human NK-1R cDNA, respectively. Orientation of the PCR product insert within the plasmid was determined by restriction mapping with *Hin*fI. Digestion at the *Bsr*GI and *Bgl*II unique restriction sites within the cloned insert resulted in the deletion of a 71-bp fragment. Sticky ends of the plasmid were then filled in by Klenow DNA polymerase (Promega), and blunt-ended recircularization of the plasmid was performed using T4 DNA ligase (Promega). Following propagation in *E. coli*, the deleted recombinant plasmid was subjected to in vitro transcription with T3 RNA polymerase (Promega) to synthesize deleted sense RNA transcripts (257 bp) for use as a competitive standard in qcRT-PCR.

   For qcRT-PCR, varying amounts of RNA standard transcripts of known concentration were spiked into aliquoted target RNA samples, and the mixtures were then subjected to RT-PCR, as described above (*see* **step 3**). As the internal standard is spiked in at the cDNA synthesis step, competition between target and standard occurs during both reverse transcription and PCR amplification. Equivalence of PCR products occurs when target and standard templates are present in equal initial concentration, permitting quantitation of the target template. Equivalence was determined as the point at which target and competitive standard PCR products were of equal band intensity **(Fig. 3)**. Results for NK-1R mRNA quantitation are expressed as number of NK-1R mRNA molecules per microgram of total RNA isolated. Total RNA was quantified using a nucleic acid quantitation kit (Invitrogen, Leek, The Netherlands). The assay was sufficiently sensitive to quantitate 100 NK-1R mRNA copies (*see* **Note 8**). This is equivalent to 10^3 NK-1R mRNA transcripts/μg RNA-isolated.

9. Immunofluorescence flow-cytometric detection of NK-1R protein. A polyclonal antibody specific for the extracellular N-terminus of the human NK-1R enabled flow cytometric analysis of NK-1R protein expression levels in PBMC and LPMC. The NK-1R antibody was raised in rabbits against a synthetic peptide (MDNVLPVDSDLSP) corresponding to the extracellular N-terminal amino acids 1–13 of human NK-1R. The IgG fraction was affinity purified on an AH-Sepharose column to which the peptide had been coupled. Antibody specificity

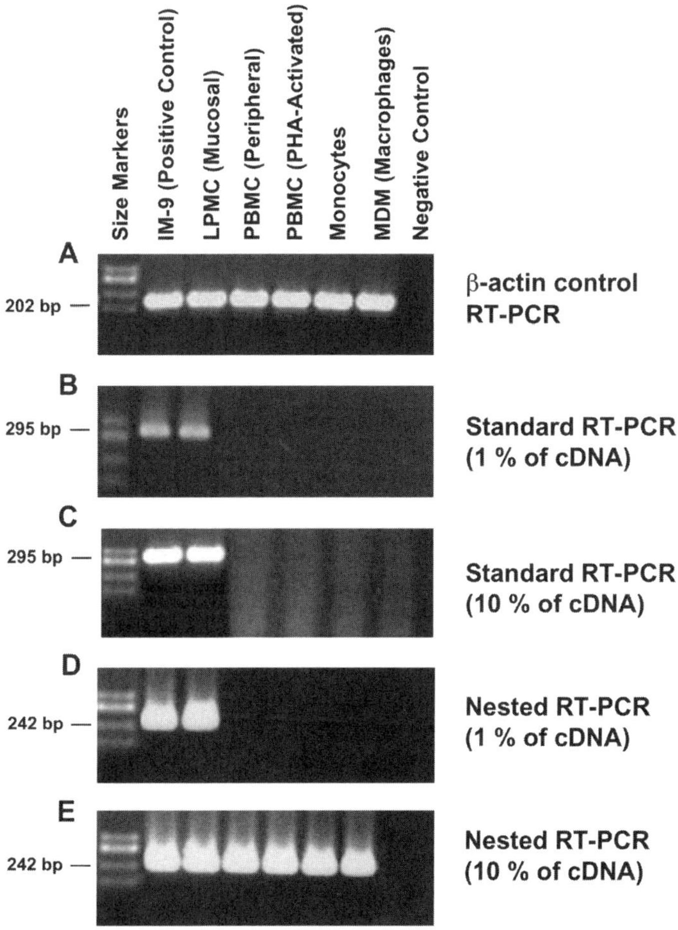

Fig. 2. NK-1R mRNA expression was analyzed by RT-PCR of equalized input RNA from IM-9 cells, LPMC, resting PBMC, PBMC activated with PHA, monocytes, MDM, and a negative control. *Hae*III-digested ϕX174 DNA size markers were used. Gel **A** shows mRNA specific amplification products (202 bp) for β-actin to control for RT-PCR amplification efficiency. Gels **B** and **C** show mRNA-specific amplification products (295 bp) from primary-round RT-PCR for NK-1R using 1% **(B)** and 10% **(C)** of cDNA. Gels **D** and **E** show mRNA-specific amplification products for NK-1R (242 bp) from secondary-round nested RT-PCR using 1% **(D)** and 10% **(E)** of cDNA. Reprinted with permission from **ref.** ***(16)***.

was confirmed by extensive radioimmunoassays and Western blotting. The fluorescence profiles of NK-1R expression in PBMC, and in LPMC are shown in **Fig. 4.** NK-1R immunofluorescence was not detected in PBMC, but was detected in LPMC preparations.

NCI-FREDERICK-SCIENTIFIC LIBRARY

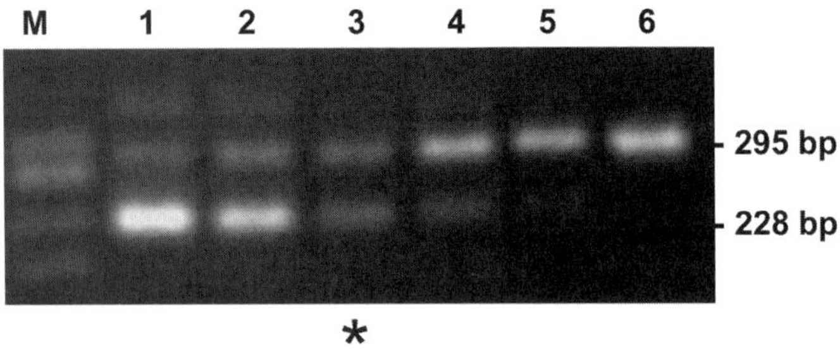

Fig. 3. Quantitation of NK-1R mRNA expression in a representative LPMC isolation. Competitive standard transcripts were spiked into the aliquoted LPMC RNA sample at concentrations ranging from 4.0×10^5 to 4.1×10^3 (2.5-fold dilution series) followed by NK-1R mRNA-specific RT-PCR (lanes 1–6). *Hae*III-digested ϕX174 DNA size markers were used (lane M). Equivalence (*) is seen at 6.4×10^4 NK-1R mRNA molecules in which target (295 bp) and competitive standard (228 bp) RT-PCR products are of equal band intensity. When adjusted for the amount of total RNA in the assay, this represents a level of expression of 7.5×10^5 NK-1R mRNA transcripts/μg total RNA isolated, or approx 7.5 NK-1R mRNA copies per cell. NK-1R mRNA was undetectable in PBMC using this assay (not shown). Reprinted with permission from **ref.** ***(16)***.

3.3. Analysis of Results

3.3.1. Quantitative Assessment of Nested RT-PCR Results

1. Using standard single-round RT-PCR, NK-1R mRNA expression was detected in the IM-9 positive control cells, and in LPMC, but not in PBMC, PBL, monocytes, or MDM **(Fig. 2B)**. When the input cDNA concentration in the PCR amplifications was increased from 1% to 10%, IM-9 cells and LPMC yielded a stronger band for the NK-1R PCR product, but PBMC, PBL, monocytes, and MDM remained negative **(Fig. 2C)**.
2. Using nested RT-PCR, when the input cDNA concentration was 1%, the results were identical to those achieved by standard RT-PCR: IM-9 cells and LPMC were positive for NK-1R mRNA, yet PBMC, PBL, monocytes, and MDM were negative **(Fig. 2D)**. The positive IM-9 and LPMC bands were stronger than for standard RT-PCR, demonstrating the greater sensitivity of the nested RT-PCR. However, when the input cDNA was increased to 10%, nested RT-PCR demonstrated NK-1R mRNA in all samples: IM-9 cells, LPMC, PBMC, PBL, monocytes, and MDM **(Fig. 2E)**.
3. Using the defined sensitivities of our RT-PCR assays, we were able to estimate the number of NK-1R mRNA transcripts expressed per cell in the different lymphoid cell isolations. The RNA preps used in these experiments were each obtained from approx 5×10^6 cells. We used one-fifth of the RNA preps for cDNA synthesis (equivalent to 1×10^6 cells), and either 1% (equivalent to 1×10^4 cells) or

10% (equivalent to 1×10^5, cells) of the cDNA for nested PCR. When the starting cDNA concentration was 1%, nested PCR did not yield a NK-1R-specific PCR product in PBMC, PBL, monocytes, or MDM. However, a NK-1R PCR product was evident in these samples when 10% cDNA was used. Since our nested RT-PCR assay can detect less than 10 NK-1R RNA transcripts of template RNA, what we are detecting is therefore less than 10 NK-1R mRNA transcripts per 10^4–10^5 cells in the PBMC, PBL, monocyte, and MDM samples. The only variable we have not accounted for is the % yield of RNA from the cells (it is difficult to estimate how much RNA a particular cell could contain). Even if the recovery of RNA was only 50%, this detection would still reflect single NK-1R mRNA transcripts in 500–5000 cells.

3.3.2. Additional Data To Support Nested RT-PCR Results

1. For the present study, we employed additional techniques to confirm our nested RT-PCR findings. We quantified NK-1R mRNA expression using a qcRT-PCR assay that we had previously developed **(Fig. 3)**. Quantitation of NK-1R mRNA expression levels in several LPMC isolations revealed a mean level of expression of $7.5 \times 10^5 \pm 2.2 \times 10^5$ NK-1R mRNA transcripts/µg total RNA ($n = 7$). Since 1 µg cellular RNA was the average RNA yield from approx 10^5 cells, this level of expression corresponds to approx 7.5 ± 2.2 NK-1R mRNA transcripts per LPMC. PBMC were found to be negative for NK-1R mRNA expression by qcRT-PCR. Since the sensitivity of the qcRT-PCR assay was calculated to be 10^3 NK-1R transcripts/µg RNA or 1 transcript/1,000 cells, we can further estimate that the level of NK-1R mRNA expression detected in PBMC by nested RT-PCR (10% cDNA) corresponds to 1 NK-1R transcript in >1000 cells.
2. Finally, we determined the level of NK-1R protein using immunofluorescence flow cytometry. LPMC showed a strong immunofluorescence peak for NK-1R, whereas NK-1R protein was undetectable in PBMC **(Fig. 4)**.

3.4. Conclusion

Our results confirm that human NK-1R is differentially expressed by the lymphoid cells of the periphery and the mucosa, and is more likely to have biological significance at the mucosal level. This finding would have been missed had we relied solely on nested RT-PCR as our method for NK-1R detection. The extremely low level of NK-1R detected in peripheral lymphoid cell preps could represent intermucosal traffic involving the transient presence of circulating mucosal homing cells ***(15)***. We conclude that NK-1R is a marker of mucosal rather than peripheral lymphoid cells ***(16)***. The finding supports the theory that SP plays a specific role in mucosal immunoregulation. SP therefore probably contributes to mucosal inflammation in pathological conditions such as inflammatory bowel disease (IBD). We have recently found increased levels of NK-1R mRNA in colonic biopsies from sites of active inflammation in IBD ***(17)***.

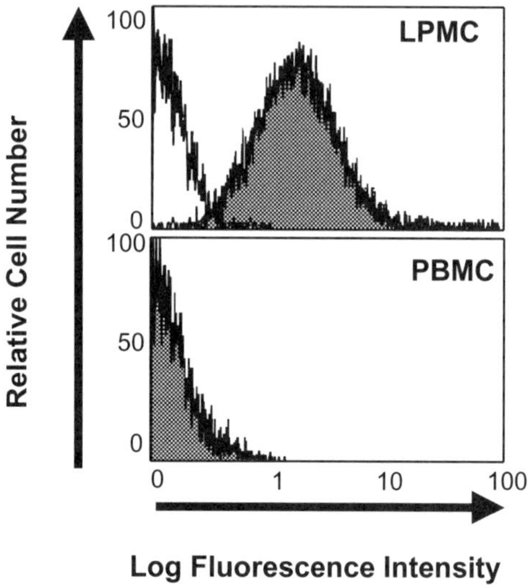

Fig. 4. Immunofluorescence profiles of NK-1R expression in PBMC and LPMC (as indicated). Strongly positive NK-1R antibody staining (shaded profile) relative to control antibody staining (open profile) is evident in LPMC, but not PBMC. This confirms our conclusions from the standard and nested RT-PCR results—that NK-1R is only expressed at a significant level in mucosal rather than peripheral lymphoid cells. Data are representative of three experiments.

Nested RT-PCR is an exquisitely sensitive technique, capable of detecting less than 10 transcripts of an RNA template. This extreme sensitivity must be factored into the interpretation of nested RT-PCR results, in order to avoid overestimation of the level of mRNA detected. Sensitivity controls are essential, and supportive data from additional techniques, such as quantitative RT-PCR or measurement of the protein level, are advisable.

4. Notes

1. Several kits are also commercially available for quick RNA isolation (e.g., RNeasy® kit from Qiagen, Crawley, UK).
2. Either random hexamers or a gene-specific primer can be used in cDNA synthesis. The advantage of random primers is that all mRNAs are copied into cDNA. This enables the detection of numerous mRNAs by RT-PCR from a single cDNA prep. However, a gene-specific primer can increase the sensitivity of the RT-PCR by generating cDNA that is enriched for the specific target. Because the RT reaction is performed at 37°C, well below the optimal annealing temperature of the specific primer, some nonspecific priming is likely to occur. When making specific-primed cDNA, it is therefore advisable to use a primer different from and downstream from the anti-sense PCR primer. This "nesting" of the anti-sense

PCR primer relative to the RT primer avoids amplification of sequences nonspecifically primed during the RT step.

3. The best number of cycles to perform for the primary and secondary PCR is somewhat arbitrary, and should probably be determined empirically. To maximize the sensitivity of the NK-1R PCR, we performed 45 cycles of the primary PCR, and 35 cycles of the secondary PCR. Usually, 25–30 cycles of both primary and secondary PCRs may be sufficient.
4. Because of the abundance of its mRNA (estimated to comprise about 10% of all mRNA in mammalian cells), fewer cycles are necessary for the control β-actin RT-PCR, i.e., 35 cycles or less.
5. Although an RNA control template is ideal for testing the sensitivity of the RT as well as the PCR phase in RT-PCR, a DNA template is more easily obtained, simply by purifying the PCR-amplified amplicon (e.g., using the QIAquick® PCR purification kit from Qiagen). After quantification, the purified amplicon can be used as a control at the PCR stage of RT-PCR. Although it will not control for variations in the RT step, this is not a serious limitation, since there is no amplification in the RT step, and the efficiency does not vary greatly between RNA samples prepared by the same isolation protocol. In most cases for which the efficiency of RT reactions has been measured by incorporation of ^{32}P-labeled nucleotides into cDNA, it is usually found that RT-catalyzed conversion of mRNA to cDNA occurs with an efficiency in the range of 10–20%. Therefore, a 10% conversion rate can be assumed as a conservative estimate.
6. If a plasmid clone of the target cDNA is not available, in vitro transcription can be used as an alternative method for generating an RNA copy of the target amplicon. To enable in vitro transcription, a promoter sequence for an RNA polymerase, such as the T7 polymerase, must be added to the 3' end of the amplicon. This can be done by synthesizing a composite primer with the T7 promoter sequence added onto the 5' end of the gene-specific anti-sense primer used in the RT-PCR. By generating a PCR amplicon with this primer and the regular sense PCR primer, the amplicon will contain the T7-promoter sequence added at its 3' end. Following purification, the amplicon serves as a template in a T7 in vitro transcription reaction, to generate an RNA copy of the sequence targeted for RT-PCR. DNA is eliminated by digestion with RNase-free DNase, and the RNA can be used as a control template for RT-PCR.
7. The following formulas are useful for calculation of the number of copies present in a particular amount of purified DNA or RNA:
   a. Spectrophotometrically, an A_{260} 1.0 is equivalent to approx 40 μg/mL of DNA or RNA.
   b. The mol wt of a DNA molecule, M.W.= number of nucleotides × 340 (average nucleotide M.W.) × 2 (number of strands).
   c. The mol wt of an RNA molecule, M.W.= number of nucleotides × 340 (average nucleotide M.W.) × (single strand).
   d. Number of moles = weight (g)/M.W.
   e. Number of molecules = number of moles × 6.023 × 10^{23} (Avogadro's number).

8. With qcRT-PCR, because two templates (target and competitor) compete for PCR amplification, the overall sensitivity is usually somewhat less than conventional single-target PCR.
   This can sometimes be compensated by performing a higher number of cycles to achieve good amplification of both templates—especially when the target copy number is very low and at the lower limit of detection.

Acknowledgments

This work was supported by the Health Research Board of Ireland, The Wellcome Trust (to Joe O'Connell), US NIH grant MH49981 (to Steven D. Douglas), and US NIH grant DA11887 (to Wen-Zhe Ho).

References

1. Payan, D.G., Brewster, D.R., Missirian-Bastian, A., and Goetzl, E.J. (1984) Substance P recognition by a subset of human T lymphocytes. *J. Clin. Invest.* **74,** 1532–1539.
2. Lucey, D.R., Novak, J.M., Polonis, V.R., Liu, Y., and Gartner, S. (1994) Characterization of substance P binding to human monocytes/macrophages. *Clin. Diagn. Lab. Immunol.* **1,** 330–335.
3. Ho, W.-Z., Lai, J.-P., Zhu, X.-H., Uvaydova, M., and Douglas, S.D. (1997) Human monocytes and macrophages express substance P and neurokinin-1 receptor. *J. Immunol.* **159,** 5654–5660.
4. Lai, J.-P., Douglas, S.D., and Ho, W.-Z. (1998) Human lymphocytes express substance P and it's receptor. *J. Neuroimmunol.* **86,** 80–86.
5. Bost, K.L. (1995) Quantification of macrophage-derived substance P receptor mRNA using competitive polymerase chain reaction, in *The Brain: Immune Axis and Substance Abuse,* Sharp, B., ed., Plenum Press, New York, pp. 219–223.
6. Roberts, A.I., Taunk, J., and Ebert, C. (1992) Human lymphocytes lack substance P receptors. *Cell. Immunol.* **141,** 457–465.
7. Kavelaars, A., Broeke, D., Jeurissen, F., Kardux, J., Meijer, A., Franklin, R., et al. (1994) Activation of human monocytes via a non-neurokinin substance P receptor that is coupled to Gi protein, calcium, phospholipase D, MAP kinase, and IL-6 production. *J. Immunol.* **153,** 3691–3699.
8. Kavelaars, A., Jeurissen, F., von Frijtag Drabbe Kunzel, J., van Roijen, J.H., Rijkers, G.T., and Heijnen, C.J. (1993) Substance P induces a rise in intracellular calcium concentration in human T lymphocytes in vitro: evidence of a receptor-independent mechanism. *J. Neuroimmunol.* **42,** 61–70.
9. Boyum, A. (1968) Isolation of mononuclear cells and granulocytes from human blood. *Scand. J. Clin. Lab. Invest.* **21,** 77–89.
10. Hassan, N.F., Campbell, D.E., and Douglas, S.D. (1986) Purification of human monocytes on gelatin-coated surfaces. *J. Immunol. Method.* **95,** 273–276.
11. Hassan, N.F., Cutilli, J.R., and Douglas, S.D. (1990) Isolation of highly purified human blood monocytes for in vitro HIV-1 infection studies of monocytes/macrophages. *J. Immunol. Methods* **130,** 283–285.

12. Bull, D.M., and Bookman, M.A. (1977) Isolation and functional characterization of human intestinal mucosal lymphoid cells. *J. Clin. Invest.* **59,** 966–974.
13. Chomczynski, P., and Saachi, N. (1987) Single-step method of RNA isolation by guanidium thiocyanate-phenol-chloroform extraction. *Anal. Biochem.* **162,** 156–159.
14. O'Connell, J., Goode, T., and Shanahan, F. (1998) Quantitative measurement of mRNA expression by competitive RT-PCR. *Methods Mol. Biol.* **92,** 183–193.
15. Shanahan, F. (1994) The intestinal immune system, in *Physiology of the Gastrointestinal Tract* Johnson, L.R., ed., Raven Press, New York, pp. 643–684.
16. Goode, T., O'Connell, J., Ho, W.-Z., O'Sullivan, G.C., Collins, J.K., Douglas, S.D., et al. (2000) Differential expression of neurokinin-1 receptor by mucosal and peripheral lymphoid cells. *Clin. Diagn. Lab. Immunol.* **7,** 371–376.
17. Goode, T., O'Connell, J., Anton, P., Wong, H., Reeve, J., O'Sullivan, G.C., et al. (2000) Neurokinin-1 receptor expression in inflammatory bowel disease: molecular quantitation and localisation. *Gut* **47,** 387–396.

III

Quantitative RT-PCR

6

Quantitative RT-PCR

A Review of Current Methodologies

Caroline Joyce

1. Introduction

The powerful amplification potential of PCR has assured its use in the detection of low-abundance mRNA in cells and tissues. RT-PCR is currently the most sensitive technique available for mRNA detection and quantification. It can accurately quantify genes present at only a few hundred copies per sample, and has become the method of choice for the examination of gene expression. The technique consists of two parts: synthesis of cDNA from RNA by reverse transcription, and amplification of a specific cDNA by polymerase chain reaction (PCR). The method requires very little RNA and differs from Northern blotting because it is somewhat tolerant of degraded RNA, as long as the RNA is intact within the region of interest.

However, quantification of mRNA by PCR is difficult because small variations in amplification efficiencies among sample tubes can lead to substantial differences in product yield, thereby rendering direct comparisons between samples invalid. Also, the kinetics of PCR product formation are only linear at relatively early cycles, tapering off to a plateau at later stages. Thus, the amount of product obtained does not necessarily reflect the amount of starting template. The development of protocols for quantitative RT-PCR has relied on internal standards to monitor the efficiency of the RT-PCR in different reaction tubes. Technically, the two most serious limitations to routine and successful application of competitive quantitative PCR are ready access to competitive internal standards and efficient methods for accurate quantitative analysis of the data.

From: *Methods in Molecular Biology, vol. 193: RT-PCR Protocols*
Edited by: J. O'Connell © Humana Press Inc., Totowa, NJ

For the detection of PCR products using either quantitative or semi-quantitative methods, several detection techniques are available. Conventional methods include gel electrophoresis followed by densitometry and radioactive labeling with detection by autoradiography, and more recently, fluorescent labeling and analysis on an automated DNA sequencer. To achieve more automated quantitative results, methods that involve capturing PCR products on solid supports followed by quantitation of the captured material have been developed. These methods achieve sensitivities in the attomole range and involve the use of enzyme-labeled probes, followed by colorimetric, chemiluminescent, or fluorescent detection.

This chapter reviews current methodologies, focusing on technical aspects of competitive PCR such as the construction of competitive templates, various PCR strategies, and modes of detecting PCR products. The principles and applications of each method are discussed, as well as the advantages and limitations associated with them.

2. Qualitative RT-PCR

Qualitative or semi-quantitative RT-PCR monitors relative changes in nucleic acid targets. It compares transcript abundance across multiple samples, using a co-amplified control gene for sample normalization. Normalization is required to account for tube-to-tube differences caused by variable RNA quality or RT efficiency, inaccurate quantification, or pipetting. Internal control and gene-specific primers are multiplexed in the same RT-PCR reaction and must be compatible (i.e., they must not hybridize to each other or produce additional products). Samples must be analyzed during the linear range of amplification for both the gene of interest and the internal control. This often occurs within 20 cycles for abundant messages. Results are then expressed as ratios of the gene-specific signal to the internal control in each sample.

2.1. External Standards

Constitutively expressed housekeeping genes of high abundance, such as β-Sctin, GAPDH, and β_2-microglobulin, are used as control genes for qualitative RT-PCR. These genes are usually expressed at moderate levels, making them easier to detect, but their expression level is not always constant. An ideal control gene is one that is easily detectable, with expression that does not vary during the cell cycle, between cell types, or in response to experimental treatments.

The expression of β-actin and GADPH is often affected by experimental treatments, stage of development, and cell type ***(1–5)***. The presence of pseudogenes for β-actin and GADPH has also limited their use as external controls. Normally, primers designed for RT-PCR span introns in the genomic

DNA sequence, so that products derived from mRNA are clearly distinguishable from the larger products derived from the genomic DNA that frequently contaminates RNA preps. The human genome contains processed pseudogenes for certain genes; these are non-expressed homologues of the authentic gene minus the introns, and therefore yield PCR products of identical size to mRNA-derived products. However, a recent publication reported the use of primers for β-actin, which do not co-amplify processed pseudogenes, and provide a more accurate measurement of β-actin levels ***(6)***.

The least variant of the potential external controls are ribosomal RNAs (rRNA). Although mRNA may vary in expression with cell type, developmental stage, or experimental conditions, cells typically maintain a constant level of rRNA. Since rRNA make up 80% of the total RNA, the use of rRNA as an external control is preferred for most experiments (excluding poly-A-selected mRNA samples, since rRNA does not contain a poly-A tail), and 18sRNA internal standards are commercially available from Ambion Ltd.

3. Quantitative RT-PCR

Quantitative competitive (qc) RT-PCR measures the absolute amount (i.e., copy number) of a specific transcript in a sample. Although the amount of product formed is easy to determine, it is difficult to determine the initial copy number of the target molecule because the overall efficiency of the RT-PCR for each sample is unknown, and the kinetics of PCR are nonlinear. This problem has been addressed by co-amplifying the molecule of interest with an internal standard, bearing identical sites for primer annealing, but distinguishable from the target via a difference either in sequence or size. The internal standard competes directly with the target template for amplification during RT-PCR. Thus, equal products are obtained from target and standard when both are present at identical initial levels. Using titration with known quantities of internal standard spiked in at the beginning of the RT-PCR, the level of target template in the sample can be determined.

In order to quantify the relative amount of PCR products derived from the target and internal standard, they are distinguished from each other, either directly according to their size (if their sizes differ), or after specifically cutting one of the two species using a restriction enzyme (a standard can be constructed with an identical size to the target by altering a restriction site in the target template). However, if the relevant PCR product is not cut completely by the restriction enzyme digestion, a false quantification will result. The amount of mRNA is quantified by titration of an unknown amount of target template against a dilution series of known amounts of the standard. An equivalence point is reached when both are amplified with the same efficiency. The size difference between target and standard should be kept as small as possible because of the inverse exponential relationship between template size and amplification efficiency. This ensures that both competitor and target template are subjected to similar PCR kinetics, and thus allows more accurate quantification.

An important consideration in RNA quantification is interference from contaminating genomic DNA, as previously mentioned. This can be eliminated by prior digestion with DNase I, or as already described, by the use of intron overlapping primers for RT-PCR. If DNA contamination is significant, the contaminating DNA template will also compete with the internal standard during amplification, so that quantification of the mRNA may be skewed.

Several protocols exist in which the RNA is first reverse-transcribed and the resulting cDNA is then co-amplified with a competitive DNA standard template. However, quantification by this method does not take into account the variability of the reverse transcription step, and such a procedure is unsuitable when exact quantification is desired. Competitive RNA standards, however, can be introduced into the test RNA sample prior to the RT step, and therefore control for variations in efficiency throughout the RT-PCR.

3.1. Internal Standards

Competitive RNA standards may be generated by cloning the endogenous target into a suitable plasmid vector, which has promoters for the T7 and the T3 RNA polymerases flanking the insertion site. Using an appropriate pair of restriction enzymes, a suitably sized fragment ($<$ 30% of the target) is then excised from the cloned target sequence and a deletion clone of the PCR target is thus generated. Following determination of the orientation of the cloned insert relative to the T7 and T3 promoters on the plasmid, the appropriate polymerase is used to transcribe sense RNA copies in vitro from the deleted insert. This RNA standard is then quantified and used as an internal competitive standard in the RT-PCR *(7)*.

Use of a DNA standard that shares the same primer template sequence, but contains a completely different intervening sequence, is termed a MIMIC, or heterologous competitor fragment. The amplification efficiencies of these competitors must be equal or very similar to that of the target. They may be generated by amplifying fragments from a different species using the human primers and a low annealing stringency PCR ***(8)*** or by ligating the primer template sequences to a non-homologous DNA fragment ***(9)***. PCR MIMICS are available commercially from Clontech Ltd. These are generated by simple PCR amplification, and involve the use of two composite primers, which result in the incorporation of the target-specific primer sequences into a heterologous DNA fragment ***(10)***. PCR MIMICs are designed to produce PCR products that

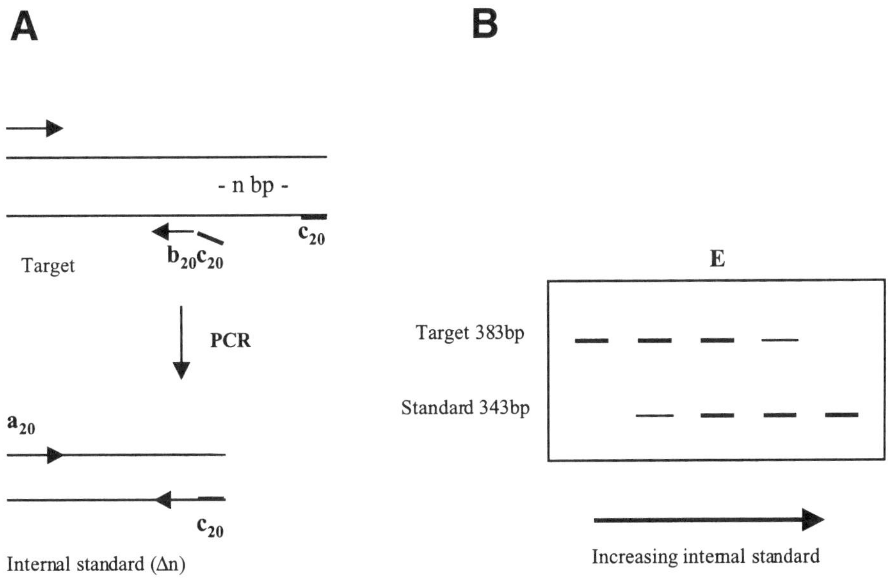

Fig. 1. Quantitative Competitive PCR. **(A)** Generation of an internal standard using composite primers. **(B)** Use of the internal standard for competitive PCR. The equivalence point (E) occurs when target and standard are present in equal starting concentration, thus permitting quantification of the target template.

are 10% larger or smaller than the endogenous target to allow efficient separation and analysis by gel electrophoresis.

Composite primers have also been used to generate homologous DNA and RNA competitor standards *(11)*. The composite primer is composed of two sections; the 5' portion is identical to one of the primers used to amplify the target template, but the 3' portion anneals to an internal site within the target template **(Fig. 1)**. After amplification of the target template with this composite primer and the "regular" primer from the opposite end, a truncated product is generated, which has identical terminal primer sites to the full-length target template. This product can be purified for subsequent use as an internal DNA standard, or cloned to facilitate the generation of an internal RNA standard. This standard will have complete sequence homology with the target template over much of its length, will have similar length, and should therefore be amplified with similar efficiency to the target template.

One potential problem with the use of competitor fragments that are homologous to the target and that differ in only a small number of base-pairs (bp) is that during later stages of PCR, when the concentration of products is high, heteroduplexes can form between the standard and target sequences. Particularly if the standard was constructed by engineering a deletion or insertion in the center of the target template, the resultant, bulky heteroduplex can gives rise to a third band of slower electrophoretic mobility than either the target or standard, and can thus complicate quantification. However, since each heteroduplex contains one strand of both the target and standard, the *ratio* of target to standard bands may not be adversely affected.

4. Detection and Quantification of RT-PCR Products

Several detection techniques are currently available for the detection of PCR products using either quantitative or semi-quantitative methods. These range from quantification of gel electrophoresis products by densitometry to colorimetric or fluorescent detection of immunologically captured PCR products.

4.1. Gel Electrophoresis and Densitometry

Quantitative detection of ethidium bromide-stained PCR product bands by densitometry after agarose gel electrophoresis is often used for product quantification. The fluorescent intensity of standards of known concentration (e.g., mol-wt markers) is used to estimate the concentration of the PCR product of interest. Image analysis software for densitometric analysis is freely available from the National Institute of Health (http:\\rsb.info.nih.gov/nih-image). However, it must be remembered that quantity of PCR product is only directly proportional to the amount of starting template during the linear phase of PCR, so a competitive standard will be needed for accurate quantification.

4.2. PCR-ELISA Digoxigenin Detection System

This PCR-enzyme-linked-immunosorbent assay (PCR-ELISA) system can be used for qualitative, semi-quantitative, and quantitative nonradioactive detection of PCR products using a microtiter-plate format. It involves the incorporation of a digoxigenin-labeled nucleotide, DIG-dUTP, during RT-PCR. The labeled PCR products are then bound to the streptavidin-coated surface of a microtitre plate by use of a biotin-labeled capture probe, which is designed to hybridize to an internal sequence of the PCR product. Alternatively, the PCR product may be immobilized by using one biotin-labeled PCR primer. The bound digoxigenin-labeled PCR products are then detected with an anti-DIG-enzyme conjugate and a colorimetric substrate such as ABTS™. The colorimetric signal allows quantitative determination of the amount of PCR product. This method is approx 100-fold more sensitive than conventional ethidium bromide-stained agarose gels, and thus frequently permits detection of product at an early cycle number in the PCR when the kinetics are still linear. Gene-specific RT-PCR ELISA kits are commercially available from Boehringer Mannheim,

Ltd. (Mannheim, Germany). These kits include primers and capture probes for specific transcripts and control genes to allow quantitative analysis.

4.3. RT-PCR and Laser-Induced Fluorescence (LIF)

In this procedure, RT-PCR products are generated by using fluorescently labeled primers, separated on polyacrylamide gels, and quantified using an automated laser fluorescent DNA sequencer. LIF is a highly sensitive methodology, and when combined with computer software, offers a means of automating post-PCR analysis *(12)*. The number of molecules in the initial sample is calculated from the ratio of individual PCR products and from the amount of internal standard as follows:

$$n\ (\text{target}) = N\ (\text{target})/N\ (\text{internal standard}) \times n\ (\text{internal standard})$$

N represents the number of molecules after amplification

and n = the initial number of molecules.

This method can use four distinct dyes, which allows either internal or external standards to be applied and analyzed within the same gel lane, thereby eliminating lane-to-lane artifacts. This approach is particularly suitable for the analysis of a large series of samples, and is applicable for clinical purposes such as the quantification of various DNA or RNA molecules in infectious diseases, leukemias, or for evaluation of gene dosage.

4.4. Real-Time Quantitative RT-PCR

Real-time PCR monitors PCR product formation continuously during the PCR reaction by means of a fluorogenic reporter. This reporter system may take the form of fluorescently labeled PCR primers or involve the use of a fluorescent dye such as Syber Green™, which interchelates double-stranded DNA. Both of these systems require prior optimization of PCR conditions to eliminate primer-dimer formation, which would interfere in subsequent analysis.

The TaqMan™ assay system uses a non-extendible fluorogenic probe to monitor PCR product formation. This probe is an oligonucleotide, dually labeled with a reporter dye covalently attached at the 5' end and a quencher dye covalently attached at the 3' end. The proximally located quencher absorbs the emission of the reporter dye as long as the probe is intact. During the exponential phase of the PCR reaction, the hybridized probe is hydrolyzed by the 3'-nuclease activity of the TaqMan DNA polymerase, separating the quencher from the reporter. This results in an increase in fluorescence emission of the reporter dye, which is quantitative for the initial amount of template *(13)*.

Because Real-Time PCR combines PCR amplification and product detection in one single step, the technique is very fast and easy to perform, compared to classical RT-PCR techniques. It allows the simultaneous quantification of a number of genes through the use of different flourophores.

4.5. Time-Resolved Fluorometry

Time-resolved fluorometry (TRF) is a rapid and sensitive alternative for detecting PCR products. TRF is based on the use of lanthanide chelates, the unique fluorescent properties of which eliminate background fluorescence in the detection phase and allow an increment in the sensitivity. Lanthanide labels used include europium (Eu), terbium (Tb^{3t}) and samarium (Sm^{3t}).

A biotin-labeled primer is used in the RT-PCR, and products are collected onto streptavidin-coated microtiter plates. After denaturation with alkali, single-stranded PCR products captured onto the wells are hybridized with a probe specific for the target sequence. Lanthanide labels are detached from the hybridized probes by incubation in enhancement solution and measured by a time-resolved fluorometer. This detection system has been used for the simultaneous detection of mRNAs for the cytokines IFN-γ and IL-4 ***(14)***. This method provides a sensitive, specific, rapid and nonisotopic alternative to conventional blotting and hybridization with radioactive probes. In addition, the multiplex fluorogenic dye detection facilitates relative quantification of various target mRNAs.

4.6. RT-PCR and Chemiluminescence

Chemiluminesence is the phenomenon observed when the vibronically excited product of a chemical reaction reverts to its ground state. The light emitted is measured in relative light units (RLUs), and is proportional to the amount of product in the sample. PCR biotinylated amplicons are captured on black streptavidin-coated microtiter plates, specifically hybridized with fluroescein-5'-isothiocyanate (FITC)-labeled oligonucleotide probes and detected by ELISA using a chemiluminescent substrate ***(15)***. The chemiluminescent substrates used in PCR detection (e.g., acridium esters) allow a wider dynamic range of measurements compared to colorimetric or fluorimetric detection. The method is also highly sensitive, and lends itself easily to automation ***(16)***.

4.7. RT-PCR and Electrochemiluminescence

Electrochemiluminescence (ECL) is achieved by an electrically initiated chemiluminescent reaction. The target is amplified using a biotinylated forward primer and a Tris (2,2'-bipyridine) ruthenium (II) (TBR)-labeled reverse primer. The amplification products are then captured on streptavidin-coated paramagnetic beads and quantified by measuring the ECL signal of the TBR label *(17)*. Results obtained are reproducible and accurate over a wide range of concentrations (3 orders of magnitude). Quantitative results can be obtained using a standard curve, which is generated with a synthetic external standard. An automated ECL-based detection system (QPCR System 5000) is commercially available from Perkin-Elmer, Ltd. This system is equivalent to radioisotopic DNA detection assays in its sensitivity, but does not require radioisotopes, and is both rapid and easy to use. Post-PCR manipulations are minimal, and easily automated; thus, this technique proves to be very useful when analyzing multiple samples in a short period of time.

5. Conclusion

It is important to avoid overestimating the accuracy of qc PCR, since standard deviations between 10% and 20% have been reported for the analysis of replicate aliquots of the same sample on different occasions. Therefore, for reliable discrimination of twofold differences in copy numbers between two samples, it is preferable to calculate a mean of replicate determinations of the same sample *(18)*. Also, when determining absolute initial amounts of mRNAs by competitive PCR using standard DNA fragments, it is important to consider that the efficiency of reverse transcription is less than 100%. The efficiency of cDNA synthesis using oligo(dT) as a primer for cDNA synthesis has been reported to be 40–50% *(19)*.

Finally, the use of microtiter plate-based detection systems for quantitative RT-PCR is becoming increasingly popular, primarily because of their affinity for automation. Also, competitors for microtiter plate-based assays do not need to have a different length to the target, since they are differentiated from wild-type amplification products by sequence. This allows the design of competitors that are more similar to the target of interest, allowing for more equivalent direct competition during RT-PCR, and thus a more accurate quantification of the gene-expression level.

References

1. Mansur, N., Meyer-Siegler, K., Wurzer, J., and Sirover, M. (1993) Cell cycle regulation of the glyceraldehyde-2-phosphate dehydrogenase/Uracil DNA glycosylase gene in normal human cells. *Nucleic Acids Rese.* **4,** 993–998
2. Finnegan, M., Goepel, J., Hancock, B., and Goyns, M. (1993) Investigation of the expression of housekeeping genes in Non-Hodgkin's Lymphoma. *Leuk. Lymphoma* **10,** 387–393.
3. Chang, T., Juan, C., Yin, P., Chi, C., and Tsay, H. (1998) Upregulation of Beta-actin, cyclophilin and GAPDH in NISI Rat *Hepatoma. Oncol. Rep.* **5,** 469–471
4. Foss, D., Baarsch, M., and Murtaugh, M. (1998) Regulation of hypoxanthine phosphoribosyltransferase, glyceraldehyde-3-phospate dehydrogenase and Beta-

actin mRNA expression in porcine immune cells and tissues. *Anim. Biotechnol.* **9,** 67–78

5. Labuhn, M., and Brack, C (1997) Age-related changes in mRNA expression of actin isoforms in Drosophilia melangaster. *Gerontology* **43,** 261–267
6. Raff, T. (1997) Design and testing of β-actin primers for RT-PCR which do not co-amplify processed pseudogenes. *Biotechniques* **23,** 456–460
7. O'Connell, J., Goode, T., and Shanahan, F. (1992) Quantitative measurement of mRNA expression by competitive RT-PCR, in *PCR in Bioanalysis,* Meltzer, S.J., ed. Humana Press, Inc., Totowa, NJ, pp. 183–195.
8. Überla, K., Platzer, C., Diamantstein, T., and Blankenstein, T. (1991) Generation of competitor DNA fragments for quantitative PCR. *PCR Methods Appl.* **1,** 136–139.
9. Siebert, P.D., and Larrick, J.W. (1993) PCR MIMICs: competitive DNA fragments for use as internal standards in quantitative PCR. *Biotechniques* **14,** 244–249
10. Siebert, P.D. (1999) Quantitaitve RT-PCR, in *Quantitative PCR Protocols,* Kochanowski, B., and Udo Reischl, eds., Humana, Totowa, NJ, pp. 61–85.
11. Celi, F.S., Zenilman, M.E., and Shuldiner, A.R. (1993) A Rapid and versatile method to synthesize internal standards for competitive PCR. *Nucleic Acids Research* **21** (4), 1047.
12. Porcher, C., Malinge, M.C., Picat, C., and Grandchamp, B. (1992) A simplified method for determination of specific DNA or RNA copy number using quantitative PCR and an automatic DNA sequencer. *Biotechniques* **13,** 106–113
13. Overbergh, L., Valckx, D., Waer, D., and Mathieu, C. (1999) Quantification of murine cytokine mRNAs using real time quantitative reverse transcriptase PCR. *Cytokine* **11** (4), 305–312
14. Halminen, M., Sjoroois, M., Makela, M.J., Waris, M., Terho, E., Lovgren, T., et al. (1999) Simultaneous detection of IFN-[g] and IL-4 mRNAs using RT-PCR and time-resolved fluorometry. *Cytokine* **11** (1), 87–93
15. Alard, P., Lantz, O., Sebagh, M., Calvo, C.F., Weill, D., Chavanel, G., et al. (1993) A versatile ELISA-PCR assay for mRNA quantification from a few cells. *Biotechniques* **15,** 730–737
16. Dufour, V., Arnauld, C., Lantz, O., Peguillet, I., Morvilliers, K., Emmanuel, A., et al. (1999) Quantification of porcine cytokine gene expression using RT-PCR, a homologous internal control and chemiluminescence for microplate detection. *J. Immunol. Methods* **229,** 49–60
17. Motmans, K., Raus, J., and Vandevyver, C. (1996) Quantification of cytokine messenger RNA in transfected human T cells by RT-PCR and an automated electrochemiluminescence-based post-PCR detection system. *J. Immunol. Methods* **190,** 107–116
18. Zimmermann, K., and Mannhalter, J. W. (1996) Technical aspects of Quantitative PCR. *Biotechniques* **21,** 268–279).
19. Bouaboula, M., Legoux, P., Pességué, B., Dumont, X., Piechaczyk, M., Casellas, P., et al. (1992) Standardisation of mRNA titration using a polymerase chain reaction method involving co-amplification with a multispecific internal control. *J. Biol. Chem.* **267,** 21,830–21,838

7

Rapid Development of a Quantitative-Competitive (qc) RT-PCR Assay Using a Composite Primer Approach

Joe O'Connell, Aileen Houston, Raymond Kelly, Darren O'Brien, Aideen Ryan, Michael W. Bennett, and Kenneth Nally

1. Introduction

The extraordinary sensitivity of reverse-transcriptase-polymerase chain reaction (RT-PCR) makes it a powerful technique for specific mRNA detection, particularly when tissue availability is limiting, and when the mRNA to be detected is present in low abundance. A disadvantage of standard RT-PCR with respect to less sensitive techniques such as Northern blot is that it is only semi-quantitative, because of the kinetics of PCR product accumulation. Exponential amplification occurs only at early cycles below the level of detection by standard ethidium bromide-stained agarose gel electrophoresis. In the later cycles, amplification efficiency decreases because of depletion of nucleotides and primers, progressive denaturation of the polymerase, and a shift in the equilibrium of template denaturation, which favors association rather than denaturation of the template DNA strands as product concentration becomes high. This results in a "plateau effect," such that there is no linearity in the relationship between product yield and initial template.

1.1. Quantitative Competitive PCR (qcPCR)

Several approaches have been employed to obtain quantitative data from PCR (*see* Chapter 5). Some of these techniques involve many manipulations following PCR, often requiring expensive equipment such as an automated DNA sequencer. One of the most widely used quantitative approaches is that of competitive PCR (**Fig. 1**) (reviewed in **ref.** ***1***). Essentially, a control PCR

From: *Methods in Molecular Biology, vol. 193: RT-PCR Protocols*
Edited by: J. O'Connell © Humana Press Inc., Totowa, NJ

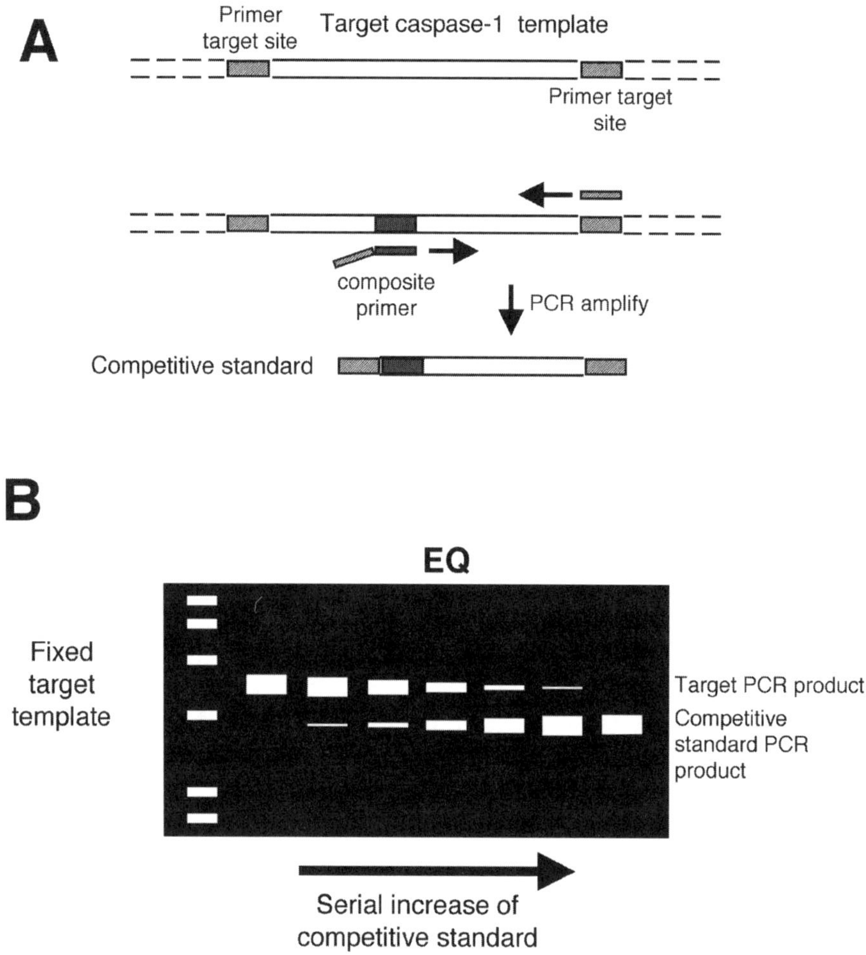

Fig. 1. Quantitative competitive (qc) PCR. **(A)** Construction of an internal standard for competitive PCR. By amplifying the target template using a composite primer approach, a truncated product is obtained, with the same primer sites as the target. This is made possible by synthesizing a sense primer for an internal priming site, which also has the regular sense primer sequence added onto its 5' end (composite primer). **(B)** Use of internal standard in qcPCR. Replicate PCRs are set up with a fixed aliquot of the unknown template. A dilution series of the competitive standard with known copy number is spiked into the PCRs. After amplification, the competitive PCRs are run on an agarose gel. Equivalence of PCR products (EQ) occurs when the target and standard templates were present in equal initial concentration, permitting quantification of the target template.

template is constructed that has identical primer sites to the target template, yet with a difference—usually simply in size—which allows amplification products from this control template to be distinguished from those of the target template. This control template is spiked in at known concentration as an internal standard during amplification of the target template. Because it shares the same primers as the target, it will compete directly with the target template during PCR amplification. If the starting amount of internal competitive standard is greater than that of the target, PCR products from the competitive standard will predominate. Similarly, if the target template is initially present at a higher concentration than the standard, the target template will out-compete during PCR. If the internal competitive-standard template is present in an equal amount to the target template, equivalent PCR products are obtained from both. In practice, multiple PCR reactions (usually 5–7) are set up containing a fixed source of target template, with serially increasing known amounts of the internal standard spiked into the reactions. Following PCR amplification, the relative amount of product from both templates in each reaction is determined from standard agarose gel electrophoresis. The equivalence point, at which there is an equal yield of target and competitive standard PCR products, is determined. The number of copies of the specific target template must be equivalent to the known number of competitive standard molecules spiked into this reaction, enabling quantification of target molecules. With qc RT-PCR, direct competition between the target sequence and the internal standard continues throughout the PCR, so the plateau effect does not interfere with quantification.

1.2. Composite Primer Strategy for Construction of Competitive Standard

A common approach to generating a competitive standard is to make a deletion to the target template, usually of less than 30%, to permit differentiation between amplification products of target and competitor on the basis of size. Unlike other types of competitive standard, no processing of the resultant PCR products, by either restriction digestion or southern hybridization, is required. No extraneous DNA sequence is introduced into the competitor, which is usually of sufficiently similar size and sequence composition to the target to result in identical efficiencies of amplification, and thus direct competition on an equal basis. Once the competitive internal standard has been constructed, no deviation is required from the standard RT-PCR protocol in order to achieve accurate quantitation of mRNA. No additional skills, equipment, reagents, or time are needed, or further manipulations of the PCR products beyond detection of the equivalence point on a standard agarose gel.

In order to generate a suitable deletion, one approach is to clone the target template into a plasmid, then use unique restriction sites to excise an appropriate fragment. After purifying the linearized plasmid from the small excised fragment by gel filtration chromatography, the sticky ends of the plasmid are filled in with Klenow and the plasmid is relegated to generate a clone containing the deleted template. This permits an RNA copy of the competitive standard to be run off the plasmid from a T7- or T3-promoter on the plasmid *(2)*. The advantage of an RNA standard is that it can actually be spiked in at the cDNA synthesis stage, so that the efficiency of the RT step is also controlled. However, there is no amplification in the RT step, and the efficiency does not vary significantly between samples. In our study, we demonstrate a much more rapid protocol for constructing a DNA standard—one that is spiked in at the PCR stage. No cloning or restriction/re-ligation are required. The standard is easily derived from the target template by PCR using the regular anti-sense primer and a composite sense primer. The composite primer primes from an internal site, downstream from the regular sense primer site **(Fig. 1)**. However, by attaching the regular sense primer sequence to the 5' end of the internal primer, PCR with this composite primer will generate a truncated product that has the regular sense and antisense primer sites at its ends. The first requirement is to select a primer pair that works well in regular RT-PCR. The competitive standard is then derived from this target template.

2. Materials

1. A program for designing primer pairs, such as the DNASTAR Lasergene Primerselect program (DNASTAR Inc., Madison, WI).
2. Access to a recent version of the European Molecular Biology Lab (EMBL) DNA sequence database (e.g., on CD-ROM from DNASTAR) to perform a homology search of the selected primers. This ensures that the primers have no significant homology to any other known gene.
3. Access to automated DNA synthesis for primer synthesis.
4. Cell lines: we obtained the cell lines in our study—HT29 and SW620—from the European Collection of Cell Cultures (ECACC), Salisbury, UK. Cells were cultured in Dulbecco's modified Eagle's medium (DMEM) supplemented with 10% fetal calf serum (FCS)and antibiotics (Gibco-BRL, Grand Island, NY). Cells were maintained at 37°C in a humidified atmosphere containing 5% CO_2.
5. Stimulus to be tested: we tested the effect of IFN-γ (R&D Systems, Abingdon, UK) on caspase-1 expression.
6. RNA isolation. Guanidine thiocyanate cell lysis buffer: 4.0 *M* guanidine thiocyanate (Sigma Chemical Co., St. Louis, MO), 25 m*M* sodium citrate, pH 7.0, 0.5% sarcosine. For extraction, use molecular-biology-grade, water-saturated phenol (Sigma). Ethanol (Sigma) is used for precipitation of the RNA, and 70% ethanol is used for washing the dried pellet. RNA pellets are re-dissolved in water that has been pretreated with diethyl pyrocarbonate (DEPC; Sigma) to inactivate any contaminating ribonucleases. DEPC, which is highly toxic, is added at 0.1% to a

bottle of distilled water in a fume-hood, and the lid is replaced. The water is left to incubate with the DEPC for several hours, with occasional vigorous shaking to disperse the DEPC throughout the water. Finally, the DEPC-treated water is autoclaved. Heat breaks down DEPC to ethanol and carbon dioxide, so autoclaved DEPC-water is not toxic. (Several kits are also commercially available for quick RNA isolation (e.g., RNeasy® kit from Qiagen, Crawley, UK). Nucleic acid quantification strips can be used to quantify RNA (e.g., nucleic acid quantitation kit, Invitrogen BV, the Netherlands).

7. cDNA synthesis: Avian myeloblastoma virus (AMV) RT and its buffer, and the ribonuclease inhibitor RNasin, are obtained from Promega (Madison, WI); random hexanucleotide primers (Boehringer Mannheim, GmbH, Mannheim, Germany) are used at 125 n*M* in the cDNA synthesis.
8. PCR: 10 m*M* deoxynucleotide-5'-triphosphate (dNTP) stocks, Taq DNA polymerase and its buffer are obtained from Promega; a proofreading, heat-stable DNA polymerase such as UlTma DNA polymerase is obtained from Perkin-Elmer (Norwalk, CT); a kit is used for purifying PCR products (e.g., QIAquick® PCR purification kit from Qiagen); densitometry software is used for determining band intensity of PCR products, e.g., National Institute of Health (NIH) Image program (available free on the internet from the National Institute of Health, USA (NIH) website at: http://rsb.info.nih.gov/nih-image).
9. Electrophoresis: molecular-biology-grade agarose (Promega); ethidium bromide solution (Sigma); *Hae* III-digested φX174 DNA-size markers (Promega).

3. Methods

3.1. RT-PCR

1. Select a PCR primer pair to amplify the cDNA of interest (caspase-1 in this instance). Our PCR primer pair was designed using the DNASTAR Lasergene Primerselect program (DNASTAR Inc.). The primer pair was chosen to span introns in the genomic sequence. This prevents co-amplification from any DNA that might contaminate the RNA prep, thus ensuring mRNA-specific amplification. Each primer was screened by a homology search of the EMBL DNA sequence database to ensure that it had no significant homology to any other known gene. The selected caspase-1 PCR primer pair had the following sense and anti-sense sequences respectively: ACCGCCCAGAGCACAAGACCT and TGTGGAAGAGCAGAAAGCGATAAA. These primers amplify all the splice variants of caspase-1 cDNA to yield a single 344-bp PCR product (*see* **Note 1**).
2. Treat isolated or cultured cells with the stimulus to be tested for its effect on the target mRNA level. In this instance, we tested the effect of IFN-γ (R&D Systems) on the level of mRNA for caspase-1 in the HT29 and SW620 cell lines. These are colonic-epithelial cell lines derived from human colonic adenocarcinoma. Cultured cells were treated with IFN-γ at 100 U/mL for 24 h prior to RNA isolation.
3. Isolate total RNA from the treated cells by first lysing in 0.5 μL of guanidine

thiocyanate lysis buffer. Add 50 µL of 2 *M* sodium acetate, pH 4, 0.5 mL of water-saturated phenol, and 0.1 mL of a mixture of chloroform and isoamyl alcohol (ratio of 49:1). Mix the tubes after addition of each reagent, and mix the final suspension vigorously. Leave the tubes on ice for 15 min, then centrifuge at 10,000*g* for 20 min in a microfuge. Remove the RNA-containing upper aqueous phase to a new tube, and add 2–2.5 vol of absolute ethanol. Leave on ice for 15 min to precipitate the RNA. Collect the precipitate by centrifugation at 10,000*g* for 30 min in a microfuge. Wash the pellet in 70% ethanol, dry, and redissolve the RNA in 10–20 µL of DEPC-treated water. Several kits are also commercially available for quick RNA isolation (e.g., RNeasy® kit from Qiagen).

4. Synthesize cDNA using the AMV RT (Promega) and random hexanucleotide primers (Boehringer Mannheim, GmbH). Perform the cDNA synthesis in a final reaction vol of 30 µL containing: 1 U of AMV reverse transcriptase (RT), 40 U of RNasin, 0.5 m*M* of each of the four dNTPs, 125 nM of random hexanucleotide primers, and approx 1 µg of total RNA. Incubate for 90 min at 42°C. Dilute the cDNA by adding 70 µLof T.E. (10 m*M* Tris-HCl, pH 8.0, 1 m*M* EDTA) and store at –20°C.
5. Test the efficiency and specificity of the selected primer pair by standard RT-PCR. We usually use PCR primers at a final concentration of 0.1 µ*M* each, dNTPs at 50 µ*M* and $MgCl_2$ at 1.5 m*M*. Use 1.0 U of Taq DNA polymerase per 50 µL reaction. We commonly use the following program of thermal cycling: denaturation at 96°C for 15 s; annealing at 55°C for 30 s, and extension at 72°C for 3 min. Perform "hot start" by heating the reaction to 80°C before adding the polymerase. Perform 35–40 cycles. Analyze PCR products by electrophoresis through 2% agarose gels and view under ultraviolet (UV) light following ethidium bromide staining. *Hae*III-digested ϕX174 DNA size markers can be used. PCR product specificity should ideally be confirmed, either by restriction mapping or DNA sequence analysis.

3.2. Construction of a Standard for qcRT-PCR

Once it has been established that the selected primer pair works well in PCR and yields a clean, specific product, the next step is to construct an internal standard for use in competitive PCR with this primer pair. The standard should be identical to the cDNA sequence targeted for RT-PCR amplification, except for an internal deletion of about one-third of the target amplicon. This permits identification of RT-PCR products derived from the target cDNA and the internal competitive standard to be distinguished from each other on the basis of size following co-amplification. The competitive standard can be derived from the target cDNA sequence by PCR using a composite primer approach (*see* **Fig. 1A**).

1. Select an internal sense primer site about one-third of the way downstream from the 5' end of the sequence amplified by the regular RT-PCR. Synthesize a sense primer that is specific for this internal site, which also has the sequence of the regular sense primer added on to its 5' end. We selected a site 96 bp downstream

from the 5' end of the caspase-1 amplicon. The composite primer had the following sequence: ACCGCCCAGAGCACAAGACCT-TCCCAGATATACTACAA CTCAATG. The hyphen indicates the joint between the regular sense primer sequence and the internal primer sequence.

2. Amplify the target cDNA with this composite sense primer and the regular antisense primer. A truncated amplicon will be derived that will lack the upper 5' one-third of the regular amplicon, but will have the sense primer site of the regular amplicon incorporated at its 5' end **(Fig.1A)**. Perform the PCR to generate the internal standard using a proofreading heat-stable DNA polymerase, such as the UlTma DNA polymerase (Perkin-Elmer), in order to minimize misincorporation leading to sequence changes in the competitive standard.
3. Purify this modified shorter amplicon, 269 basepairs (bp) for our caspase-1 construct, for use as the internal competitive standard for qcRT-PCR. A number of kits are commercially available for purifying PCR products (e.g., QIAquick® PCR purification kit from Qiagen). Its important to purify the construct to eliminate primers and primer artifacts, which could contribute to background in subsequent qcRT-PCR.
4. Determine the concentration of the standard spectrophotometrically, or from an ethidium bromide-stained agarose gel by comparison with a standard. For example, if 100 ng of *Hae*III-digested ϕX174 DNA size markers are run on the gel, the 600-bp band will contain 11 ng of DNA. From the concentration of the competitor, calculate the copy number per mL (*see* **Note 2**).

3.3. qcRT-PCR

1. Prepare a dilution series of the competitor for use as a set of standards. We usually use a five-fold or 2.5-fold series. For the caspase-1 qcRT-PCR, we used a five-fold series of the competitor, with six dilutions ranging from 8.0×10^3 down to 2.6 copies per µL.
2. Prepare a set of PCR reactions (as per the regular PCR in **Subheading 3.1., step 5**) all containing the same amount of the test cDNA template (1 µL). Add 1 µL of each dilution of the competitive standard into one reaction, and also perform a control with no added competitor, as well as a PCR-negative control as usual. For the caspase-1 qcRT-PCR, we had six tubes containing the six different amounts of the competitor, one tube without competitor, and a template-free PCR-negative control. Subject the tubes to thermal cycling, as already described (*see* **Note 3**).
3. Following amplification, run all PCRs on an agarose gel as usual. Equivalence of RT-PCR products occurs when target and competitor cDNA templates are present in equal initial concentration, permitting quantitation of the target template. Equivalence is determined as the point at which target and competitive standard PCR products are of equal band intensity. For precise quantitation, densitometry can be performed using the NIH Image program (available free on the internet at: http://rsb.info.nih.gov/nih-image) to determine the exact ratio of target:standard products in the PCR closest to equivalence (*see* **Note 4**). The concentration of the target is determined by multiplying this ratio by the known standard concentra-

tion of that PCR. Express results as the number of mRNA transcripts per μg of total RNA (*see* **Note 5**), assuming a 10% conversion of mRNA to cDNA (*see* **Note 6**).

Using our qcRT-PCR for caspase-1, we were able to quantify and compare the upregulation of caspase-1 between HT29 and SW620 cells. From the qcRT-PCR data shown in **Fig. 2**, IFN-γ (100 U) treatment for 24 h resulted in a massive upregulation of the basal level of caspase-1 mRNA in HT29 (from 8.2×10^3, to 7.9×10^5, transcripts/μg of RNA), whereas a more modest upregulation was achieved in SW620 (from 1.7×10^2 to 5.8×10^4 transcripts/μg of RNA). The constitutive, basal level of caspase-1 mRNA was 48-fold greater in HT29 than in SW620, and the final, IFN-γ upregulated level was 14-fold greater in HT29 than in SW620. We had already found that IFN-γ caused profound sensitization of HT29 cells to programmed cell death, or apoptosis, but had a lesser sensitizing effect on SW620 cells ***(3)***. The differential upregulation of caspase-1—which is involved in apoptosis signaling—correlated well with the differential apoptosis-sensitizing effects of IFN-γ in the two cell lines. Thus, caspase-1 was identified as a key regulator of apoptosis in these colonic-epithelial cell lines.

4. Notes

1. Many human mRNAs are expressed as two or more splice variants that differ in size. Primers that span site(s) of differential splicing will generate different sized-PCR products for the different splice variants. For qcRT-PCR, these multiple-sized products would complicate the competition with the internal standard, and also give rise to various heteroduplexes. For this reason, for qcRT-PCR, choose a primer pair that does not span site(s) of differential splicing, so that a single RT-PCR product is obtained from all mRNA variants. The primer pair should span at least one intron to avoid amplification of genomic DNA that can contaminate RNA preps.
2. The following formulae are useful for calculation of the number of DNA copies present in a given amount of purified DNA.
   a. Spectrophotometrically, an $A_{260} = 1.0$ is equivalent to approx 40 μg/mL of DNA.
   b. The mol wt of a DNA molecule, M.W. = number of nucleotides × 340 (average nucleotide M.W.) × 2 (number of strands).
   c. Number of moles = weight (g)/M.W.
   d. Number of molecules = number of moles × 6.023×10^{23} (Avogadro's number).
3. For standard RT-PCR, we commonly perform 35–40 cycles. However, with qcRT-PCR, because two templates (target and competitor) compete for PCR amplification, it may be necessary to perform more cycles to achieve good samplification of both templates—especially when the target copy number is very low and at the lower limit of detection. For this reason, we sometimes increase the cycles to 45.
4. In theory, direct competition between target and competitor occurs on an equal basis throughout the PCR when initially present at approximately equal

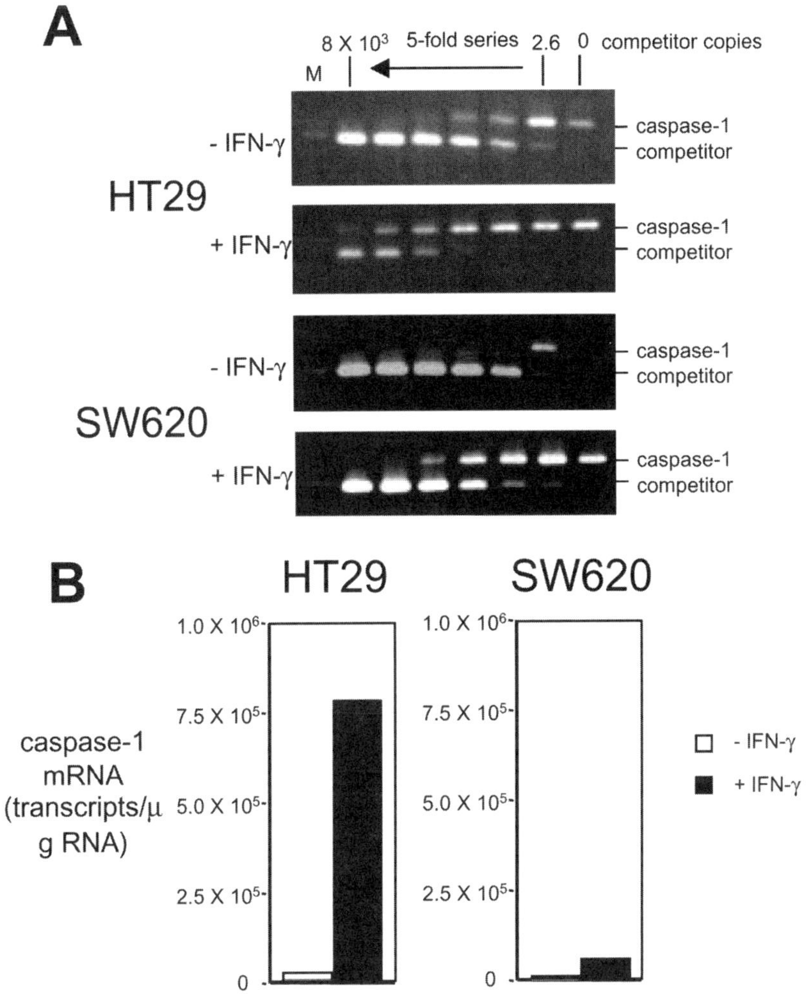

Fig. 2. IFN-γ upregulates caspase-1 in colonic-epithelial cell lines. **(A)** Caspase-1-specific qcRT-PCR was performed on equalized amounts of RNA from SW620 and HT29 cells with (+) or without (−) IFN-γ pretreatment (100 U) for 24 h. A five-fold dilution series of the competitive standard was spiked into a fixed aliquot of cDNA from each sample, and competitive PCR was performed. Upregulation of caspase-1 is evident from the shift in equivalence to higher competitor levels (i.e., left) following IFN-γ treatment. **(B)** After densitometry of bands, the amount of caspase-1 cDNA was calculated from the ratio of target:competitor PCR products in the reactions closest to equivalence. IFN-γ was found to induce a massive upregulation of caspase-1 mRNA in HT29 cells, and a lesser upregulation in SW620. The basal level of caspase-1 was higher in HT29 than SW620 cells, and IFN-γ-treated HT29 cells had 14-fold more caspase-1 mRNA than IFN-γ-treated SW620. The extent of caspase-1 upregulation correlated well with the extent of IFN-γ-mediated sensitization of both cell lines to apoptosis. Reprinted with permission from **ref.** ***(3)***.

concentrations. Therefore, calculations are probably most accurate using the target:template ratio closest to equivalence. Occasionally in an experiment, the dilution series of the standard may not include the precise quantity to compete with the amount of target in a particular sample, so that the point of equivalence is missed. In this event, the qcRT-PCR can be repeated with a different set of standards to focus on the range where equivalence should occur. Alternatively, the target concentration can be estimated by calculating the concentration from the product ratios at either side of the "missed" equivalence point, and obtaining the average from these two estimates.

5. The total RNA concentration can be determined spectrophotometrically. However, this method requires a lot of material. Also, spectrophotometric measurements often overestimate nucleic acid concentrations, since many preps will contain high amounts of free nucleotides, which also absorb at 260 nm. Methods based on staining of intact nucleic acids are more sensitive, and are not affected by free nucleotides. Detection strips are commercially available for rapid estimation of nucleic acid concentration (e.g., nucleic acid quantitation kit, Invitrogen BV).
6. As previously mentioned, the RNA form of competitive standards can be used to control for the RT as well as the PCR phase in qcRT-PCR, but this is not possible with DNA competitive standards. However, this is not a serious limitation, since there is no amplification in the RT step, and the efficiency does not vary greatly between RNA samples prepared by the same isolation protocol. In most cases in which the efficiency of RT reactions has been measured by incorporation of ^{32}P-labeled nucleotides into cDNA, it is usually found that RT-catalyzed conversion of mRNA to cDNA occurs with an efficiency in the range of 10–20%. Therefore, we suggest assuming a 10% conversion as a conservative estimate.

Acknowledgments

This work was supported by grants from the Health Research Board of Ireland (HRB) and a "New Blood" fellowship award from the Wellcome Trust-HRB to Joe O'Connell.

References

1. Zimmermann, K., and Mannhalter, J.W. (1996). Technical aspects of quantitative competitive PCR. *Biotechniques* **21,** 268–279.
2. O'Connell, J., Goode, T., and Shanahan, F. (1998). Quantitative measurement of mRNA expression by competitive RT-PCR. *Methods Mol. Biol.* **92,** 183–193.
3. O'Connell, J., Bennett, M.W., Nally, K., O'Sullivan, G.C., Collins, J.K., and Shanahan, F. (2000). Interferon-γ sensitizes colonic epithelial cell lines to physiological and therapeutic inducers of colonocyte apoptosis. *J. Cell. Physiol.* **185,** 331–338.

8

Quantitation of Gene Expression by RT-PCR and HPLC Analysis of PCR Products

Franz Bachmair, Christian G. Huber, and Guenter Daxenbichler

1. Introduction

The quantitative measurement of specific mRNA species is of major importance for studies on gene expression. Northern blotting is a relatively insensitive method requiring microgram amounts of RNA. It is time consuming and semi-quantitative at best. Because of the limitations of Northern blotting, various strategies have been developed for quantitation of cDNA by polymerase chain reaction (PCR)-based methods *(1–4)*. Competitive PCR, in which a synthetic segment is co-amplified together with the target segment, is one of these approaches. Besides adding more steps to the experimental protocol, competitive PCR is limited because the competitor and target sequences are not necessarily amplified with the same efficiency, which impairs reliable quantitation *(5)*. Recently, real-time PCR, in which the generated PCR-products are quantified fluorimetrically after each cycle, has become widely used *(6)*. However, determination of fragment size for positive fragment identification is not—at least not directly—possible with this method.

We were interested in comparing expression of the human retinoic acid receptor (RAR) genes, particularly RAR-γ, in very small samples of human tumor material. The RT-PCR method uses RNA as a template to produce cDNA, which can be used to reproduce and amplify specific cDNA target sequences by PCR. Either Moloney murine leukemia virus (MMLV) or Avian myeloblastoma virus (AMV) reverse transcriptases (RTases) are used for this procedure. However, a significant problem with using RNA as a template is the inability of mesophilic viral RTases to synthesize cDNA through stable

From: *Methods in Molecular Biology, vol. 193: RT-PCR Protocols*
Edited by: J. O'Connell © Humana Press Inc., Totowa, NJ

secondary RNA structures. Several methods have been described to destabilize regions of complementarity, such as the use of methylmercury, dimethyl sulfoxide (DMSO) and increased reaction temperatures. The recombinant DNA polymerase derived from the thermophilic eubacterium Thermus thermophilus was found to possess efficient RT activity at elevated temperatures in the presence of manganese ions. The increased reaction temperatures for the RT reaction allow for the synthesis of cDNA, even through stable secondary RNA structures. In addition, both reverse transcription and cDNA amplification occur in a single reaction tube, thereby minimizing the risk of contamination ***(7–10)***.

The standard method for the separation of DNA fragments is electrophoresis in polyacrylamide and agarose gels, in which visualization of the nucleic acid of interest usually requires ethidium bromide staining or autoradiography. Despite impressive advances, however, modern electrophoresis remains a laborious methodology that cannot be run unattended or automatically, and is prone to limited reproducibility and accuracy. And although capillary electrophoresis has put electrophoresis on the same instrumental footing as high-performance liquid chromatography (HPLC), the separation is still prone to poor reproducibility caused by matrix effects, which precludes its application in the routine analysis of PCR products without prior sample purification ***(11)***. Several reports have demonstrated the suitability of ion-pair reversed-phase high-performance liquid chromatography (IP-RP-HPLC) with alkylated, micropellicular polystyrene—divinylbenzene (PS–DVB-C18) particles for the rapid and high-resolution separation of DNA restriction fragments and PCR products without prior sample preparation ***(10,13–17)***. Compared to conventional gel electrophoresis, HPLC offers the advantage of full automation from sample injection to data analysis. Moreover, quantitation of the separated DNA is achieved directly by measurement of ultraviolet (UV)-absorbance at 254 nm ***(10)***.

Here, we present a method that combines an optimized RT-PCR procedure with the efficient separation and precise quantification of PCR products by HPLC with ultraviolet (UV) detection.

2. Materials

2.1. Tissue Pulverization

1. To pulverize fresh-frozen tissue into a frozen powder, use a robust, steel pulverizer. The pulverizer we use consists of a stainless-steel cylinder (inner diameter 4 cm, height 5 cm) and a loosely fitting steel piston driven by a rotary hammer (UBH 2/20, Bosch, Germany). This size is appropriate for tissue samples ranging from 0.5 to 5 g in weight. For smaller samples, we use a smaller type of pulverizer.
2. To resuspend and homogenize tissue powders in buffer for extraction of RNA, use a homogenizer such as the Ultra Turrax (Janke and Kunkel, Germany).

2.2. RNA Preparation by the Guanidiumthiocyanate Method

1. Denaturing solution: 4.0 *M* guanidinium thiocyanate, 10 nm EDTA, 10 μ*M* Tris-HCl, 1% n-lauroylsarcosine, and 1% β-mercaptoethanol (all from Sigma Chemical Co., St. Louis, MO).
2. Prepare a solution of 2 *M* sodium acetate (pH 4.0).
3. Water-saturated phenol, chloroform-isoamyl alcohol (24/1, v/v), 2-propanol and transfer RNA solution (0.1 mg/mL) are all obtained from Sigma.
4. A solution of proteinase K (10 mg/mL) is obtained from Roche Diagnostics (Vienna, Austria)
5. Prepare a solution of 75% ethanol.

2.3. RT-PCR

1. Master mix: 50 μ*M* bicine, 115 μ*M* potassium acetate, 8% glycerol (pH 8.2), 300 μ*M* of each dNTP, 2.5 m*M* manganese acetate, 0.45 m*M* of each primer and 100 U/mL of rTth DNA polymerase (Applied Biosystems, Foster City,USA).
2. PCR primer sequences for the retinoic acid receptor-γ (RAR-γ) mRNA were as follows: forward: 5'-GGAAGAAGGGTCACCTGACAGC-3' and reverse: 5'-TGCACTTGGTAGCCAGCTCAC-3'.

2.4. IP-RP-HPLC Analysis of RT-PCR Products

1. Stainless steel column (50 × 4.6 mm I.D.) packed with 2.1 μ*M* PS-DVB-C18 beads (DNASep^{TM} from Transgenomic, San Jose, CA, USA).
2. Gradient pump (Model LKB 2150, Pharmacia, Uppsala, Sweden).
3. Vacuum degasser (Knauer, Berlin, Germany).
4. Column thermostat (Haake FE2, Karlsruhe, Germany).
5. Variable wavelength detector (Model LKB 2151, Pharmacia).
6. Autosampler with a 50-mL biocompatible sample loop (Model AS 100 autosampler, Bio-Rad, Richmond, CA, USA).
7. PC-based data system (Borwin, JMBS Developments, Biolab, Vienna, Austria).
8. 0.1 *M* triethylammoniumacetate, pH 7.0 (TEAA), prepared from a 1 *M* stock solution (Fluka, Buchs, Switzerland).
9. Acetonitrile (ACN; Riedel-de Haen, Seelze, Germany).
10. DNA mol-wt markers: pBR322-*Hae*III digest, and a mixture of a pBR328-*Bgl*I and pBR328-*Hinf*I digests (Boehringer Mannheim); ϕX-174-*Hind*II digest and pBR322-*Msp*I digest (United States Biochemical, Cleveland, OH, USA); pBR322-*Alu*I digest (MBI Fermentas, Vilnius, Lithuania).

3. Methods

3.1. Tissue Pulverization

1. Ensure that the member of theater staff responsible for collecting the surgical tissue specimen washes the specimen immediately after surgery in buffered saline solution, cuts it into cubes of about 1 cm, and freezes it in liquid nitrogen.

2. Place the frozen sample into the prechilled cylinder of the pulverizer. Attach the prechilled piston to the rotary hammer and grind the tissue at gentle pressure and approx 500 revolutions/min Only a short exposure (approx 20 s) to this procedure should be necessary to obtain fine frozen tissue powder.

3.2. RNA Preparation

Extract total cellular RNA from either frozen powdered tumor tissue samples or dispersed cells by the following guanidinium thiocyanate procedure *(12)*. Briefly:

1. Add denaturing solution to the powdered tumor tissues (1 mL/ 50–100 mg) in 4.5 mL cryotubes (Falcon). Homogenize using an Ultra-Turrax homogenizer at full speed for a few seconds.
2. Add 0.1 mL of sodium acetate, 1 mL of water-saturated phenol, and 0.2 mL of chloroform-isoamyl alcohol. After extensive shaking, keep the samples on ice for 20 min.
3. Centrifuge for 30 min at 4000*g* at 4°C and remove the aqueous phase.
4. Precipitate the RNA from the aqueous phase for at least 12 h at 0°C after adding 2 vol of 2-propanol.
5. Collect the precipitate by centrifugation, wash the pellet once with 2 mL of ice-cold 75% ethanol, dry, and redissolve in distilled water.
6. Repeat the phenol extraction and ethanol precipitation as described in **steps 2–5**, but first treat the re-dissolved RNA sample with proteinase K (Roche Diagnostics, Vienna, Austria) in a final concentration of 200 mg/mL for 5 min at 55°C (*see* **Note 1**). Re-dissolve the final RNA pellet in 50 µL of distilled water.
7. Determine the RNA concentration by measuring the optical density (OD) at 260 nm in a UV-Vis spectrophotometer and dilute all samples to 200 µg total RNA/mL with deionized water. Dilute all RNA preparations with an aliquot of transfer RNA solution (100 µg/mL) (*see* **Note 2**).

3.3. RT-PCR

1. Perform RT-PCR using a suitable reagent kit. For our example, we used the GeneAmp Ez rTth RNA PCR Kit (Applied Biosystems). Add total RNA (100 ng in 5 µL) to 45 L of master mix. Add PCR primers. The primers used for our amplification of RAR-γ mRNA were flanking an intron in the genomic sequence to avoid misleading results caused by contamination with genomic DNA. A single mRNA-specific fragment of 198 base-pairs (bp) was produced *(17)*.
2. Perform thermal cycling. We used a Perkin-Elmer 9600 GeneAmp PCR thermocycling system using 0.2-mL reaction tubes. First, incubate the tubes to perform the RT reaction as follows: 40 min at 60°C followed by a 90-s denaturation step at 94°C.
3. Next, perform thermal cycling for the PCR step as follows: 18 amplification cycles of 30 s at 94°C, 30 s at 60°C for primer annealing, and 1 min at 72°C for primer extension. Next, perform 5 further cycles but with the DNA synthesis step

elongated for 15 s, followed by another 5 cycles, but with the DNA synthesis step elongated for another 15 s to 1 min 30 s. Finally, cool the samples to 4°C and maintain at this temperature until HPLC analysis.

3.4. IP-RP-HPLC Analysis of DNA Mol-Wt Markers and RT-PCR Products

1. Perform HPLC analysis of the PCR products using the following chromatographic conditions: 5 min linear gradient from 10–17% ACN in 0.1 *M* triethylammonium acetate, pH 7.0 (TEAA), followed by 2 min at 17% ACN in TEAA, and finally, 2 min at 10% ACN in TEAA for re-equilibration of the column; use a column temperature of 50°C.
2. Transfer the PCR reaction tubes to the autosampler rack, which is equipped with suitable adaptors; empty 1.5-mL reaction tubes serve as convenient holders for the 0.2-mL PCR tubes.
3. When the run is set up, a 25 L volume of each sample, standard, or DNA mol-wt marker, respectively, is injected into the column automatically at 9-min intervals.
4. Important considerations for optimization of the HPLC run. **Figure 1** illustrates the separation of DNA fragments from a restriction digest of pBR322 DNA with *Hae*III which we normally use for size determination of the PCR products. The gradient slope was optimized in this run to allow high resolution of DNA fragments up to lengths of 200 base-pairs (bp) in minimal analysis time. Since the difference in retention times of the eluting DNA fragments critically depends on the steepness of the ACN gradient, higher resolution could be obtained with a shallower gradient of 7–19% ACN in 30 min at the cost of longer analysis time, as demonstrated in **Fig. 2** by the separation of five different DNA restriction digests with fragments ranging in size from 46–2176 bp. Another important factor that influences the resolution of the fragments is the concentration of TEAA in the mobile phase. The resolution increased with increasing TEAA concentration in the range of 0.025–0.125 M TEAA ***(14)***. In IP-RP-HPLC, the retention of nucleic acids is determined by the number of negative charges and, hence, the length of the nucleic acid molecules; the longer the fragment, the stronger the interaction with the positively charged triethylammonium ions adsorbed at the surface of the nonpolar stationary phase. Because the amount of triethylammonium ions adsorbed at the stationary phase depends on the concentration of ACN in the mobile phase, nucleic acids are retained less when the ACN concentration is increased ***(14)***. At temperatures up to 50°C, IP-RP-HPLC separates double-stranded DNA molecules according to their chain length and not their base composition. This is in contrast to anion-exchange chromatography, in which the base composition has a significant impact on retention of DNA molecules ***(18)***. The high resolution, together with the short analysis time achievable with IP-RP-HPLC, are a consequence of the nonporosity of the PS-DVB-C18 particles and their narrow size distribution (2.1–0.12 m). These characteristics of the chromatographic packing result in minimized diffusional path lengths and high chromatographic efficiency.

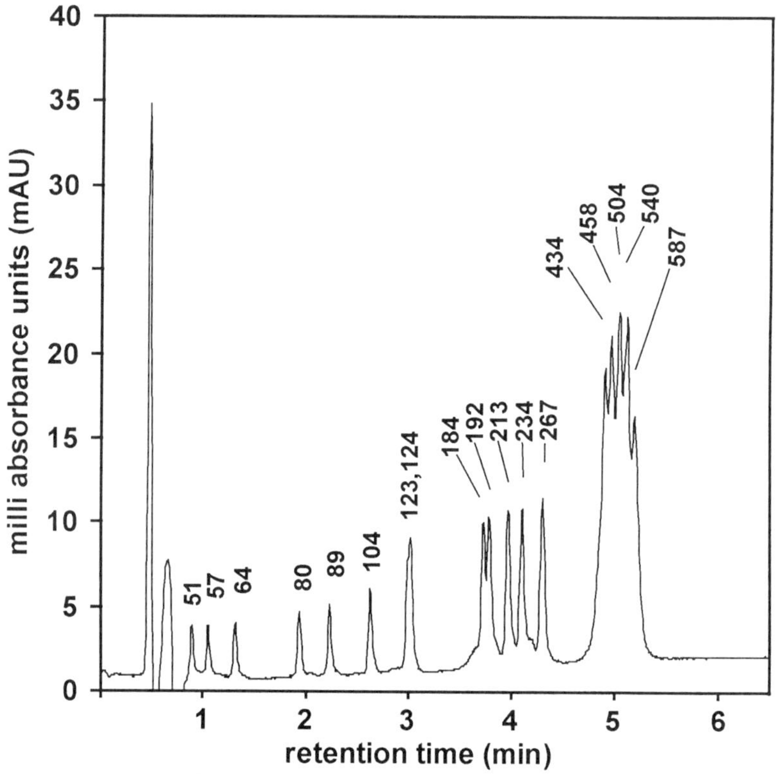

Fig. 1. IP-RP-HPLC separation of DNA restriction fragments (pBR322-*Hae*III digest) ranging in length from 8–587 bp, as indicated by the peak labels. The column used was: PS-DVB-C_{18}, 2.1 m, 50 4.6 mm I.D.; a linear gradient was applied, from 10–17% acetonitrile in 0.1 *M* TEAA, pH 7.0, for 5 min, followed by 17% ACN for 2 min; the flow rate was 0.8 mL/min; the temperature was 50°C; detection was at UV 254 nm.

3.5. Determination of Fragment Length

In IP-RP-HPLC, the dependence of retention time on size can be utilized for fragment size evaluation. The strict dependence of retention time upon size at temperatures of 30°C, 40°C, and 50°C is illustrated in **Fig. 3**, where retention times of 73 DNA fragments from five different mol-wt markers, ranging in size from 46–2176 bp, are plotted against the logarithm of fragment length (retention data from **Fig. 2**). Usually, the separation of one set of mol-wt markers with fragments in the same size range as the fragment to be characterized is

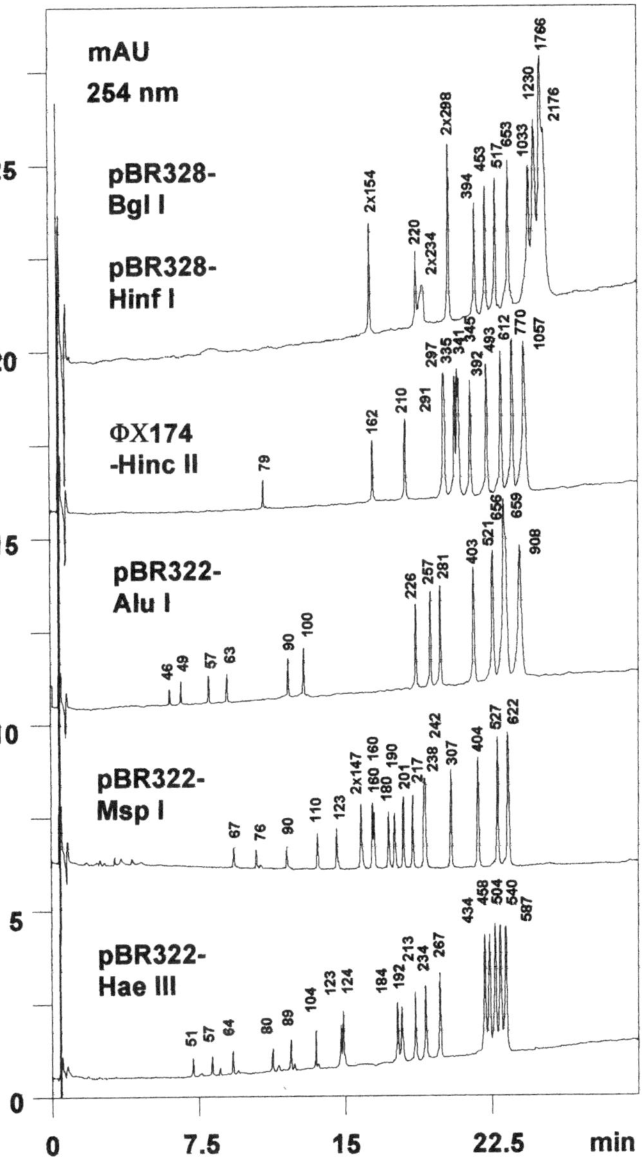

Fig 2. Chromatographic separation of five different sets of DNA mol-wt markers (pBR322-*Hae*III, pBR322-*Msp*I, pBR322-*Alu*I, ϕX174-*Hind*II, pBR328-*Bgl*I and pBR328-*Hinf*I digests, 1 μg each) under identical gradient conditions. The column used was PS-DVB-C_{18}, 2.1 m, 50 4.6 mm I.D.; a linear gradient was applied, from 7–19% ACNe in 0.1 *M* TEAA in 30 min; the flow-rate was 0.75 mL/min.; the temperature was 50°C; detection was at UV 254 nM.

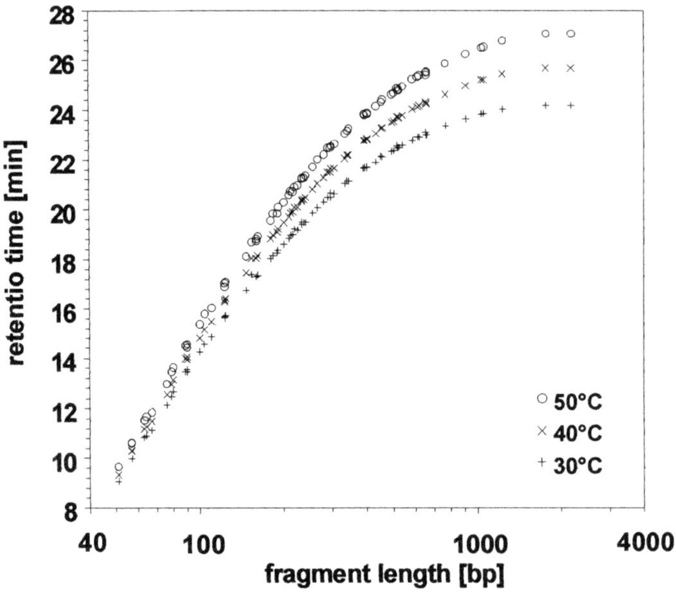

Fig. 3. Semilogarithmic size calibration plot of retention times vs length of DNA fragments ranging from 46–2176 bp.

sufficient for reliable size characterization. The relationship between length of the DNA fragments and retention time is given by the following equation: log bp = a tr + b , where t_r is the retention time, bp is the fragment length in basepairs, a is the slope, and b is the Y intercept. There are different ways to calculate fragment length by interpolation using retention times of product and standards.

1. In the local method **(Fig. 4A)**, a linear regression curve is created by using two standard points below and one point above the fragment, and the size is determined. A second curve is created from one standard point below and two standard points above, and a second value is assigned. The two size values are averaged to determine the unknown fragment length (*see* **Note 3**).
2. For the global method **(Fig. 4B)**, a set of five data points below and five above the unknown fragment are used for linear regression analysis. The global method includes a greater number of standards, and thus may allow a better correction for eventual local irregularities in retention. It is appropriate if standards are in a range where there is a linear relationship between log fragment size and retention time approx (up to 500 bp, **Fig. 3**).

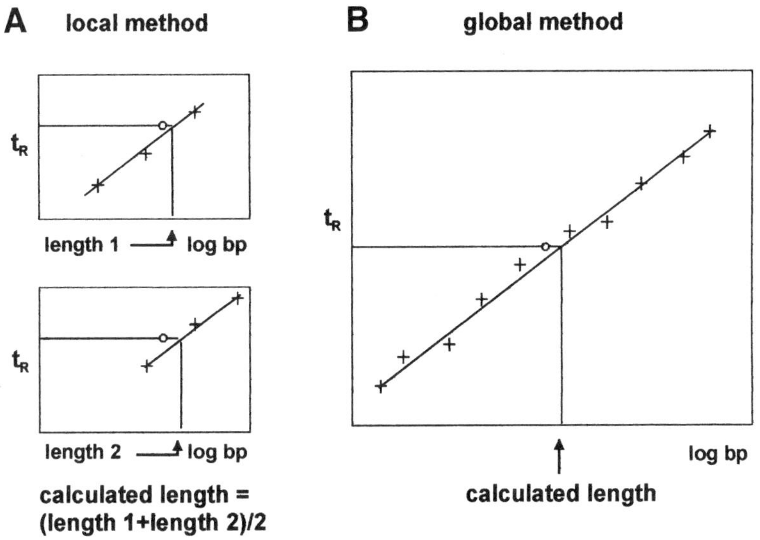

Fig. 4. Calculation of fragment size from chromatographic retention time data applying the local (**A**) and global (**B**) methods.

Results obtained with the two methods are usually not significantly different *(14)*. A third calculation method uses a very broad range of standards to construct a size calibration curve of higher order, e.g., by a polynom of second order (**Fig. 3**).

3.6. Quantitation of RT-PCR Products Using Standard RNA Preparations

1. Detect and quantify the amplified fragments by ultraviolet (UV) absorption at 254 nm and automatic peak area integration. **Figure 5** shows a chromatogram of the 198-bp PCR amplified RAR-γ fragment.
2. For each set of samples for RT-PCR-HPLC, it is best to run several dilutions of a standard RNA preparation for analysis. In our example, standard RNA had been prepared, as described, from 5 g of a breast cancer specimen and had been stored at –80°C in multiple aliquots to have sufficient material for at least 2 yr. This served for checking the linearity of the RT-PCR kinetics and for identifying samples that gave a signal in the plateau phase and thus had to be diluted. In addition, by relating individual sample quantification results to standards, with results given as arbitrary units, may increase the inter-assay reproducibility. However, in our experience, measurement of the sample absorption alone is usually sufficient to obtain reproducible results for different samples (*see* **Subheading 3.7.**). This is because absorption at 254 nm is an intrinsic property, and does not result from a color reaction (*see* **Note 4**).

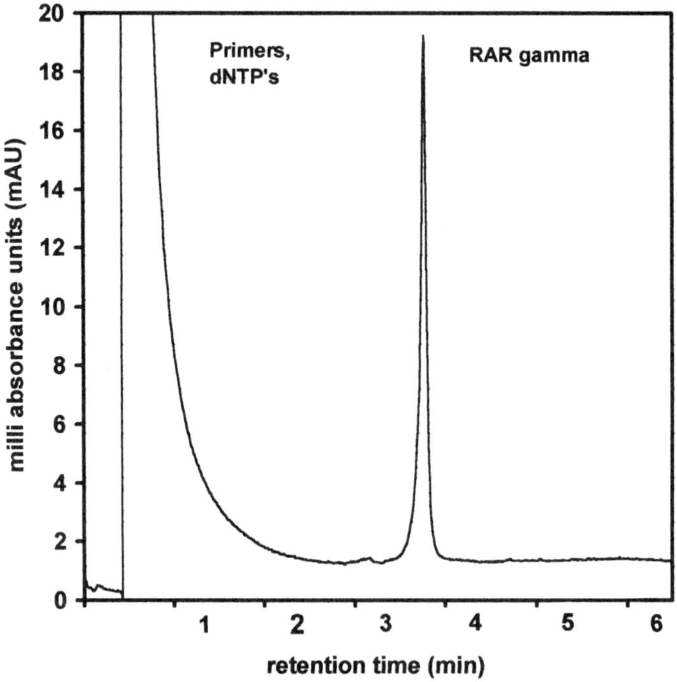

Fig. 5. HPLC and UV detection of the RT-PCR product for RAR-γ mRNA. Chromatographic conditions are as described for **Fig. 1**. Expected product length: 198 bp. Experimentally determined product length: 192 bp, derived from standard curves generated by the local method **(Fig. 4A)**.

3.7. Quality Criteria of the Method

1. Accuracy of product length determination was evaluated by analyzing chromatograms of ten RT-PCR reactions of one RNA sample. The expected 198-bp DNA fragment of RAR eluted at 3.73 ± 0.12 min, under the chromatographic conditions indicated in **Fig. 5**, from which an average product length of 192-bp was calculated by the local method. This compares well to the expected length of 198 bp (3% relative deviation). Intra- and inter-assay coefficients of variation (CVs) of 3% and 2%, respectively, for retention time, and 3% and 4%, respectively, for calculated product length, indicate good reproducibility of product size estimation (**Table 1**).
2. The peak area was used to quantitate RT-PCR products, since it is proportional to the amount of DNA injected. The CV for the peak area obtained when performing ten simultaneous RAR-RT-PCRs from the same RNA preparation was 6%. At different concentrations of this RNA preparation, the intra-assay precision was only slightly changed, and remained below 7%. When performing the reaction on seven different days, the CV was 8% (**Table 1**).

Table 1
Quality Criteria of the Method Described (RT-PCR and HPLC)

	Retention time (min)	Product length	Area[a]
Intra-assay precision (n = 10)			
Mean value	3.77	192	31.6
Coefficient of variation	3%	3%	6%
Inter-assay precision (n = 7)			
Mean value	3.85	196	25.3
Coefficient of variation	2%	4%	8%
Inter-assay precision including RNA extraction (n = 6)			
Mean value	3.65	186	68.1
Coefficient of variation	0.7%	1%	6%

[a] Different peak area values result from different initial amounts of mRNA used.

3. To determine the overall precision, RNA extraction was included in the analysis—i.e. six RNA extractions from the same tumor powder were performed and the RNA samples were processed as described in **step 2**. Not surprisingly, variations in the retention time and calculated product size were again very small (0.7% and 1%, respectively). CV for RNA quantity was only 6%; thus, potential variations in the RNA extraction procedure did not diminish the method's precision (**Table 1**).
4. Linearity of the method was tested with an RNA preparation that contained a high amount of the specific mRNA. Serial dilutions of this preparation were subjected to RT-PCR with subsequent HPLC analysis. Linearity within a relatively broad range (10–125 ng of total RNA) was excellent (r = 0.999), as can be seen from **Fig. 6**. Besides linearity testing, a preparation with medium mRNA content was used to determine the appearance of the plateau phase for amplification in relation to the number of cycles **(Fig. 7)**. The Figure shows 28 cycles seem to be suitable, as, at a medium concentration of mRNA, the plateau phase of amplification is not reached. Obviously, the beginning of the plateau phase was also dependent on the starting quantity of RNA.
5. Practicability of the method is good because it requires only two pipetting steps, not including the RNA isolation procedure. When using an autosampler for HPLC, about 100 samples can be analyzed within 12 h.

   In summary, the method presented here is easy to perform and quantitative. It also allows the evaluation of qualitative features of the amplified mRNA sequence—i.e., size and uniformity.

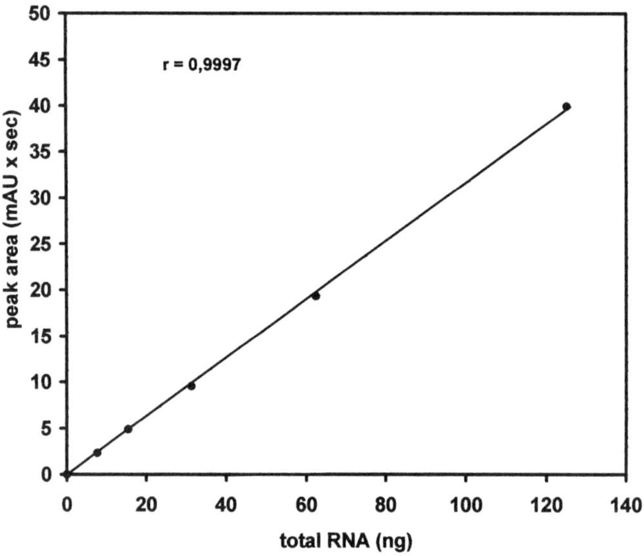

Fig. 6. Typical standard curve for RAR-γ mRNA RT-PCR. A total RNA preparation of a mammary carcinoma sample was aliquoted and diluted with tRNA solution (100 μg/mL). After RT-PCR and HPLC, the RAR-γ peak area was determined for each dilution, and plotted vs the amount of starting RNA.

4. Notes

1. Proteinase K treatment of the redissolved RNA eliminates contaminating protein. This allows better separation of the organic and aqueous phases during the subsequent re-extraction. This procedure minimizes DNA contamination of the final RNA preparation.
2. The addition of transfer RNA solution to the purified RNA generates a standardized matrix for the subsequent RT-PCR reactions, and also prevents nonspecific binding of RNA to the walls of the sample tubes.
3. With Method 1, only four data points of the mol-wt marker near the fragment of unknown length are needed for calibration. The averaging of two different calibration curves helps to minimize the effects of any possible abnormalities in retention of the calibration standards.
4. Because of the high reproducibility of the RT-PCR combined with IP-RP-HPLC, the addition of an internal standard and the co-amplification of a housekeeping gene are not necessary to measure precisely the expression of specific genes in cultured cells in which RNA quality is normally good. For tissues, where the handling and processing of different surgical specimens prior to freezing is not rigorously controlled, and thus RNA degradation may occur in some samples, co-amplification of two different areas of the specific mRNA is recommended. If

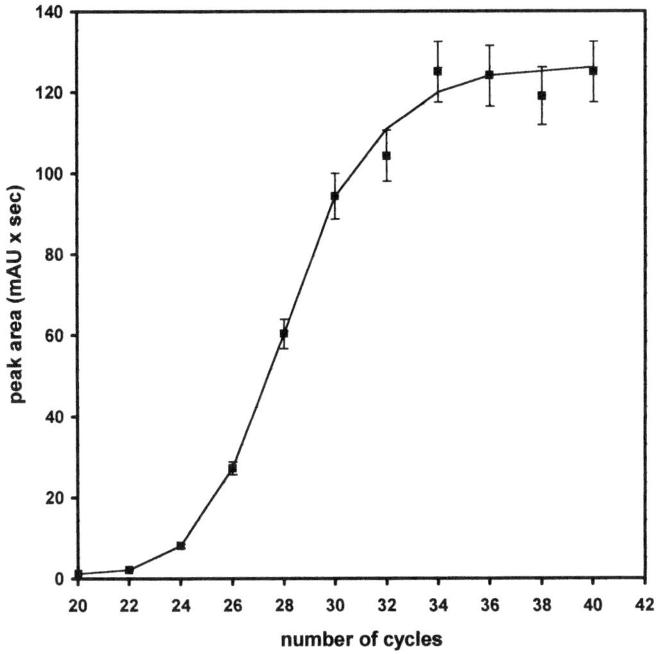

Fig. 7. Dependence of the HPLC peak area on the number of PCR cycles performed. Equal amounts (100 ng) of total RNA were subjected to RT-PCR with the indicated number of PCR cycles. Peak area for each product was then determined by HPLC, and plotted vs cycle number.

quantities of both amplified products are not comparable within a predefined range, results are unacceptable. We find this to be a more reliable approach than the co-amplification of a housekeeping gene, as stabilities of different mRNA species may differ drastically.

References

1. Horikoshi, T., Danenberg, K., and Volkenandt, M. (1993) Quantitative measurement of relative gene expression in human tumors. *Methods Mol. Biol.* **5,** 177-188.
2. Becker-Andre, M., and Hahlbrock, K. (1989) Absolute quantitation using the polymerase chain reaction (PCR). A novel approach by a PCR aided transcript titration assay (PATTY). *Nucleic Acids Res.* **17,** 9437–9446.
3. Gilliland, G., Perrinn, S., and Blanchard, K., Bunn. F. (1990) Analysis of cytokine mRNA and DNA: Detection and quantitation by competitive polymerase chain reaction. *Proc. Nat'l Acad. Sci. USA* **87,** 2725–2729.
4. Wang, A.M., Doyle, M.V., and Mark, D.F. (1989) Quantitation of mRNA by the polymerase chain reaction. *Proc. Nat'l Acad. Sci. USA.* **86,** 9717–9721.
5. Schweitzer, B., Horikoshi, T., et al. (1992) Quantitation of gene copy number and

mRNA using the polymerase chain reaction. *Proc. Soc. Exp. Biol. Med.* **200,** 1–6.
6. Heid, C.A., Stevens, J., Livak, K.J., and Williams, P.M. (1996) Real time quantitative PCR. *Genome Res.* **6,** 986–994.
7. Young, K., Resnick, R.M., and Myers, T. W. (1993) Detection of hepatitis C-virus RNA by a combined reverse transcription-polymerase chain reaction assay. *J. Clin. Microbiol.* **31,** 882–886.
8. Mulder, J., McKinney, M. , Christopherson, C., Sninsky, J., Greenfield, L., and Kwok, S. (1994) Rapid and simple PCR assay for quantitation of human immunodeficiency virus type I RNA in plasma: Application to acute retroviral infection. *J. Clin. Microbiol.* **32,** 292–300.
9. Myers, T.W., and Gelfand, D.H. (1991) Reverse transcription and DNA amplification by a Thermus thermophilus DNA polymerase. *Biochemistry* **30,** 7661–7665.
10. Huber, C.G., Oefner, P.J., Preuss, E., and Bonn, G.K. (1993) High resolution liquid chromatography of DNA fragments on non-porous poly(styrene-divinylbenzene) particles. *Nucleic Acids Res.* **21,** 1061–1066.
11. Oefner, P.J., Bonn, G.K., Huber, C.G., and Nathakarnkitkool, S. (1992) Comparative study of capillary zone electrophoresis and high-performance liquid chromatography in the analysis of oligonucleotides and DNA. *J. Chromatogr.* **625,** 331–340.
12. Chirgwin, J.M., Przybyla, A.E., MacDonald, R.J., and Rutter, W.J. (1979) Isolation of biologically active ribonucleic acid from sources enriched in ribonuclease. *Biochemistry* **18,** 5294–5299.
13. Oefner, P.J., Huber, C.G., Puchhammer-Stockl, E. Umlauft, F.,Gruenewald, K., Bonn, G.K.,et al. (1994) High performance liquid chromatography for routine analysis of hepatitis C virus cDNA/ PCR products. *BioTechniques* **16,** 898–908.
14. Huber, C.G., Oefner, P.J., and Bonn, G.K. (1995) Rapid and accurate sizing of DNA fragments by ion pair chromatography on alkylated non-porous poly(styrene-divinylbenzene) particles. *Anal. Chem.* **67,** 578–585.
15. Huber, C.G., Oefner, P.J., and Bonn, G.K. (1993) High-resolution liquid chromatography of oligonucleotides on nonporous alkylated styrene divinylbenzene co-polymers. *Anal. Biochem.* **212,** 351–358
16. Huber, C.G., Oefner, P.J., and Bonn, G.K. (1993) Rapid analysis of biopolymers on modified non porous polystyrene divinylbenzene particles. *Chromatographía* **37,** 653–658.
17. Lehmann, J.M., Hoffmann, B., and Pfahl, M. (1991) Genomic organisation of the retinoic acid receptor gamma gene. *Nucleic Acids Res.* **19,** 573–578.
18. Huber, C.G. (1998) Micropellicular stationary phases for high-performance liquid chromatography of double-stranded DNA. *J. Chromatogr.* **806,** 3–30.

9

Time-Resolved Fluorometric Detection of Cytokine mRNAs Amplified by RT-PCR

Kaisa Nieminen, Markus Halminen, Matti Waris, Mika Mäkelä, Johannes Savolainen, Minna Sjöroos, and Jorma Ilonen

1. Introduction

PCR (polymerase chain reaction) products are conventionally detected by agarose gel electrophoresis and ethidium bromide staining. However, the specificity and sensitivity of detection can be increased by hybridization with a labeled probe that is complementary to an internal sequence in the amplified target. Time-resolved fluorometry (TRF) utilizes nonradioactive lanthanide chelates with large Stoke's shifts as the labels *(1)*. For the time-resolved measurement of lanthanide fluorescence, a fluorometer is used (e.g., those manufactured by Wallac) that has a xenon flash lamp at 340-nm wavelength with 1-s intervals between flashes. After the excitation light flash and a delay time of 0.4 s, the emission from the lanthanides is measured at longer wavelengths (613, 643, and 545 nm for Eu, Sm^{3+}, and Tb^{3+}, respectively) for another 0.4 s. The excitation-emission measurement cycles are repeated 1000× during a total measurement time of 1 s per sample. The lanthanides, e.g., europium (Eu^{3+}), samarium (Sm^{3+}) and terbium (Tb^{3+}), have manifold longer fluorescence lifetimes than most biological compounds. A time delay after the excitation pulse eliminates nonspecific background by allowing the short-lived autofluores-cence of the biological material to decline to an insignificant level before measuring the specific signals from the lanthanides *(2)*. This delay time is the time-resolving principle and combined together with large Stoke's shifts, it is the critical element that makes TRF such a sensitive technique. Substances with the same fluorescence characteristics as lanthanides are rarely, if ever,

From: *Methods in Molecular Biology, vol. 193: RT-PCR Protocols*
Edited by: J. O'Connell © Humana Press Inc., Totowa, NJ

found in natural samples, and background from rare earth metals is also very low. The fluorescence signals obtained are directly proportional to the amount of probe bound and captured on a solid phase in assays that are conveniently performed in a microtiter-plate format. Furthermore, the different emission wavelengths of the lanthanide chelates have permitted the development of multi-label assays. By using different fluorescence filters for detection, the labels can be easily distinguished from each other in an individual hybridization reaction.

In this chapter we describe three assays that apply TRF technology to the detection of cytokine gene-expression levels in peripheral-blood mononuclear cells (PBMC). As outlined in **Fig. 1**, total cellular RNA is extracted, and the mRNAs are reverse-transcribed into cDNAs by poly(dT)-primed reverse transcription using Moloney murine leukemia virus (MMLV) reverse transcriptase (RT). Specific biotinylated primer pairs are used to simultaneously PCR-amplify cDNAs for interleukin (IL)-4 and interferon (IFN)-γ, IL-12 p40 subunit and IL-13, or transforming growth factor-beta 1 (TGF-β1), with the constantly-expressed beta-actin (β-actin) as an internal control in each reaction. The primer concentrations are optimized to limit the amplification of β-actin sequences, since cytokines are expressed at a much lower level. The biotinylated PCR products are collected on streptavidin-coated microtiter plates by affinity capture, denatured, and hybridized in solution with Eu^{3+}, Sm^{3+} and Tb^{3+} labeled sequence-specific probes. After washings, the lanthanides are dissociated from the chelates used to label the probes by adding beta-diketone-based acidic enhancement solution(s), in which the lanthanides rapidly form new, highly fluorescent chelates. The fluorescence emitted by the detached labels is measured by time-resolved fluorometer, and the signals from the cytokine RT-PCR products are standardized relative to the β-actin signals to compensate for variations in initial cell numbers or reverse-transcription efficiency, or handling losses. In the assays for IFN-γ and IL-4 *(3)*, and for IL-12 p40 and IL-13, each pair consists of cytokines with opposing biological functions. Thus, the ratio of the two counteracting cytokines can be calculated directly, providing useful biological information regardless of the level of the β-actin control. However, the β-actin control still serves as a useful marker of successful RNA extraction and reverse transcription reaction.

The specificity and sensitivity of these assays are advantageous when the quantity of initial mRNA is limited. The simultaneous detection of different targets saves assay time and reagents, and is especially beneficial when applied to a large series of samples. The methods described here can be used for semi-quantitative detection of cytokine expression levels in PBMC, but may also be adapted for precise quantitation; for example, by creating external standard curves.

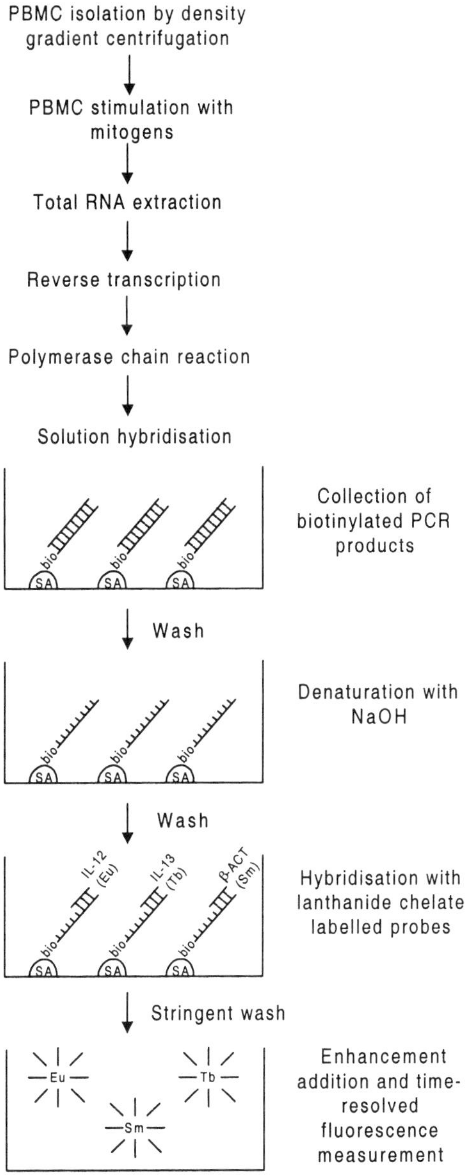

Fig. 1. General outline of the assay. Human PBMC are isolated by Ficoll-density gradient centrifugation and stimulated with mitogens to produce cytokines. After stimulation, total RNA is extracted from the harvested cells, and the mRNAs are 0reverse-transcribed into cDNAs. Subsequently, IL-12 p40, IL-13 and β-actin cDNAs are amplified by PCR with primer pairs—including biotinylated 5'-primers and collected onto streptavidin-coated microtiter wells. The immobilized and denatured PCR products are hybridized in a solution with lanthanide chelate-labeled probes and the fluorescence of the labels is measured with time-resolved fluorometer after the addition of enhancement solution.

2. Materials

2.1. Total RNA Extraction

1. Use a commercially available kit to isolate RNA, such as the Ultraspec™' RNA isolation system (Biotecx Laboratories Inc., Texas, USA). Store at the kit at 4°C and protect from light during handling.
2. Chloroform and isopropanol are obtained from Sigma Chemical Co. (St. Louis, MO).
3. A solution of glycogen (20 mg/mL) is obtained from Boehringer Mannheim (Mannheim, Germany), and stored at –20°C.
4. Prepare a solution of 75% ethanol and store at –20°C.
5. Nuclease-free water is obtained from Promega (Madison, WI) and stored at –20°C.

2.2. Reverse Transcription

1. For cDNA synthesis, use poly(dT)12–18 primers (33 U/mL stock) (Pharmacia Biotech, Uppsala, Sweden).
2. 5X RT buffer: 5X MMLV buffer (Promega): 250 m*M* Tris-HCl, pH 8.3, 375 m*M* KCl, 15 m*M* $MgCl_2$, 50 m*M* dithiothreitol (DTT).
3. Deoxynucleotide-5'-triphosphate (dNTP) mix (Pharmacia Biotech, Uppsala, Sweden): 25 m*M* each dATP, dTTP, dCTP, dGTP.
4. MMLV RT, RNase H minus (200 U/µL), is obtained from Pharmacia Biotech.
5. RNase inhibitor: recombinant RNasin (40 U/µL) is obtained from Promega.
   Store the reagents for reverse transcription at –20°C.

2.3. Polymerase Chain Reaction

1. 10X PCR buffer: 10X DyNAZyme buffer (Finnzymes, Espoo, Finland): 100 m*M* Tris-HCl, pH 8.8, 15 m*M* $MgCl_2$, 500 m*M* KCl, 1% Triton X-100.
2. dNTP mix (Pharmacia Biotech): 20 m*M* each dATP, dTTP, dCTP, dGTP.
3. Our PCR primer stocks had the following concentrations: IFN-γ 5 pmol/µL, IL-4 5 pmol/µL, IL-12 p40 10 pmol/µL, IL-13 10 pmol/µL, β-actin 5 pmol/µL, TGF-β1 5 pmol/µL. The sequences of the primers are listed in **Table 1**.
4. 25 m*M* $MgCl_2$; used in PCR mixtures A and B.
5. 500 m*M* $(NH_4)2SO_4$; used only in PCR mixture B.
6. Dimethylsulfoxide (DMSO); used only in PCR mixture B.
7. Thermostable DNA polymerase (2 U/µL) (DyNAZyme DNA polymerase, native enzyme (Finnzymes).
8. Nuclease-free water (Promega).
   Store the reagents for PCR at –20°C.

2.4. Solution Hybridization

1. DELFIA® assay buffer (Wallac, Turku, Finland). Store at 4°C.
2. DELFIA® 25X Wash Concentrate (Wallac). Store at 4°C. Prepare the 1X Wash Concentrate in distilled water. Approximately 2 L of 1X Wash Concentrate is needed per plate.
3. Prepare solutions of 50 m*M* NaOH and 5 *M* NaCl.

Table 1
Primers for PCR

	Primer sequence 5'-3'	5'-Label	Reference
IFN-γ (**1**)	ATG AAA TAT ACA AGT TAT ATC TTG GCT TT	biotin	Butch et al., 1993 ***(4)***
IFN-γ (**2**)	GAT GCT CTT CGA CCT CGA AAC AGC AT		Butch et al., 1993 ***(4)***
IL-4 (**1**)	ATG GGT CTC ACC TCC CAA CTG		Brenner at al., 1991 ***(5)***
IL-4 (**2**)	TCA GCT CGA ACA CTT TGA ATA T	biotin	Brenner at al., 1991 ***(5)***
IL-12 p40 (**1**)	GGC CCG CAC GCT AAT GCT		CDC*
IL-12 p40 (**2**)	GGG TGA CGT GCG GAG CTG	biotin	CDC*
IL-13 (**1**)	TTT ACC CCT CCC TAA CCC TC		CDC*
IL-13 (**2**)	ACA TCA CCC AGA ACC AGA AG	biotin	CDC*
TGF-β1 (**1**)	TCC CGG CAC CGC CGA GCC C	biotin	Leung et al., 1992 ***(6)***
TGF-β1 (**2**)	AGG CTC CAA ATG TAG GGG CAG G		Leung et al., 1992 ***(6)***
β-actin (**1**)	TGA AGT CTG ACG TGG ACA TC	biotin	Melby et al., 1993 ***(7)***
β-actin (**2**)	ACT CGT CAT ACT CCT GCT TG		Melby et al., 1993 ***(7)***

*Sequences obtained from CDC (Centers for Disease Control, Atlanta) by personal communication with Larry Andersson. (1=forward; 2=reverse).

Table 2
Probes for Solution Hybridization

	Sequence 5'–3'	5'-Label	Reference
IFN-γ	ACT TGA CAT TCA TGT CTT CC	Eu	Boyle et al., 1993 *(8)*
IL-4	AAG AAC ACA ACT GAG AAG GA	Tb	
IL-12 p40	AAT GGC TTC AGC TGC AAG	Eu	CDC*
IL-13	GTT GAT CAG GGA TTC CAG	Tb	CDC*
TGF-β1	TGG GCT CGT GGA TC	Eu	
β-actin	GGA GCA ATG ATC TTG ATC TT	Sm	

*Sequences obtained from CDC (Centers for Disease Control, Atlanta) by personal communication with Larry Andersson.

4. Tween-100.
5. Lanthanide chelate-labeled probes (Wallac). Store at -20°C. The sequences and labels of the probes are listed in **Table 2**.
6. DELFIA® enhancement solution, enhancer, and streptavidin-microtiter strips (Wallac). Store at 4°C.

2.5. Fluorometry

1. In our case, we used a Wallac 1234 time-resolved fluorometer (Perkin-Elmer Life Sciences; Wallac).

3. Methods

3.1. Total RNA Extraction

1. Human PBMC are isolated by standard Ficoll-density-gradient centrifugation, and are stimulated with mitogens to produce cytokines. After stimulation, transfer the culture of 5–10 × 10^6 PBMC into a sterile 1.5-mL Eppendorf tube and pellet the cells by centrifugation at 800*g* for 5 min. Remove the supernatant and add 1 mL of Ultraspec™ RNA solution. Resuspend by a brief vortexing and store at –70°C until further processing (*see* **Note 1**).
2. Add 200 µL of chloroform and vortex for 15 s. Incubate on ice for 5 min and centrifuge at 12,000*g* (4°C) for 15 min to separate the phases.
3. Transfer the upper aqueous phase (450–500 µL) containing the total cellular RNA into a new tube. Avoid taking the proteins from the interphase. Add an equal volume of isopropanol and 2 µL of glycogen to serve as a carrier for RNA. Mix and incubate at –20°C for 10 min before centrifuging at 12,000*g* (4°C) for 20 min.
4. Remove the supernatant and wash the pellet with 1 mL of 75% ethanol by briefly vortexing and subsequently centrifuging at 8,000*g* (4°C) for 5 min. Remove the supernatant and repeat the wash with 0.2 mL of 75% ethanol.
5. After the washes, remove ethanol carefully and let the total RNA pellet air-dry for 5-10 min. Do not use a vacuum drier.

6. Dissolve the pellet in 20 µL of nuclease-free water by vortexing for 30 s, and subsequently incubating in a water bath at 55°C for 15 min. After incubation, vortex briefly and spin down the condensate. Store at -70°C until further processing.

3.2. Reverse Transcription

1. To prime the mRNAs, mix an 18.5-µL aliquot of total RNA extract (1-10 µg) with 0.7 µL of poly$(dT)_{12-18}$ primers (0.5 µg poly$(dT)_{12-18}$ primers/µg RNA) in a 0.5-mL tube (*see* **Note 2**). Incubate in a heat block at 70°C for 5 min. Cool the tubes on ice for 3 min and spin down the condensed drops.
2. Add 5.8 µL of a master mix containing the following reagents, so that their concentrations in the final reaction vol of 25 µL are as indicated in parentheses: 5 µL of 5X MMLV RT buffer (1×), 0.5 µL of dNTP mix (0.5 m*M* each), 0.1 µL of RNase inhibitor (4 U) and 0.2 µL of MMLV RT, RNase H minus (40 U) (*see* **Note 3**). Mix gently and incubate at 37°C for 60 min. Stop the reaction by heating at 90°C for 5 min. Spin down the condensate and store at -70°C until further processing.

3.3. Polymerase Chain Reaction

The PCRs are performed in a final vol of 50 µL containing 5 µL of template cDNA and 45 µL of the PCR mastermix. The PCR mastermix is firstly prepared in a large tube and then aliquoted into 0.5-mL tubes. Finally, a drop of mineral oil is added to prevent evaporation. Negative and positive PCR controls are included in each PCR run (*see* **Note 4**).

1. Prepare three different PCR master mixes (A, B, and C) for each of the three different multiplex PCRs, so that the final concentrations are as indicated in parentheses (*see* **Note 5**):
   a. Reaction mixture for IL-4/IFN-γ/β-actin triplex-PCR (for one reaction): Add 5 µL of 10X DyNAZyme buffer (1×), 0.5 µL of dNTP mix (200 µM each), 0.7 µL of IFN-γ primers (3.5 pmol), 0.7 µL of IL-4 primers (3.5 pmol), 0.3 µL of β-actin primers (1.5 pmol), 6 µL of $MgCl_2$ (3 m*M*) and 29.6 µL of nuclease-free water into a 0.5-mL tube. Mix and add 0.5 µL of DyNAZyme DNA-polymerase (1 U).
   b. Reaction mixture for IL-12 p40/IL-13/β-actin triplex-PCR (for one reaction): Add 5 µL of 10X DyNAZyme buffer (1×), 0.5 µL of dNTP mix (200 µM each), 0.75 µL of IL-12 p40 primers (7.5 pmol), 0.75 µL of IL-13 primers (7.5 pmol), 0.3 µL of β-actin primers (1.5 pmol), 6 µL of $MgCl_2$ (3 m*M*), 1 µL of $(NH_4)_2SO_4$ (10 m*M*), 2 µL DMSO (4%), and 26.4 µL of nuclease-free water into a 0.5-mL tube. Mix and add 0.5 µL of DyNAZyme DNA-polymerase (1 U).
   c. Reaction mixture for TGF-β1/β-actin duplex-PCR (for one reaction): Add 5 µL of 10X DyNAZyme buffer (1×), 0.5 µL of dNTP mix (200 µM), 1.7 µL of TGF-β1 primers (8.33 pmol), 1.7 µL of β-actin primers (8.33 pmol) and 35.6 µL of nuclease-free water into a 0.5-mL tube. Mix and add 0.5 µL of DyNAZyme DNA-polymerase (1 U).

Table 3
The PCR Programs

IL-4/IFN-γ/β-actin	IL-12/IL-13/β-actin	TGF-β1/β-actin
1× the initial denaturation step: 98°C; 10 s 95°C; 2 min	1× the initial denaturation step: 94°C; 4 min	1× the initial denaturation step: 94°C; 2 min
32× the amplification step: 95°C; 1 min 50°C; 1 min 72°C; 1.25 min	32× the amplification step: 94°C; 1.25 min 60°C; 1.25 min 72°C; 1.25 min	33× the amplification step: 94°C; 1 min 62°C; 1 min 72°C; 1 min
The final extension step: 72°C; 7 min	The final extension step: 72°C ; 6.25 min	The final extension step: 72°C; 10 min

2. Add 5 μL of template cDNA, 5 μL of nuclease-free water (negative PCR control) or 5 μL of a known amplifiable template—positive PCR control, e.g., cDNA from phytohemagglutinin (PHA) stimulated PBMC—under the oil layer. Spin down briefly.
3. Perform the PCR reaction in a Perkin-Elmer Thermal Cycler with the programs listed in **Table 3** (*see* **Note 6**).
4. Store the PCR products at –20°C until analysis.

3.4. Solution Hybridization

1. Apply 10 μL of PCR products into duplicate wells of a DELFIA®microtiter plate coated with streptavidin (*see* **Note 7**). Include two wells of sterile distilled water (10 μL each) to serve as negative controls for the hybridization.
2. Add 50 μL of DELFIA®assay buffer (pre-equilibrated to room temperature) per well. Shake the plate in a DELFIA®, Plateshake at room temperature for 30 min. to bind the biotinylated PCR products to the wells.
3. Remove unbound material by washing the wells in a DELFIA® Platewash 4× with 1X DELFIA®Wash Concentrate.
4. Add 150 μL of 50 m*M* NaOH per well and shake the plate at room temperature for 5 min to denature the bound DNAs.
5. Wash 4× with 1X DELFIA® Wash Concentrate to remove the non-biotinylated strands.
6. Prepare separate hybridization solutions (A, B, and C) which are specific for each of the multiplex PCR products. Add 100 μL of freshly prepared hybridization solution per well (*see* **Note 8**). The final concentrations in each hybridization solution (100 μL final vol) are as indicated in parentheses:
   a. Hybridization solution for IFN-γ/IL-4/β-actin detection (per well): 2.5 ng of Eu-labeled IFN-γ probe, 2.5 ng of Tb-labeled IL-4 probe, 7.5 ng of

Sm-labeled β-actin probe, 17 μL of 5 *M* NaCl (0.85 *M*), 0.1 μL Tween-100 (0.1%), and DELFIA® assay buffer up to a final vol of 100 μL.

b. Hybridization solution for IL-12 p40/IL-13/β-actin detection (per well): 2.5 ng of Eu-labeled IL-12 p40 probe, 5 ng of Tb-labeled IL-13 probe, 5 ng of Sm-labeled β-actin probe, 17 μL of 5 *M* NaCl (0.85 *M*), 0.1 μL Tween-100 (0.1%), and DELFIA® assay buffer up to a final vol of 100 μL.

c. Hybridization solution for TGF-β1/β-actin detection (per well): 1.5 ng of Eu-labeled TGF-β1 probe, 1.5 ng of Sm-labeled β-actin probe, 17 μL of 5 *M* NaCl (0.85 *M*), 0.1 μL Tween-100 (0.1%), and DELFIA® assay buffer up to a final vol of 100 μL.

7. Seal the plate with adhesive tape and incubate at the following conditions (*see* **Note 9**):
   a. Hybridization conditions for IFN-γ/IL-4/β-actin detection: 48°C, 4 h.
   b. Hybridization conditions for IL-12 p40/IL-13/β-actin detection: 48°C, 1 h
   c. Hybridization conditions for TGF-β1/β-actin detection: 46°C, 2 h.
8. After hybridization, remove the unbound probes by washing the wells 6× with 1X DELFIA®wash concentrate preheated to 42°C for the IFN-γ/IL-4/β-actin and IL-12 p40/IL-13/β-actin detections, or to 44°C for the TGF-β1/β-actin detection.
9. Add 100 μL of DELFIA®enhancement solution per well and shake the plate at room temperature for 25 min to dissociate the Eu^{3+} and Sm^{3+} ions.
10. Measure Eu^{3+} and Sm^{3+} fluorescence signals with the time-resolved fluorometer (*see* **Note 11**).
11. To dissociate the Tb^{3+} ions, add 25 μL of DELFIA® enhancer and shake at room temperature for 10 min. This is applied only for hybridizations A and B.
12. Measure Tb^{3+} fluorescence signals with the time-resolved fluorometer.

4. Notes

1. Prior to and during RNA isolation, take adequate precautions to avoid contamination by RNase, or previously amplified PCR products. The pipets, filtered pipet tips, tube trays and the flowhood should be ultraviolet (UV)-irradiated for approx 30 min prior to the procedure. Only RNase-free solutions, reagents, and glassware should be used, and sterile disposable plasticware should be used whenever possible. To avoid RNase contamination from skin, disposable gloves should be worn and frequently changed.
2. Prior to and during setup of the cDNA synthesis reaction, take adequate precautions to avoid contamination by RNase, or previously amplified PCR products. The reaction is performed in a laminar hood after UV-irradiating it and the pipets, filtered pipet tips and tube trays for approx 30 min before starting the procedure. 70% ethanol can be used for cleaning the working space. Thawed reagents should be kept on ice to minimize the effects of RNases and RNA degradation. PCR products should not be brought into the same hood. Gloves should be worn at all times and frequently changed during the procedure, because "common areas" such as door handles must be touched in between the steps.

3. RT-buffer may contain some precipitates after freezing. Heat the tube to 70°C for 1–2 min, and then vortex vigorously to dissolve these. The RT enzyme should be added to the master mix immediately prior to dispensing the aliquots, in order to minimize inactivation of the enzyme by a higher temperature.
4. Contamination can be prevented by keeping separate areas for preparing the PCR mixture, pipetting the template, and performing the actual PCR reaction, and for detecting the PCR products. Filtered pipet tips and disposable gloves should be used, and the tubes should be kept closed during transfer.
5. During the PCR setup, all the reagents, master mixes, and templates should be kept on ice after thawing. The DNA polymerase should be added immediately before aliquoting the master mix. The water should be added first, followed by the PCR buffer, primers, and other reagents.
6. A drop of mineral oil should be added into the wells of the heat block in the thermal cycler to improve the transfer of heat between the steps.
7. The lot numbers (and dates) of reagents and plates should be recorded to enable possible trouble shooting.
8. The probes should be thawed and added into the hybridization solution just prior to aliquoting the hybridization solution. In a diluted concentration, probes can bind to plasticware. Thus, for long-term storage, probes should be at high concentrations.
9. In-solution hybridization, the specific binding of the probes is highly dependent on the hybridization and wash temperatures. Thus, it is important to keep these temperatures constant to avoid losing the signal or obtaining a false result.
10. If the fluorescence signals are very high, the template can be diluted or the number of the PCR cycles reduced, and vice-versa if the signals are very low.
11. If the background signals (i.e., the signals of the hybridization controls) are very high, the amount of the probe used in hybridization should be reduced or the temperature of the hybridization oven and the washing solution should be checked. To avoid high background signals it is also important to rinse the forks of the platewash with distilled water between washes. The forks should be left in 75% ethanol over night and rinsed carefully with distilled water before use. Careful attention should be given to the labeling degree of the probe whenever the probe batch is changed, and the probe concentration used per well should be optimized once again. When the labeling degree of the new probe is higher than the previous one, smaller amounts of the new probe should be used, and vice-versa.

References

1. Soini, E., and Lövgren, T. (1987) Time-resolved fluorescence of lanthanide probes and application in biotechnology. *CRC Crit. Rev. Anal. Chem.* **18,** 105– 154.
2. Lövgren, T., Hurskainen, P., Dahlén, P., and Iitiä, A. (1995) Detection of lanthanide chelates by time-resolved fluorescence, in *Nonisotopic Probing, Blotting and Sequencing* Kricka, L., ed., Academic Press Inc., San Diego, pp. 331–376.
3. Halminen, M., Sjöroos, M., Mäkelä, M.J., Waris, M., Terho, E., Lövgren, T., and

Ilonen, J. (1999) Simultaneous detection of IFN-γ and IL-4 mRNAs using RT-PCR and time-resolved fluorometry. *Cytokine* **11,** 87–93.

4. Butch, A., Chung, G.-H., Hoffmann, J., and Nahm, M. (1993) Cytokine expression by germinal center cells. *J. Immunol.* **150,** 39–47.
5. Brenner, C.A., Daniel, S.L., and Adler, R. (1991) Cytokine mapping: observation and quantification of cytokine mRNA in small number of cells using the polymerase chain reaction, in *Cytokines, a Practical Approach* Balkwill, F.R., ed., Oxford University Press, New York, pp.51–59.
6. Melby, P.C., Darnell, B.J., and Tryon, V.V. (1993) Quantitative measurement of human cytokine gene expression by polymerase chain reaction. *J. Immunol. Methods* **159,** 235–244.
7. Boyle, M.J., Berger, M.F., Tschuchnigg, M., Valentine, J.E., Kennedy, B.G., Divjak, M., et al. (1993) Increased expression of interferon-gamma in hyperplastic lymph nodes from HIV-infected patients. *Clin. Exp. Immunol.* **92,** 100–105.
8. Leung, J.C.K., Lai, C.K.W., Chui, Y.L., Ho, R.T.H., Chan, C.H.S., and Lai K.N. (1992) Characterization of cytokine gene expression in CD4+ and CD8+ T cells after activation with phorbol myristate acetate and phytohaemagglutinin. *Clin. Exp. Immunol.* **90,** 147–153.

10

Mimic-Based RT-PCR Quantitation of Substance P mRNA in Human Mononuclear Phagocytes and Lymphocytes

Jian-Ping Lai, Steven D. Douglas, and Wen-Zhe Ho

1. Introduction

Substance P (SP), the most extensively studied and most potent member of the tachykinin family, is a modulator in neuroimmunoregulation *(1–4)*. SP has been described as a peptide that is almost exclusively of neural origin *(5,6)*. More recently, SP has been identified in non-neuronal cell types, including murine macrophages *(7,8)*, human endothelial cells *(9,10)*, eosinophils *(11)*, and Leydig cells derived from human and mouse testis *(12)*. We have demonstrated that human monocytes and lymphocytes express the SP gene mRNA transcripts and produce SP protein *(13–15)*.

SP is synthesized from four preprotachykinin-A (PPT-A, SP precursor) mRNAs (α, β, γ, and δ) *(15–17)*. These four isoforms of PPT-A mRNAs are the result of alternative RNA splicing of the primary transcript of the PPT-A gene *(17)*. The sequence in exon 3 encodes SP, and the sequence in exon 6 encodes neurokinin A (NKA). The SP sequence is encoded by all four PPT-A mRNAs (α, β, γ, and δ), whereas the NKA precursor sequence is present only in β- and γ-PPT-A mRNAs. The four isoforms of PPT-A mRNAs differ in their exon combinations (**Fig.1**). Although β-PPT-A contains all 7 exons of the corresponding gene, α-PPT-A lacks exon 6, γ-PPT-A lacks exon 4 *(18, 19)*, and δ-PPT-A lacks both exon 4 and 6 *(15, 17, 20)*.

In an attempt to identify PPT-A gene expression in human immune cells, we have recently characterized splicing products of the PPT-A gene transcript in human monocytes/macrophages *(13)* and lymphocytes *(14, 15)* using reverse-transcriptase-polymerase chain reaction (RT-PCR) with several human SP (HSP) primer pairs. We have also established a strategy of designing a specific

From: *Methods in Molecular Biology, vol. 193: RT-PCR Protocols*
Edited by: J. O'Connell © Humana Press Inc., Totowa, NJ

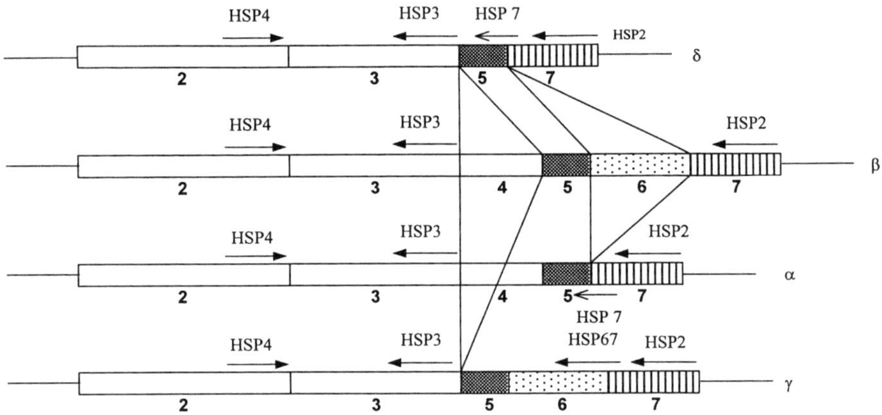

Fig. 1. Alternative splicing products of PPT-A mRNA transcripts. Exons 2, 3, 4, 5, 6, and 7 are indicated. Exon 3 encodes for SP, and exon 6 encodes for neurokinin-A (NKA). Locations and orientations (arrows) of the primer sites for primers HSP2, HSP3, HSP4, HSP67, and HSP7 are indicated *(21)*. HSP2 is located at the 3' end of the mRNAs for all four PPT-A transcripts, and is used as a specific primer in the simultaneous cDNA synthesis of all four isoforms. The HSP4 primer site is in exon 2, which is present in all the four isoforms. The HSP3 primer site is within exon 3; the HSP67 primer sequence is derived from the exon 6–7 junction , which occurs only in the β- and γ-isoforms; and the HSP7 primer sequence is derived from the exon 5–7 junction, which occurs only in the α and δ isoforms. Thus, in the isoform-specific PCR strategy (**Fig. 3**), the primer pair of HSP3/HSP4 amplifies all four isoforms of PPT-A mRNA; the primer-pair of HSP4/HSP67 is used to amplify only the β and γ isoforms, and the primer-pair of HSP4/HSP7 amplifies only the α and δ isoforms. Thus, for a particular unknown sample, one can use the HSP4/HSP3 primer-pair to analyze total mRNAs of the PPT-A gene, and the primer pairs of HSP4/HSP7 and HSP4/HSP67 to analyze the expression patterns of PPT-A mRNA isoforms (**Fig. 4**).

primer pair (HSP4/HSP3) to amplify all four isoforms (α, β, γ, and δ) of PPT-A mRNA transcripts, resulting in a single 121 basepair (bp) fragment that can be quantitatively measured by the SP-mimic-based PCR assay *(21)* (**Fig. 2**). We have also established a primer pairing strategy to differentiate the four isoforms of PPT-A mRNA transcripts in human monocytes and lymphocytes *(21)* (**Fig. 1** for details of the isoform-specific primer combinations, **Fig. 3** for a flow-chart of the isoform-specific PCR strategy, and **Fig. 4** for actual results of isoform-specific PCRs).

RT-PCR has become an important tool to examine levels of mRNA transcripts because of its high sensitivity *(22)*. However, quantification of mRNA levels can be problematic because of the exponential nature of PCR amplification. Small variations in amplification efficiency may lead to dramatic changes

Prepare SP cDNA
Isolate RNA

↓

Synthesize 1st–strand cDNA
Using SP Specific Primer

↓

Conduct Competitive PCR
Make Preliminary Ten-Fold Serial
Dilutions of SP Mimic

↓

Add Portion of Each SP Mimic Dilution
to 1st–strand cDNA Samples

↓

Perform PCR Using Primer Pair
HSP4/HSP3

↓

Make Fine-Tuned Two-Fold Serial
Dilutionsof SP Mimic Starting with a
Selected Ten-Fold dilution

↓

Add Dilutions to cDNA, Perform PCR
Using Primer pair HSP4/HSP3

↓

SP Mimic Dilution 1:1 with
SP cDNA Identifies the Amount
of SP cDNA in PCR sample

Fig. 2. Outline for using a SP-mimic in competitive PCR to quantify SP mRNA levels. The SP cDNA is prepared by reverse transcription of RNA using a SP-specific primer (HSP2). To quantitate SP cDNA, decreasing amounts of SP-mimics are added to PCR reactions containing a constant amount of SP cDNA. Following PCR amplification using primer pair HSP4/HSP3, the products derived from the SP-mimic and SP cDNA are resolved and amounts compared on an agarose gel.

in product yields, which obscures differences in the levels of the target mRNA during amplification. This is particularly true for determinating levels of SP mRNA, because human immune cells contain very low levels of SP mRNA.

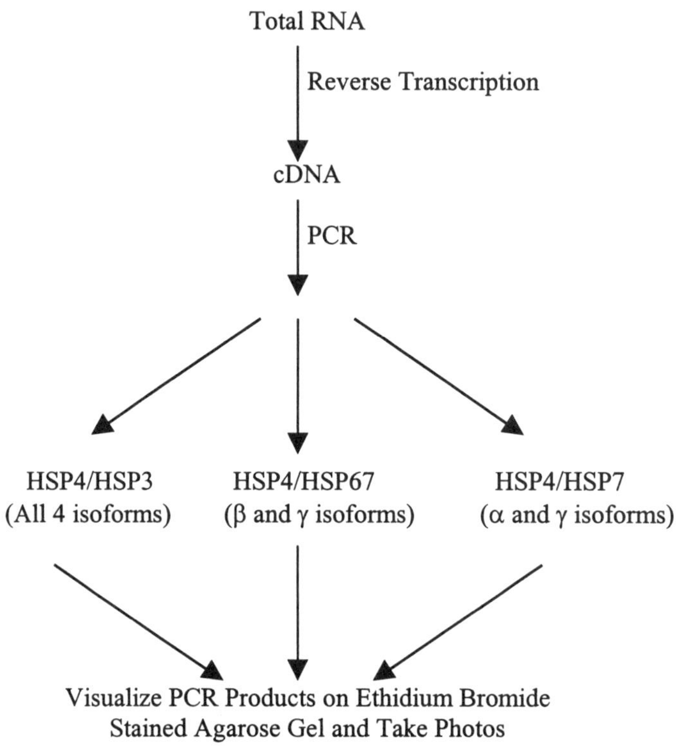

Fig. 3. Flowchart showing the RT-PCR procedure of detecting and differentiating the four isoforms of PPT-A (SP precursor) mRNA. The SP cDNA is prepared by reverse transcription of RNA using a SP-specific primer (HSP2), and the cDNA is amplified by the three primer pairs as indicated.

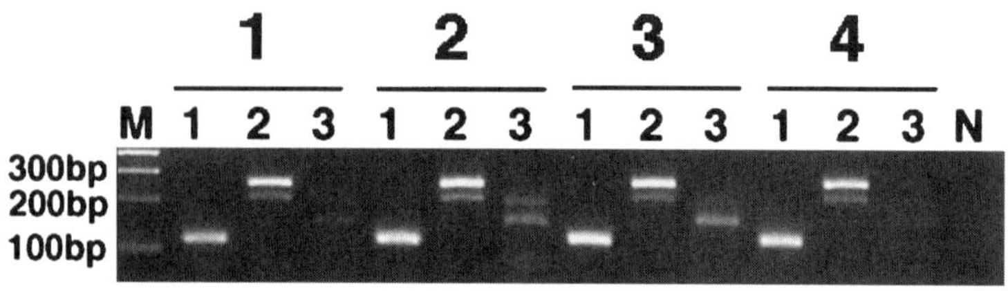

Fig. 4. RT-PCR amplification of PPT-A gene transcripts from human macrophages and PBL. The upper numbers, 1–4, represent four different samples: samples 1 and 2 were macrophages isolated from two donors, and samples 3 and 4 were PBL isolated from two donors. The three different PPT-A PCRs were performed on each sample (lanes 1–3 respectively, lower numbers). For each sample, lane 1: total PPT-A transcripts were amplified with the primer-pair HSP4/HSP3; lane 2: only β and γ transcripts were amplified with HSP4/HSP67; lane 3: only α and δ transcripts were amplified with HSP4/HSP 7. Total PPT-A transcripts were detected in all four samples. The β and γ isoforms of PPT-A were strongly expressed in both of the macrophage and

Presently, competitive RT-PCR is considered to be a reliable and reproducible assay, since this approach has the potential to control for variations in amplification efficiency between individual samples. An exogenous control PCR template is created with identical primer sites to the target template, but with a difference in size that enables amplification products from this control template to be distinguished from those of the target template. The control template is used as an internal standard and is co-amplified with the target template. Because target and control templates share the same primers, they compete for PCR amplification. Thus, if the internal competitive standard is present in equal starting amount to the target template, equal PCR products are obtained from both. By co-amplifying serial dilutions of the standard with fixed aliquots of the sample cDNA, the dilution that produces equal product bands will reveal the concentration of the unknown starting cDNA target. This method was first described by Gilliland et al. *(23)* and Becker-Andre et al. *(24)*.

The internal standard can be a homologous DNA fragment that has the same primer sites as the target DNA, but is designed to generate a PCR product of a different size. Alternatively, a non-homologous DNA fragment of a desired size may be engineered to contain primer sites. It may be obtained by PCR amplification using the target-specific primers and cross-species genomic DNA under reduced annealing stringency, or it may be obtained by amplification of a DNA fragment using composite primers. We use the latter method to generate SP-mimics by a mimic construct kit (Clontech Laboratories Inc., Palo Alto, CA). The SP-mimic-based PCR utilizes an exogenous internal standard (SP-mimic) in a competitive PCR. In our SP-mimic-based PCR, the primer pair HSP4/HSP3 is used to amplify both SP cDNA and the SP-mimic (the internal standard), and the SP-mimic competes with SP cDNA for PCR amplification. The amount of SP cDNA can be calculated based on the amount of the mimic added in each reaction. The initial calculation of SP mRNA level is based on the assumption that the efficiency of cDNA synthesis is 100%. However, the actual efficiency of reverse transcription is less than 100%. This calculation provides the minimum number of SP mRNA molecules in a particular sample.

In **Subheading 3.3.2.**, we describe a protocol for mimic-based RT-PCR to quantify total SP mRNA transcripts, and an RT-PCR protocol to differentiate between some of the four isoforms of PPT-A mRNA transcripts, using several

---

*(continued)*PBL isolations. The α and δ isoforms of PPT-A were weakly expressed in both of the macrophage and PBL isolations, and were negligible in one PBL isolation (sample 4). Lane N: negative control, which was an RT-PCR performed with omission of RT, using 1.2×10^9 molecules of RNA transcript derived from the PPT-A β-isoform plasmid clone *(21)*. For the negative control, PCR was performed using the primer-pair HSP4/HSP3, and no PCR products were obtained.

SP primer-pair combinations. The SP-mimic-based PCR exhibits good reproducibility and accuracy. As little as a two-fold difference in levels of SP mRNA in human immune cells can be distinguished on ethidium bromide-stained gels with the aid of the Doc 1000 device and Multi-Analyst software.

2. Materials

Purchase all reagents as molecular-biology-grade, i.e., RNase-and DNase-free.

2.1. Sample Preparation

1. Human blood monocytes/macrophages and lymphocytes.
2. 15-mL plastic centrifuge tubes (e.g., Falcon).
3. Dulbecco's Modified Eagle's medium (DMEM).
4. Ficoll-Plaque, research-grade (Amersham Pharmacia Biotech AB, Uppsala, Sweden).
5. Gelatin-coated T75 tissue-culture flasks. Autoclave 100 mL of 2% gelatin solution (in deionized water) at 121°C for 15 min. Dispense 10 mL of the autoclaved gelatin solution into each T75 flask when it cools to about 60°C, and coat T75 flasks for 2 h at 37°C, keeping them horizontal with loosened caps at all times. Aspirate the excess gelatin solution from each flask, date, and store horizontally in an incubator at 37°C for at least 2 d before use. The gelatin-coated flasks are stored in the incubator at 37°C until use. They are good for 1 mo.

2.2. RNA Extraction

1. Sterile 1.5-mL capped tubes.
2. TRI REAGENT (Molecular Research Center Inc. , Cincinnati, OH): a reagent for the single-step isolation of total RNA from tissues, cells cultured in monolayer, or cell pellets. 1 mL of TRI REAGENT processes up to 10^7 cells.
3. BCP (Molecular Research Center Inc.): It is a molecular-biology-grade 1-bromo-3-chloropropane.
4. Isopropanol (Sigma, St. Louis, MO).
5. Prepare a solution of 75% ethanol.
6. DEPC-treated sterile water (BIOTECX Laboratories, Inc., Houston, TX).
7. Equipment:
   a. Timer
   b. Heating block for incubating extracted RNA at 55–60°C.

2.3. RT-PCR

To avoid RNase contamination, wear gloves when working with RNA, preparing solutions and master mixes. Keep hands clean, and prepare all solutions and reagents using autoclaved or sterile glassware, microcentrifuge tubes, reaction tubes, and pipet tips. Be sure to keep instruments clean. In general, molecular-biology-grade (RNase- and DNase-free) water should be used for all RT-PCR assays. Solutions for RNA dilution should be prepared with molecular-biology-grade or DEPC-treated water. All RT-PCR reagents should be aliquoted to smaller volumes to minimize damage from repeated freeze/thaw.

1. GeneAmp PCR System 2400; sterile 0.2-mL PCR reaction tubes; AmpliTaq Gold (5 U/µL) and its 10× PCR buffer; $MgCl_2$ (25 m*M*) (Perkin-Elmer, Norwalk, CT).
2. High-pressure liquid chromatography (HPLC) grade water (Fisher Scientific, Fair Lawn, NJ).
3. Molecular-biology-grade water (DEPC-treated) (BIOTECX Laboratories, Inc.).
4. Sterile 1.5-mL microcentrifuge tubes and pipet tips.
5. Total RNA samples.
6. Avian myeloblastoma virus (AMV) reverse transcriptase (RT) (10 U/µL) and RNasin recombinant ribonuclease inhibitor (40 U/µL) (Promega, Madison, WI).
7. Prepare a 10-m*M* mix of the four deoxynucleotide triphosphates (dNTPs) using a dNTP set containing 100 m*M* stock solutions of the sodium salts of each dNTP, pH 8.3 (Boehringer Mannheim GmbH, Mannheim, Germany). Allow each dNTP solution to thaw completely on ice, and vortex each thoroughly before mixing together. Mix an equal vol of 100 m*M* dATP, dCTP, dGTP, and dTTP (total 4 vol) in a sterile 1.5-mL microfuge tube and add 6 vol of DEPC water to the tube so that the final concentration of each dNTP is 10 m*M* (10 × dilution). Vortex and briefly spin down the solution, and then aliquot 50 µL into sterile 1.5-mL tubes clearly labeled with the contents and date. Store at −20°C.
8. PCR primers: sense and anti-sense primers are synthesized by Integrated DNA Technologies, Inc. (Coralville, IA). The following are the primer sequences used for RT-PCR amplification of SP mRNA from human monocytes/macrophages and lymphocytes (*see* **Fig. 1**): HSP2: 5'-TGCATTGCACTCCTTTCA-3' (anti-sense, used for the first strand cDNA synthesis of PPT-A mRNA); HSP4: 5-GACAGC- GACCAGATCAAGGAGGAA-3' (sense); HSP3: 5'-CAGCATCCC GTTT- GCC-3' (anti-sense); HSP7: 5'-CTTTCAT AAGCCATTTTGTGAG AGA-3' (anti-sense) and HSP67: 5'-CTTTCATAAGC CACAGAATTTAAA-3' (anti-sense). The RT-PCR products of PPT-A mRNA are α-isoform (197 bp), β-isoform (251 bp), γ-isoform (206 bp), and δ-isoform (152 bp). All the primers are solubilized in HPLC-grade water to a final concentration of 20 µ*M*.
9. PTT-A cDNA-containing plasmid at known concentration (*see* **Note 1**).
10. SP-mimic dilution series (*see* **Note 2**): Working on ice, dilute the most concentrated stock mimic 10-fold serially using FPLC-grade water to a concentration of 12 molecules per 2 µL (**Table 1**). Distribute them as 50-µL aliquots and store at −70°C.
11. Ethidium bromide solution (10 mg/mL) and gel-loading buffer (Sigma).
12. 100-bp ladder, Promega (Cat. No. G695A).

2.4. Quantification

1. The agarose electrophoresis results are recorded and analyzed with a Gel Doc 1000 imaging device and Multi-Analyst software (BIO-RAD, Hercules, CA).

Table 1
Preparation of SP-Mimic Dilution Series

SP-mimic	Amount	+	Molecular biology H_2O	Copies/2 µL
M1				1.2×10^7/2 µL
M2	20 µL M1	+	180 µL	1.2×10^6/2 µL
M3	20 µL M2	+	180 µL	1.2×10^5/2 µL
M4	20 µL M3	+	180 µL	1.2×10^4/2 µL
M5	20 µL M4	+	180 µL	1.2×10^3/2 µL
M6	20 µL M5	+	180 µL	1.2×10^2/2 µL
M7	20 µL M6	+	180 µL	1.2×10/2 µL
M8	20 µL M7	+	180 µL	1.2 / 2 µL
Negative control	–		200 µL	0

3. Methods

3.1. Sample Preparation

1. Add 15-mL of Ficoll-Plaque into 50-mL plastic tubes.
2. Dispense 10 mL of DMEM into each of another set of new 50-mL plastic tubes, and then expel 30 mL of heparinized peripheral-blood samples into each of these tubes. Mix gently with a pipet. Load 30 mL of the diluted blood samples onto the top of the Ficoll-Plaque solution in **step 1** and centrifuge for 45 min at 800*g* at room temperature.
3. Prime the gelatin-coated T75 flasks (*see* **Subheading 2.1.5**) by adding 10 mL of DMEM to each.
4. Aspirate the top orange layer (plasma) from the Ficoll-Plaque separating tubes, collect the cloudy layer of cells (PBMC) (approx 10 mL per tube) with a pipet and transfer the cells into a new sterile 50-mL tube.
5. Aspirate the DMEM from the gelatin-coated flasks and add 10 mL of fresh DMEM to each flask again.
6. Dispense 10-mL of the cells (PBMC) to each of the gelatin-coated T75 flask (thus 20 mL of vol total per flask now).
7. Place flasks with loosened caps at 37°C for 1 h in order for monocytes to adhere to the gelatin surface.
8. Collect the supernatants that contain peripheral-blood lymphocytes (PBL) from each of the T75 flasks and place into new 50-mL tubes for later experimental use.
9. Wash the flasks 5–6 × with DMEM until no floating cells (PBL) are observed by microscopy.
10. Aspirate the DMEM after the last wash and add 5 mL of DMEM plus 20% fetal calf serum (FCS), and 5 mL of 10 m*M* ethylenediaminetetraacetic acid (EDTA) to each flask to detach monocytes from the gelatin-coated surface.
11. Incubate the flasks at 37°C for up to 15 min and then strike the flasks hard with

the hands in order to detach the monocytes from the gelatin surface.

12. Transfer the cell suspension into a 50 mL tube and then wash the flasks with 10 mL DMEM by "piggy-backing" the liquid from one flask to the next.
13. Wash both the isolated monocytes and PBL at 209*g* for 10 min at room temperature once.
14. Resuspend monocytes with 5 mL of DMEM plus 10% FCS, and PBL with 5 mL of RPMI-1640 plus 10% FCS.
15. Count the cells and seed monocytes in 24-well plates at a concentration of 1×10^6 cells/well. Monocytes become macrophages after 7–10 d in culture. PBL are cultured in RPMI-1640 plus 10% FCS medium containing 1μg/mL phytohemagglutinin (PHA) for 72 h. The PBL are then treated with interleukin-2 (IL-2) (50 μg/mL) and maintained in culture in 24-well plates at a density of 1×10^6 cells /mL.

3.2. RNA Extraction (see *Note 3)*

To minimize contamination with previously amplified PCR products, the RNA extraction is performed in a laminar flow-hood that is separated from the RT-PCR setup laboratory and post-PCR processing area.

1. Homogenization: lyse monocytes/macrophages (1×10^6cells/well) in the 24-well plates with 1 mL of TRI REAGENT. Collect the PHA-stimulated PBL (1×10^6cells/mL) by centrifugation at 800*g* for 5 min and then lyse. Make sure the cells are lysed by pipeting the cell lysate up and down several times. Store the homogenate for 5 min at room temperature to permit the complete dissociation of nucleoprotein complexes.
2. Phase separation: to the homogenate, add 0.1 mL (BCP), cover the samples tightly, and shake vigorously for 15 s. Store the resulting mixture at room temperature for 5–15 min. and centrifuge at 13,400*g* for 15 min at 4°C. Following centrifugation, the mixture separates into a lower red, phenol-chloroform phase, interphase, and the colorless upper-aqueous phase. RNA remains exclusively in the aqueous phase. The volume of the aqueous phase is approx 50% of the volume of the TRI REAGENT used for homogenization.
3. RNA precipitation: transfer the aqueous phase to a new 1.5-mL tube and precipitate the RNA from the aqueous phase by mixing with 0.5 mL of isopropanol. Store samples at room temperature for 5–10 min and centrifuge at 13,400*g* for 8 min at 4–25°C.
4. RNA wash: remove the supernatant carefully and wash the RNA pellet (by vortexing) with 75% ethanol (at least 1 mL) and subsequent centrifugation at 13,400*g* for 5 min at 4–25°C.
5. RNA solubilization: remove the ethanol wash as much as possible, and as carefully as possible. Briefly air-dry the RNA pellet for 5–8 min. and dissolve the RNA in 30 μLof DEPC-treated water. Incubate the RNA solution at 60°C for 10 min and then store the RNA at −70°C.

Table 2
Master Mix Guide for Reverse Transcription

Reagents	For one tube	For ten tubes	Final concentration
10X PCR buffer ($MgCl_2$ free)	2 µL	20 µL	1X PCR buffer
$MgCl_2$ 25 m*M*	5 µL	50 µL	6.5 m*M*
dNTPs 10 m*M* each	1 µL	10 µL	0.5 m*M* each (A, C, G,T)
Primer HSP2 (2 µ*M*)	1 µL	10 µL	0.1 µ*M*
RNasin	0.5 µL	5 µL	20 U
AMV reverse transcriptase	1.5 µL	15 µL	15 U
H_2O	5 µL	50 µL	—
Total master mix	16 µL	160 µL	—

3.3. RT-PCR Amplification (see Note 4)

During the set-up of RT-PCR reactions, keep reagents on ice until they are returned to storage at −20°C. Use fresh pipette tips between all reagents, standards, and samples. Precision of pipetting is important in this procedure.

3.3.1. Reverse Transcription

1. Clearly label a 0.2-mL PCR reaction tube for each amplification to be performed.
2. Make a master mix to ensure that each tube has an equal concentration of reagents. Make the master mix for more tubes than are actually needed, to compensate for loss during pipetting. **Table 2** is provided as a guide. Place the specified amount of each reagent in a sterile 1.5-mL microcentrifuge tube.
3. Gently mix the master mix by inverting the tube several times and then spinning down the mix, so that the reagents are well mixed and all the reagents on the walls will be brought down to the bottom of the tube.
4. Distribute 16 µL of the RT master mix to each of the reaction tubes, and then pipet 4 µL of RNA (1 µg) into each of the reaction tubes. Briefly tap each reaction tube to mix the template and reagents. Centrifuge each of the reaction tubes briefly to eliminate any bubbles and to clear liquid from the side walls of the tubes.
5. Perform the reverse transcription at 42°C for 1 h, and terminate by incubating the reaction tubes at 99°C for 5 min, and then cool them to 4°C. The samples are now ready for PCR amplification of SP cDNA.

3.3.2. PCR Detection and Differentiation of Four Isoforms of PPT-A mRNA

In order to analyze the expression patterns of PPT-A mRNA transcripts in human monocytes/macrophages and PBL, the cDNA samples are amplified by PCR using three primer-pairs as follows: HSP4/HSP3 (for all four isoforms),

Table 3
Master Mix Guide for the PCR Amplification of SP cDNA

Reagents	For one tube	For ten tubes	Final concentration
10X PCR buffer ($MgCl_2$-free)	5 µL	50 µL	1X PCR buffer
$MgCl_2$ 25 m*M*	3 µL	15 µL	1.5 m*M*
dNTPs 10 m*M* each	1 µL	10 µL	200 µ*M* each (A, C, G, T)
Primer HSP4 (20 u*M*)	1 µL	1 0 µL	0.4 µ*M*
PrimerHSP3 or HSP67 or HSP7 (20 µ*M*)	1 µL	10 µL	0.4 µ*M*
AmpliTaqGold	0.3 µL	3 µL	1.5 U
H_2O	36.7 µL	367 µL	To 48 µL
Total master mix	48 µL	480 µL	—

HSP4/HSP7 (for α and δ isoforms only), and HSP4/HSP67 (for β and γ isoforms only) (**Figs. 1** and **3**).

1. Clearly label three 0.2-mL PCR reaction tubes for each amplification to be run—one for all four isoforms of PPT-A mRNA , one for β and γ isoforms, and the last for α and δ isoforms.
2. Make the three master mixes to ensure that each tube has an equal concentration of reagents. **Table 3** is provided as a guide to creating the master mixes (*see* **Note 5**). Always make master mixes for more tubes than are actually needed. Place the specified amount of each reagent in a sterile 1.5-mL microcentrifuge tube.
3. Gently mix the master mix by inverting the tube several times, and then briefly centrifuge the tube to eliminate any bubbles and to clear liquid from the side walls so that reagents are equally distributed.
4. To each PCR reaction tube, add 48 µL of the master mix and 2 µL of cDNA from the reverse transcription reaction.
5. Mix the reagents by tapping the tubes, and then briefly centrifuge the tubes to bring the reagents and cDNA down to the bottom of the tube.
6. Begin the PCR using the programs in **Table 4** for these three primer pairs.
7. Add 15 µLof gel loading buffer to each reaction tube, mix well, and centrifuge briefly to bring the contents to the bottom of the tube.
8. Separate 15 µLof each PCR product on a 4% Tris-acetate EDTA (TAE) agarose gel until the loading dye is about halfway down the gel.
9. Turn off the power supply and remove the gel and gel tray. Photograph the gels in the gel tray on an ultraviolet (UV)-light box.

Table 4
PCR Cycle Programs for Amplification of SP cDNA

Primer-pair	PCR cycle program
HSP4/3 and HSP4/HSP67	95°C for 9 min followed by 45 cycles of 95°C for 30 s, 48°C for 30 s, and 72°C for 30 s; a final elongation at 72°C for 7 min, then cool to 4°C
HSP4/HSP7	95°C for 9 min followed by 45 cycles of 95°C for 30 s, 55°C for 30 s, and 72°C for 30 s; a final elongation at 72°C for 7 min, then cool to 4°C

3.3.3. Preliminary Competitive PCR Amplification

First, titrate a constant amount of your experimental SP cDNA against serial dilutions (10-fold) of the SP PCR mimic (1200, 120, and 12 copies). Based on the results, we will set up a fine-tuned SP-mimic serial dilutions (twofold) for the quantitative PCR.

1. Clearly label three 0.2-mL PCR reaction tubes for each sample to be amplified (one for 12, one for 120, and one for 1200 copies of the SP-mimic).
2. Make a master mix (**Table 5** can be used as a guide) to ensure that each tube has an equal concentration of reagents. Make the master mix for more tubes than are actually needed. Place the specified amount of each reagent in a sterile 1.5-mL microcentrifuge tube.
3. Gently mix the master mix by inverting the tube several times, and then briefly centrifuge the tube to eliminate any bubbles and to clear liquid from the side walls so that reagents are equally distributed.
4. To each of three tubes for each cDNA sample, add the an aliquot of the cDNA, the PCR master mix, and the SP-mimic as follows:
   a. 2 µL cDNA from the reverse-transcription reaction.
   b. 2 µL one dilution (M5, M6, or M7, *see* **Table 1**) of the SP-mimic.
   c. 46 µL PCR Master mix (*see* **Table 5**).
   d. Total 50 µL final reaction vol.
5. Begin PCR using the cycling parameters for the primer pair HSP4/HSP3 (*see* **Table 4**).
6. Electrophorese 12.5 µL of each PCR sample on a 4% agarose gel.

3.3.4. Fine-Tuned Competitive PCR Amplification

1. Determine which 10-fold mimic dilution produces SP-mimic and target cDNA template bands of similar intensity. Then use the mimic dilution tube 10-fold less dilute to start making your twofold serial dilutions. For example, if you determine that the M6 dilution (120 copies/2 µL) gives an SP-mimic band of similar intensity to the target band, begin the twofold serial dilution series with the M5 dilution (1200 copies/2 µL).

Table 5
Master Mix Guide for the Mimic-Based PCR

Reagents	For one tube	For ten tubes	Final concentration
10X PCR buffer ($MgCl_2$-free)	5 µL	50 µL	1X PCR buffer
$MgCl_2$ 25 m*M*	3 µL	15 µL	1.5 m*M*
dNTPs 10 m*M* each	1 µL	10 µL	200 µ*M* each (A, C, G, T)
Primer HSP4 (20 µM)	1 µL	10 µL	0.4 µ*M*
Primer HSP3 (20 µM)	1 µL	10 µL	0.4 µ*M*
AmpliTaqGold	0.3 µL	3 µL	1.5 U
H_20	34.7 µL	47 µL	To 46 µL
Total master mix	46 µL	460 µL	—

2. Clearly label four 0.2-mL PCR reaction tubes 2M1–2M4.
3. To make the twofold serial dilution series, place 5 µLof HPLC water into each tube. Then, make the dilution series as indicated below:

Concentration (copies/2 mL)	Tube label	
1200 (M5)	2M1	SP-mimic solution M5 (**Table 1**)
600	2M2	Add 5 mLof 2M1, mix, and change pipet tip
300	2M3	Add 5 mLof 2M2, mix, and change pipet tip
150	2M4	Add 5 mLof 2M3, mix, and change pipet tip

4. Note: If multiple samples are to be quantitated, the volume of the two-fold dilutions should be increased accordingly.
5. Set up four new 0.2 mL PCR tubes for each sample to be quantitated by the mimic-based PCR amplification.
6. Add to each tube the following reagents:
   a. 2 µL of cDNA from the reverse-transcription reaction.
   b. 2 µL of one dilution of the twofold series of SP-mimic (2M1, through 2M4).
   c. 46 µL of the PCR master mix.
   d. 50 µL final reaction volume.
7. Begin PCR amplification using the cycling parameters for the primer pair HSP4/HSP3 (*see* **Table 4**).
8. Electrophorese 12.5 µL of each sample on a 4% ethidium bromide-containing agarose gel.

Using the Gel Doc 1000 imaging device and Multi-Analyst software, determine which two-fold serial dilution gives a PCR band for the SP cDNA of

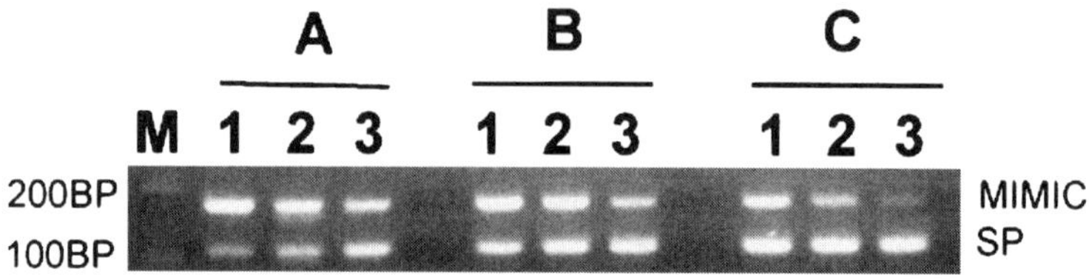

Fig. 5. Quantification of SP mRNA from human PBMC. **A**, **B**, and **C**: Three RNA samples isolated from human PBMC are reverse-transcribed with primer HSP2, and one-tenth of each cDNA sample in the presence of known molecules of the SP-mimic are amplified with the primer pair HSP4/HSP3. Lanes 1, 2, and 3: copy numbers of the SP-mimic (120, 60, and 30 copies) added to each PCR reaction. The ratio of SP to mimic in sample A, B, and C is as follows: 1.05 (lane 3), 0.97 (lane 2), and 1.12 (lane 1). These results indicated that sample A, B, and C have 30, 60, and 120 copies of SP cDNA molecules, respectively ***(21)***.

Table 6
Ratios of Substance P cDNA to SP-Mimic

SP MIMIC	Sample A	Sample B	Sample C
120 copies	0.34	0.79	1.12
60 copies	0.64	0.97	1.5
30 copies	1.05	1.3	2.74

Sample A, B, and C: three total RNA samples isolated from human PBMC were amplified using the primer pair HSP4/HSP3. In order to quantitatively measure SP cDNA, the gel image was recorded and analyzed using Gel Doc 100 image devices and Multi-Analyst software. The values in this table are the ratios of SP cDNA to SP-mimic.

equal band intensity to the mimic band. **Figure 5** and **Table 6** show examples of quantification of SP mRNA from human PBMC (*see* **Note 6**).

4. Notes

1. PPT-A cDNA-containing plasmids were obtained by cloning a PCR-amplified PPT-A cDNA in our laboratory ***(21)***. Four isoforms of PPT-A mRNA transcripts were amplified with primer pairs of HSP4/HSP5 and HSP4/HSP7, as previously described in detail ***(14, 15)***. The PCR products amplified by these primers were separated on a 4% agarose gel and then purified with the Wizard PCR Prep DNA Purification System (Promega). The purified cDNA of PPT-A isoforms (α, β, γ, δ) were then cloned into a plasmid using the Eukaryotic TA Cloning Kit (Invitrogen Corporation, San Diego, CA). For each PPT-A isoform, at least 10 clones were chosen and grown overnight in 5 mL of Lunia Bertani (LB) broth

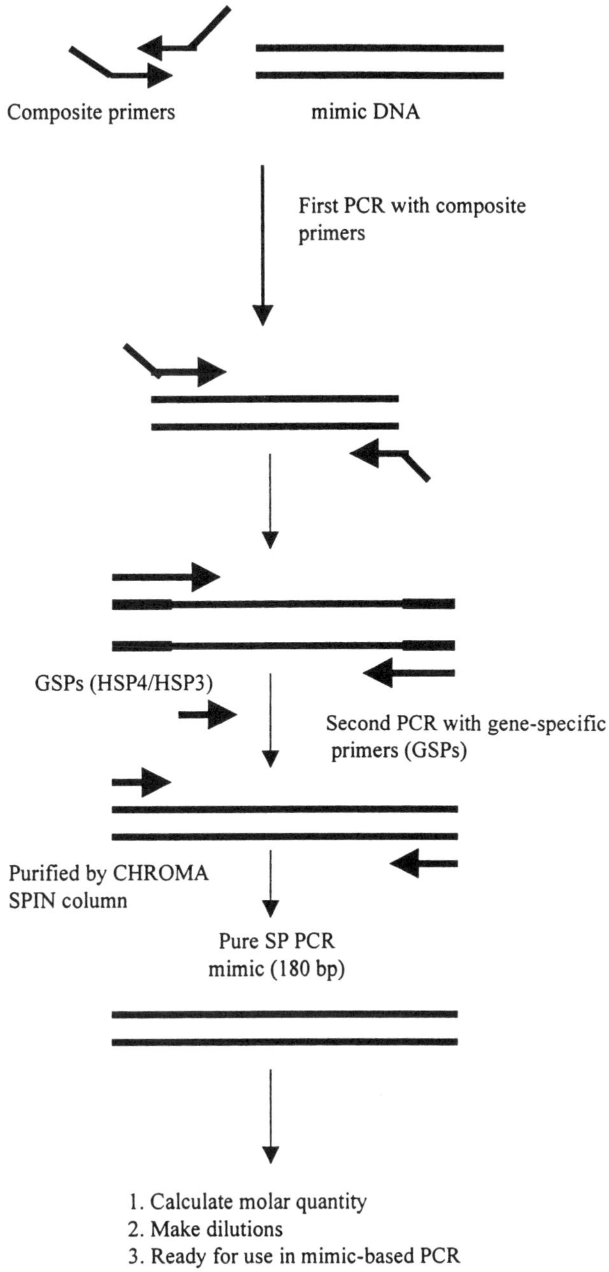

Fig. 6. SP PCR-mimic construction strategy.

containing ampicillin (50 µg/mL). The cloned plasmids containing cDNA of PPT-A isoforms were then purified with Wizard Plus Miniprep DNA Purification System (Promega). The presence and orientation of the PPT-A isoform inserts were determined by restriction analysis, using *Eco*RV and DNA sequencing. The purified plasmids were linearized by *Eco*R1 restriction digestion and purified by phenol-chloroform extraction and alcohol precipitation. These cloned PPT-A isoform-containing plasmids were used as standard controls for studying the primer-pairing strategy and as a template to synthesize mRNA in vitro to evaluate the sensitivity of the RT-PCR and determine the accuracy of the quantitative mimic-based PCR for SP mRNA transcripts.

2. In order to quantitatively measure SP cDNA, an SP-mimic was constructed with the PCR MIMIC Construction Kit (Clontech Laboratories, Inc.). **Figure 6** shows the SP-mimic construction strategy. The SP-mimic has a size of 180 bp, with primer sites for the HSP4 and HSP3 primers at its ends. Serial dilutions of known amounts of the SP-mimic were added to the PCR reactions containing constant amounts of SP cDNA. Since the SP-mimic fragment contains the same sequences of HSP4/HSP3 primer sites as SP cDNA, it competes with target SP cDNA during the amplification reaction. The amount of SP cDNA could then be determined based on the SP-mimic dilution, at which the PCR products for the SP-mimic and the SP cDNA appeared the same. The PCR-amplified products were then resolved on a 4% agarose gel. The results were recorded and analyzed with a Gel Doc 1000 imaging device and Multi-Analyst software.
3. To minimize RNase contamination, wear gloves and keep tubes closed throughout the procedure. Complete evaporation of ethanol from the washed RNA pellet is especially important, because the RNA will be redissolved in a small volume (5–20 µL), which, if not dried sufficiently, may contain a relatively high level of ethanol. After homogenization, samples can be stored at −70°C for at least 1 mo. The RNA precipitate can be stored in 75% ethanol at 4°C for at least 1 wk, or at least 1 yr at −20°C.
4. Rigorous steps are taken to avoid risk of contamination of samples, or cross-contamination between samples, at all stages of the RT-PCR protocol. Reaction mixtures for the reverse transcription and PCR stages are always prepared in a laminar flow-hood; the remainder of aliquots of the PCR master mixes are disposed after the first use; reverse transcription and PCR are performed in a room separated from the laminar flow hoods where RT-PCR reactions are set up; and the subsequent gel electrophoresis of amplified PCR products are performed on a designated bench in a separate room from the room where RNA extraction, RT-PCR setup, and RT-PCR amplification are performed.
5. We routinely make master mixes of SP PCR reagents without AmpliTaq Gold in a quantity of 300 reactions, divide them into 0.5-mL aliquots, and store at −20°C. Thaw one aliquot of the master mix just before PCR and make the working master mix depending on how many samples will be amplified. Add AmpliTaq Gold to the mix, mix well, and then dispense them into PCR reaction tubes. In this

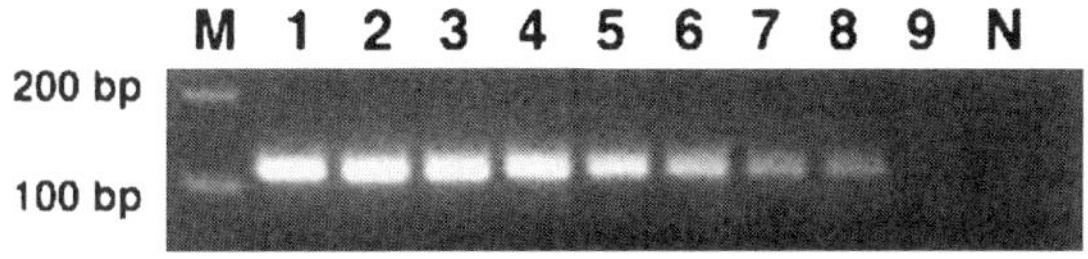

Fig. 7. RT-PCR sensitivity analysis with RNA transcripts derived from a plasmid clone of the PPT-A β-isoform. Primer HSP2 was used for reverse transcription, and primer pair HSP4/HSP3 was used for PCR. Lanes 1 to 9: 10-fold serial dilutions of RNA starting from 1.2×10^9 (lane 1) to 12 (lane 9) molecules per reaction; Lane M: 100-bp ladders; Lane N: negative control, which lacked PCR-amplified product when RT was omitted from the RT reaction using 1.2×10^9 molecules of RNA. A sensitivity of 120 molecules was attained (lane 8) ***(21)***.

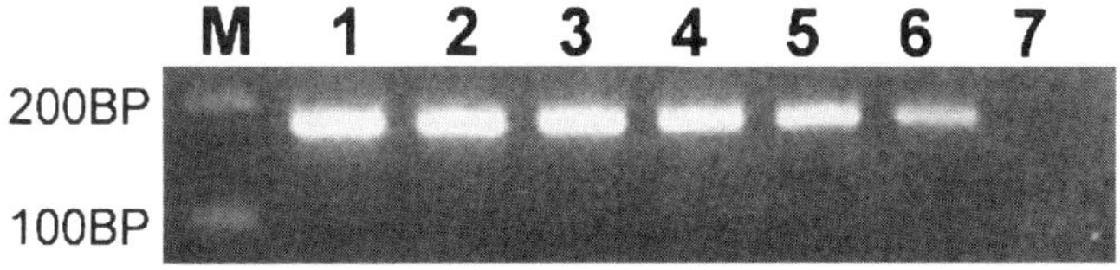

Fig. 8. PCR sensitivity analysis with SP-MIMIC. Primer pair HSP4/HSP3 was used for PCR. The SP-MIMIC was amplified for 45 cycles with an expected product size of 180 bp. Lanes 1 to 7: 10-fold serial dilutions of SP-MIMIC starting from 1.2×10^6 to 1.2 molecules per reaction. A sensitivity of 12 copies was attained (lane 6) ***(21)***.

way, the variation of the PCR reagents between several PCR amplifications will be minimized.

6. The sensitivity of the RT-PCR described is 120 copies, and the sensitivity of PCR amplification of the SP-mimic is 12 copies (**Figs. 7** and **8**).

References

1. McGillis, J.P., Mitsuhashi M., and D. G., Payan, D. G. (1990) Immunomodulation by tachykinin neuropeptides. *Ann. NY Acad. Sci.* **594,** 85–94.
2. Gilbert, M.S., Bunnett N.W., and D.G., Payan. (1992) Antibodies to the rat substance P receptor: production and characterization. *Cell. Mol. Neurobiol.* **12,** 529–545.
3. Black, P.H. (1994) Immune system-central nervous system interactions: effect and immunomodulatory consequences of immune system mediators on the brain. *Antimicrob. Agents and Chemother.* **38,** 7–12.
4. Black, P.H. (1994) Central nervous system-immune system interactions: psychoneuroendocrinology of stress and its immune consequences. *Antimicrob. Agents and Chemother.* **38,** 1–6.
5. Pernow, B. (1983) Substance P. *Pharmacol. Rev.* **35,** 85–141.

6. Nakanishi, S. (1987) Substance P precursor and kininogen: their structures, gene organizations, and regulation. *Physiol. Rev.* **67,** 1117–1142.
7. Pascual, D.W., and K. L., Bost. (1990) Substance P production by P388D1 macrophages: a possible autocrine function for this neuropeptide. *Immunology* **71,** 52–56.
8. Bost, K.L., S.A., Breeding, and D.W., Pascual. (1992) Modulation of the mRNAs encoding substance P and its receptor in rat macrophages by LPS. *Reg'l Immunol.* **4,** 105–112.
9. Linnik, M.D., and Moskowitz, M.A. (1989) Identification of immunoreactive substance P in human and other mammalian endothelial cells. *Peptides* **10,** 957–62.
10. Milner, P., Kirkpatrick, K.A., Ralevic V., Toothill, V., Pearson, J., and.Burnstock, G. (1990) Endothelial cells cultured from human umbilical vein release ATP, substance P and acetylcholine in response to increased flow. *Proc. Royal Soc. London - Series B: Biol. Sci.* **241,** 245–248.
11. Aliakbari, J., Sreedharan, S.P., Turck, C.W., and Goetzl, E.J. (1987) Selective localization of vasoactive intestinal peptide and substance P in human eosinophils. *Biochem. Biophys. Res. Commun.* **148,** 1440–1445.
12. Chiwakata, C., Brackmann, B., Hunt, N., Davidoff, M., Schulze, W., and Ivell, R. (1991) Tachykinin (substance-P) gene expression in Leydig cells of the human and mouse testis. *Endocrinol.* **128,** 2441–2448.
13. Ho, W.Z., Lai, J.P., Zhu, X.H., Uvaydova, M., and Douglas, S.D. (1997) Human monocytes and macrophages express substance P and neurokinin-1 receptor. *J. Immunol.* **159,** 5654–5660.
14. Lai, J.P., Douglas, S.D., and Ho, W.Z. (1998) Human lymphocytes express substance P and its receptor. *J. Neuroimmunol.* **86,** 80–86.
15. Lai, J.P., Douglas, S.D., Rappaport, E., Wu, J.M., and Ho, W.Z.. (1998) Identification of a delta isoform of preprotachykinin mRNA in human mononuclear phagocytes and lymphocytes. *J. Neuroimmunol.* **91,** 121–128.
16. MacDonald, M.R., McCourt, D.W., and Krause, J.E. (1988) Posttranslational processing of alpha-, beta-, and gamma-preprotachykinins. Cell-free translation and early posttranslational processing events. *J. Biol. Chem.* **263,** 15,176–15,183.
17. Harmar, A.J., Hyde, V., and Chapman, K. (1990) Identification and cDNA sequence of delta-preprotachykinin, a fourth splicing variant of the rat substance P precursor. *FEBS Lett.* **275,** 22–24.
18. Nawa, H., Kotani H., and. Nakanishi S. (1984) Tissue-specific generation of two preprotachykinin mRNAs from one gene by alternative RNA splicing. *Nature* **312,** 729–734.
19. Krause, J.E., Chirgwin, J.M., Carter, M.S., Xu, Z.S., and Hershey, A.D. (1987) Three rat preprotachykinin mRNAs encode the neuropeptides substance P and neurokinin A. *Proc. Natl. Acad. Sci. USA* **84,** 881–885.
20. Khan, I. and Collins S.M. (1994) Fourth isoform of preprotachykinin messenger RNA encoding for substance P in the rat intestine. *Biochem. Biophys. Res. Commun.* **202,** 796–802.

21. Lai, J.-P., Douglas, S.D., Zhao, M., and Ho, W.-Z. (1999) Quantification of substance P mRNA in human mononuclear phagocytes and lymphocytes using a mimic-based RT-PCR. *J. Immunol. Methods* **230,** 149–157.
22. Larrick, J.W. (1992) Message amplification phenotyping (MAPPing)—principles, practice and potential. *Trends Biotech.* **10,** 146–52.
23. Gilliland, G., Perrin, S., Blanchard, K., and Bunn, H.F. (1990) Analysis of cytokine mRNA and DNA: detection and quantitation by competitive polymerase chain reaction. *Proc. Natl. Acad. Sci. USA* **87,** 2725–2729.
24. Becker-Andre, M., and Hahlbrock, K. (1989) Absolute mRNA quantification using the polymerase chain reaction (PCR). A novel approach by a PCR aided transcript titration assay (PATTY). *Nucleic Acids Res.* **17,** 9437–9446.

IV

Detection and Analysis of RNA Viruses

11

Detection and Quantification of the Hepatitis C Viral Genome

Liam J. Fanning

1. Introduction

The use of molecular-based methodologies for the identification of viral nucleic acid has evolved into a clinically useful adjunct to the biochemical, serological, and histological assays used to evaluate patient health. The clinical application of viral-load assays for monitoring viral infections has been the impetus for the development of standardized tests for the hepatitis C virus (HCV). Standardization and precision of quantification have been instrumental in assessing the natural history of HCV, in which viral RNA levels have a large dynamic range. Assessment of the HCV viral load is now a routine part of the algorithm for the diagnosis of chronic hepatitis C *(1–5)*. Patient's response to anti-viral therapy can be assessed, in part, by quantification of serum viral loads *(6,7)*.

1.1. Detection of HCV

In essence, there are two types of assays for the detection of HCV. The first test in the diagnostic algorithm is the qualitative assessment of viral absence or presence. The second part of this algorithm is quantification of the amount of the virus present in the specimen. The extensive availability of the reagents and enzymes needed to perform reverse transcription and the polymerase chain reaction (PCR) has resulted in an explosive development of assays for the detection of HCV RNA in many laboratories worldwide. These in-house assays can suffer from a lack of standardization, with variable sensitivity. The in-house reverse-transcriptase-polymerase chain reaction (RT-PCR) assays are

From: *Methods in Molecular Biology, vol. 193: RT-PCR Protocols*
Edited by: J. O'Connell © Humana Press Inc., Totowa, NJ

usually nested, and if not controlled properly, can be associated with increased identification of false-positive reactions. The commercially available assays offer the advantage of being validated, have in-built anti-contamination systems, and show less variation between assays than the in-house varieties. The commercially available assays are limited because they are expensive, and thus, find greater utility in clinical virology laboratories and in the clinical trial setting. A comprehensive discussion of the large numbers of protocols available for in-house RT-PCR detection of HCV is beyond the scope of this chapter.

1.2. Specimen Storage

Specimen processing and storage conditions can influence the stability, and thus the detectability, of HCV RNA. Well-controlled sample processing and storage conditions are critical to the quantitative analysis of HCV RNA *(8,9)*. Cryopreservation of specimens at −80°C appears to offer the most stable condition for long-term storage. Where possible, prompt measurement of the viral load without long-term cryopreservation is the option of choice *(10)*. Long-term cryopreservation for several years with subsequent batch analysis may yield artefactual changes in viral load and thus affect accurate quantification of the viral RNA. HCV RNA is not stable when stored at an ambient temperature or 4°C for longer than a few days. Multiple freeze-thaw cycles can decrease the integrity of the specimens, resulting in artefactual modifications of viral load.

1.3. Qualitative Assessment of HCV

The Roche AMPLICOR™ system is a combined RT-PCR-enzyme-linked immunosorbent assay [RT-PCR-ELISA] assay in which HCV nucleic acid can be converted from RNA to cDNA and then amplified to generate sufficient quantity of amplicons to detect using enzyme-linked immunosorbent assay (ELISA). The RT-PCR is performed in a single tube, reducing contamination and facilitating automation. There have been several versions of the AMPLICOR system. v2.0 is the most sensitive to date, with a lower limit of detection of 100 viral copies per mL.

HCV RNA is isolated from 200 μL of serum using a guanidine-thiocyanate lysis buffer. An internal control is added to the lysis solution to serve as an extraction and amplification control, thus ensuring result integrity of results. After heat treatment at 60°C for 10 min, an equal volume of isopropanol is added. Following incubation at room temperature for 2 min the RNA is precipitated by centrifugation at 12,500*g*. The pellet is washed in 70% ethanol and resuspended in a bicine buffer. Reverse transcription and PCR of the 5'UTR (untranslated region) is performed using the Tth DNA polymerase from Thermus thermophilus. The anti-sense primer is 5' biotinylated. The addition of a biotin molecule facilitates detection of the amplicons by ELISA. The system also incorporates a protection mechanism against contamination from previous amplification products. Amplification of the target nucleic acid incorporates a uracil instead of a thymidine. Tracking of amplified product into new reactions is prevented by the activity of the enzyme uracil-N-glycosylase (UNG). UNG hydrolyzes DNA which contain a uracil instead of a thymidine. This selective destruction of potential carryover from previous amplification reactions leaves thymidine-containing DNA and uracil-containing RNA templates intact for selective amplification. Incubation at 50°C prior to initiation of the reverse transcription reaction destroys any contaminating products. Analysis of the amplicons is conducted through ELISA.

1.4. Quantitative Measurement of Serum HCV Load

Quantitative assessment of HCV-RNA levels via signal amplification and quantitative PCR are valuable tools in the clinical management of patients before, during, and after therapy. Many of the commercially available quantitative PCR assays have clinically relevant detection thresholds of approx 100 viral copies/mL of serum. Ideally, assays for the quantification of HCV viral load should span the dynamic range of viremia observed in serum (**Fig. 1**). The dynamic range of HCV is extensive, and viral loads range from 3 $\log_{10}$ viral copies per mL to more than 7.7 $\log_{10}$ viral copies per mL in most clinical settings. This implies that an ideal quantification assay should have a linear range of amplification that flanks these values. The most recent version of the Roche MONITOR™ quantitative HCV assay has a linear range of amplification from 3.3 to 5.7 log10 viral copies per mL ***(11)***. With appropriate dilution of patient sera prior to extraction/amplification, the linear range of this latter assay can be extended to 7.7 $\log_{10}$ viral copies per mL. Earlier versions of the Roche MONITOR™ assay suffered from a biased amplification of HCV genotype 1. Non-uniformity of amplification of non-genotype 1 isolates resulted in the underestimation of viral load from these specimens. However, reformulation of the enzymatic mix used in the amplification of the target nucleic acid ensured more complete denaturation and more constant amplification across all genotypes. The SuperQuant assay (National Genetics Institute, Culver City, CA, USA) is linear between 2 and 8 $\log_{10}$ viral copies per mL, although samples greater than 6.7 $\log_{10}$ viral copies per mL require dilution (**Fig. 1**).

Manufacturers of the Roche MONITOR assay assign a cut-off of 3.3 $\log_{10}$ viral copies per mL and manufacturers of the Chiron branched-chain DNA (bDNA) -2 assay have established a detection limit of 5.3 $\log_{10}$ equivalent genomes/mL. This suggests that the Roche product is the more sensitive of the two assays. In reality, the Roche copy and the Chiron genome do not represent the same amount of HCV RNA in a clinical sample. The greatest difficulty with the interpretation of quantitative information accruing from the viral

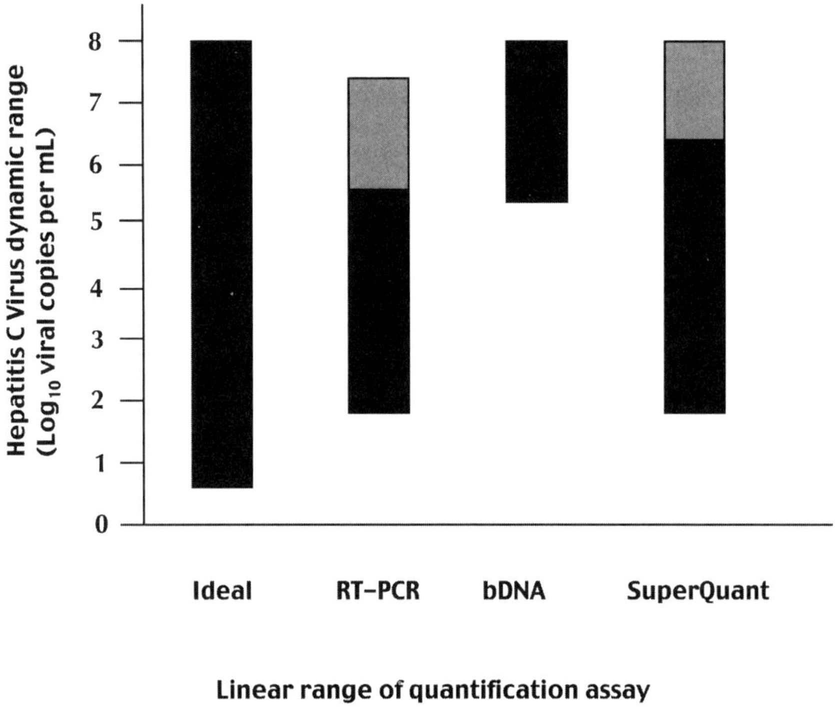

Fig. 1. Comparison of ideal linear range of quantification with three commercially available assays for the hepatitis C virus (HCV): RT-PCR (AMPLICOR/MONITOR versions v2.0 and 2.0, respectively, Roche, USA), bDNA (HCV 2.0, Bayer Diagnostics, USA: unit of measurement $\log_{10}$ genome equivalents per mL), SuperQuant (National Genetics Institute, USA). Grey boxed area represents extended range of quantification with appropriate dilution. The HCV viral loads per mL as assessed by the MONITOR, the bDNA, and the SuperQuant HCV quantification assays, are not equivalent. This relationship has been investigated by several groups ***(15, 16)***.

nucleic acid isolation/detection methodologies available is that comparison between technologies and units of reporting viral load did not until recently have an international standard against which all quantitative values can be directly compared ***(12)***. A collaborative study was organized by National Institute for Biological Standards, as well as the Control and Central Laboratory of the Netherlands Red Cross Blood Transfusion Service to assess the suitability of three candidate preparations as a potential International Standard, and to determine the RNA content of the candidate materials. The concentration was expressed in International Units (IU), based on the mean titer, and the concentration assigned is 105 International Units/mL (IU/mL). The WHO Expert

Committee on Biological Standardization accepted the standard, and in October 1997 this material became the First International Standard for HCV RNA NAT *(13)*. Recently, the Roche MONITOR system has converted the units of quantification from viral copies per mL to IU/mL, with a more precisely defined linear range of amplification—i.e., 2.8 to 5.9 $\log_{10}$ IU/mL (600–8.5 × 10^5 viral copies per mL). Calibration against this international standard provides a reference for all viral-load determinations.

1.5. Quantification of Serum HCV Load

1. Quantitative HCV RT-PCR by the Roche MONITOR test (F. Hoffmann-La Roche Ltd., CH-4070 Basel, CH) is based on a denaturation of serum nucleases by a cathotropic agent and an alcohol precipitation. Briefly, HCV RNA is isolated from 100 µL of serum using a guanidine thiocyanate lysis buffer. After heat treatment at 60°C for 10 min, an equal volume of isopropanol is added. Following incubation at room temperature for 2 min the RNA is precipitated by centrifugation. The pellet is washed once in 70% ethanol and resuspended in a bicine buffer. A known quantity of internal standard is included in each preparation of HCV RNA. RT-PCR of the 5'untranslated region generates the amplicons, which are then subsequently quantified by ELISA.
2. The bDNA assay (Chrion Corporation, USA) for measuring the amount of HCV RNA in a sample of plasma does not use the classical amplification of target by use of PCR (**Fig. 2**). Instead the RNA extracted from HCV is quantified directly. The target nucleic acid is captured by use of a molecular anchor oligonucleotide, which is complementary to the viral target and to the molecular tag that is attached to the microwell plate. Using a molecular adjunct, which hybridizes to the target nucleic acid, a bridge molecule is hybridized to the adjunct nucleic acid. Finally, "amplifier" sequences are added to the molecular bridge. The attachment of these "amplifier" sequences gives a branched structure. A labeled probe is hybridized to the "amplifier", the substrate dioxetane is added to the reaction mixture, and light is emitted. Measurement of the chemiluminescence produced is used to quantify the amount of virus in the starting material.

   RT-PCR based quantification (AMPLICOR/MONITOR & SuperQuant) and branched-chain methodologies have proven to be extremely valuable and effective in determining the natural history of HCV infections and the efficacy of the appropriate anti-viral therapies *(2,6,7,10,14)*. Initially, there were substantial differences between the RT-PCR procedure and the bDNA assay, and the RT-PCR based assay was more sensitive. The bDNA assay is the more reproducible of the two methodologies. The new versions of both assays are comparable in sensitivity and accuracy. However, patients who are nonreactive in the bDNA assay must be retested by RT-PCR because low viral titers are not detected by the bDNA assay *(15)*. In contrast to quantitative PCR, the bDNA assay is highly standardized, although reference to the international standard is not currently available on this platform. However, this is anticipated in the near future. The bDNA assay does

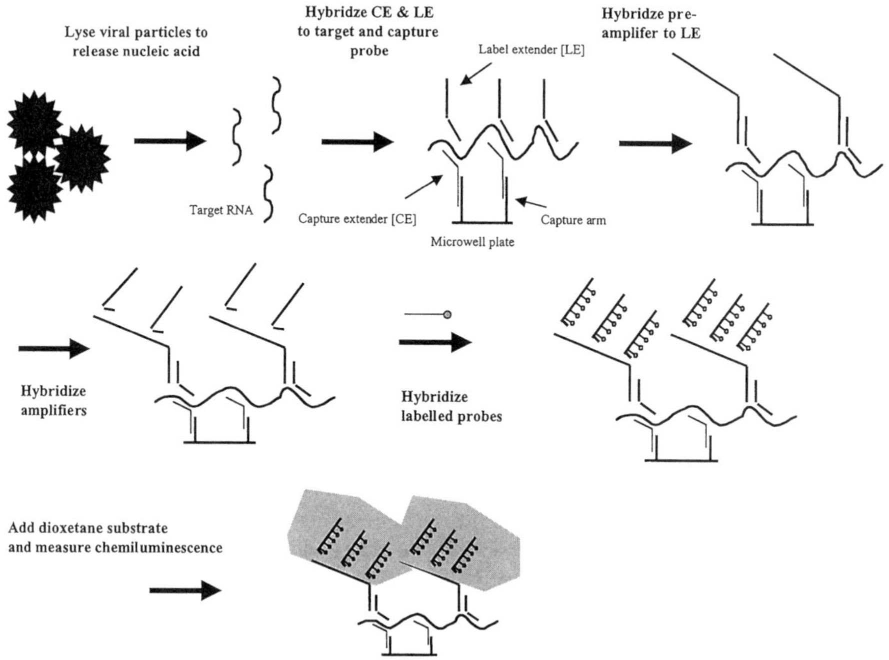

Fig. 2. Quantiplex branched-chain quantification assay for the hepatitis C virus (HCV). The range of quantification for this assay is from 5.3–8.0 $\log_{10}$ genome equivalents per mL. Abbreviations: CE, capture extender; LE, label extender.

not suffer from the genotype bias that earlier versions of the Roche MONITOR system exhibited.

In a study using the HCV RNA Assay (bDNA-1; Chiron Corporation), the HCV RNA 2.0 assay (bDNA-2; Chiron Corporation), and the Roche MONITOR assay, highly reproducible results were observed upon repeat testing of samples by both the bDNA-1 and the bDNA-2. Higher variability was observed in the Roche MONITOR assay (correlation coefficient of 0.537, compared with 0.942 for the bDNA-1 assay and 0.964 for the bDNA-2 assay) ***(16)***. Differences in the efficiency of detection of genotypes 1, 2, and 3 were observed for the bDNA-1 and Roche MONITOR assays. These results suggest that many of the previously published studies evaluating the effect of genotype and virus load on the response to alpha interferon and disease progression, which used older versions of the ROCHE MONITOR assay, may require reinvestigation. However, as stated previously, the more recent versions of these technologies have pan-genotype equalization of sensitivity and accuracy ***(11,12)***.

3. Transcription-mediated amplification (TMA) assesses RNA levels via an RNA amplification step followed by a dual kinetic assay in which bound probes to (a) target and (b) internal controls emit differing wavelengths of light (Bayer diag-

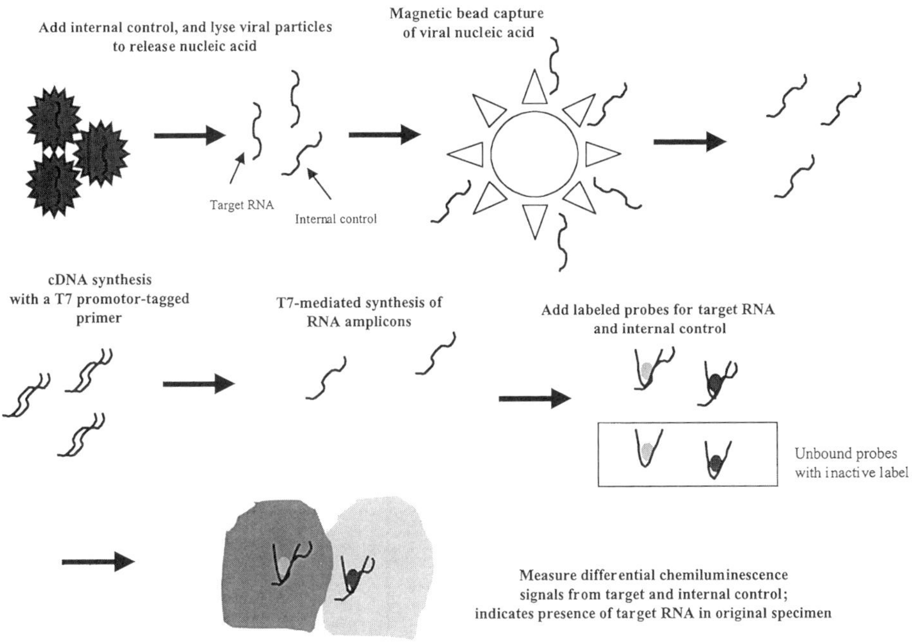

Fig. 3. Transcription-mediated amplification assay for the detection for the hepatitis C virus (HCV). The detection threshold of this assay is 0.7 $\log_{10}$ 5 IU/mL.

nostics Laboratories, Emeryville, CA). The methodology is similar in format to that of the branched-chain technology, because TMA utilizes a capture step for the target and internal-control nucleic acids **(Fig. 3)**. The next stage is conceptually similar to that of the Roche MONITOR assay. Amplification of the target RNA is performed by an autocatalytic, isothermal synthesis of RNA using reverse transcriptase (RT) and T7 RNA polymerase. However, with TMA, the amplicons generated are RNA and not DNA, as in the RT-PCR-based AMPLICOR/MONITOR system. Two primers are used; one containing the T7 RNA-promoter sequence in addition to sequences complementary to the target nucleic acid, the second primer contains sequences complementary to part of the first strand of synthesized cDNA. In essence, the T7 promoter-containing primer binds to the target RNA, and the RT then synthesizes cDNA. The RNA component of the RNA-cDNA hybrid is destroyed by the RNase H activity of the RT. The second primer then binds to the single-stranded cDNA, and a complementary DNA strand is synthesized. The T7 RNA polymerase then synthesizes RNA transcripts from the T7-promoter sequence from this newly generated double-stranded DNA. Each newly generated RNA re-enters the TMA process and serves as a template for a new round of replication. The end result of the cyclic events of synthesis of cDNA—hydrolysis of RNA, synthesis of second-strand DNA and synthesis of RNA—results in the exponential amplification of the target RNA. This methodology is a relatively new development, and experience with this technology is still limited. However, TMA has a stated detection threshold for HCV RNA of less than 50 IU/mL. Recently, evidence of the clinical utility of this methodology has become apparent. In one study, Sarrazin et al. demonstrated that 36% of treated patients who tested HCV-negative using the Roche AMPLICOR RT-PCR system (v2.0) tested positive by TMA ***(17)***. If validated in other prospective studies, TMA has the potential to assist in the clinical management of HCV-infected patients.

4. Individuals occasionally present with varying combinations of co-infection with HCV, Hepatitis B virus (HBV) and/or HIV-1. Simultaneous isolation of these viruses is advantageous when sample availability is limited. In addition, it offers the advantage of reducing the expenditure of time and money that is involved in the isolation of viral nucleic acid for downstream manipulations. Currently, there are few validated protocols for the simultaneous extraction of HCV, HBV, and HIV. In essence, these methodologies depend upon the concentration of viral particles by ultracentrifugation before application of the isolation procedures described above in **Subheadings 1.3.** and **1.4.**; (1) Viral particles can be concentrated by ultracentrifugation at 23,000*g* for 160 min at 16°C; alternatively ***(18)***, (2) addition of NaCl and polyethyleneglycol to the sera, and incubation at 4°C prior to ultracentrifugation at 40,000*g* for 1 h at 4°C ***(19)***, and (3) ultracentriugation at 24,000*g* for 1 h at 4°C, followed by removal of 90% of the supernatant, with viral nucleic acid being extracted from the pellet and remaining 10% of the supernatant ***(20)***. The comparative efficiency in concentrating viral particles of these three methodologies has not yet been determined. However, addition of this simple ultracentrifugation step is likely to be a valuable adjunct to the current range of methodologies and technologies used to isolate viral nucleic acid.

In summary, the precision of HCV diagnostic testing has direct clinical relevance in the provision of patient care, as it can identify those patients who are recently infected, identify the type of the infecting virus, allow monitoring of patient viral load and disease progression, and enable optimization of therapeutic intervention. Most importantly, hepatitis C virology testing provides patients with the information that can lead to realistic expectations about their prognosis.

References

1. Alter, M.J., Kruszon-Moran, D., Nainan, O.V., McQuillan, G.M., Gao, F., Moyer, L.A., et al. (1999) The prevalence of hepatitis C virus infection in the United States 1988 through 1994. *N. Engl. J. Med.* **341,** 556–562.
2. Fanning, L., Kenny, E., Sheehan, M., Cannon, B., Whelton, M., O'Connell, J., et al. (1999) Viral load and clinicopathological features of chronic hepatitis C (1b) in a homogeneous patient population. *Hepatol.* **29,** 904–907.
3. Kenny-Walsh, E. (1999) Clinical outcomes after hepatitis C infection from contaminated anti-D immune globulin. Irish Hepatology Research Group. *N. Engl. J. Med.* **340,** 1228–1233.

4. Lu, R.H., Hwang, S.J., Chan, C.Y., Chang, F.Y., and Lee, S.D. (1998) Quantitative measurement of serum HCV RNA in patients with chronic hepatitis C: comparison between Amplicor HCV monitor system and branched DNA signal amplification assay. *J. Clin. Lab. Anal.* **12,** 121–125.
5. Nomura, H., Kimura, Y., Rikimaru, N., Tada, H., Okamoto, O., Shiraishi, G., et al. (1997) Usefulness of HCV-RNA assays in efficacy evaluation of interferon treatment for chronic hepatitis C: Amplicor HCV assay and branched DNA probe assay. *J. Infect.* **34,** 249–255.
6. McHutchison, J.G., Gordon, S.C., Schiff, E.R., Shiffman, M.L., Lee, W.M., Rustgi, V.K., et al. (1998) Interferon alpha-2b alone or in combination with ribavirin as initial treatment for chronic hepatitis C. Hepatitis Interventional Therapy Group. *N. Engl. J. Med.* **339,** 1485–1492.
7. Poynard, T., Marcellin, P., Lee, S.S., Niederau, C., Minuk, G.S., Ideo, G., et al. (1998) Randomised trial of interferon alpha 2b plus ribavirin for 48 weeks or for 24 weeks versus interferon alpha 2b plus placebo for 48 weeks for treatment of chronic infection with hepatitis C virus. International Hepatitis Interventional Therapy Group (IHIT). *Lancet* **352,** 1426–1432.
8. Gordon, S.C., Dailey, P.J., Silverman, A.L., Khan, B.A., Kodali, V.P., and Wilber, J.C. (1998) Sequential serum hepatitis C viral RNA levels longitudinally assessed by branched DNA signal amplification. *Hepatol.* **28,** 1702–1706.
9. Halfon, P., Khiri, H., Gerolami, V., Bourliere, M., Feryn, J.M., Reynier, P., et al. (1996) Impact of various handling and storage conditions on quantitative detection of hepatitis C virus RNA. *J. Hepatol.* **25,** 307–311.
10. Fanning, L., Kenny-Walsh, E., Levis, J., Choudhury, K.R., Cannon, B., Sheehan, M., et al. (2000) Natural fluctuations of hepatitis C viral load in a homogeneous patient population: a prospective study. *Hepatol.* **31,** 225–229.
11. AMPLICOR HCV MONITOR TEST, Package insert 13–06–83100–003. Branchburg, NJ, USA.
12. Martinot-Peignoux, M., Boyer, N., Le Breton, V., Le Guludec, G., Castelnau, C., Akremi, R., et al. (2000) A new step toward standardization of serum hepatitis C virus-RNA quantification in patients with chronic hepatitis *C. Hepatology* **31,** 726–729.
13. Saldanha, J., Lelie, N., and Heath, A. (1999) Establishment of the first international standard for nucleic acid amplification technology (NAT) assays for HCV RNA. WHO Collaborative Study Group. *Vox Sang* **76,** 149–158.
14. Davis, G.L., Esteban-Mur, R., Rustgi, V., Hoefs, J., Gordon, S.C., Trepo, C., et al. (1998) Interferon alpha-2b alone or in combination with ribavirin for the treatment of relapse of chronic hepatitis C. International Hepatitis Interventional Therapy Group. *N. Engl. J. Med.* **339,** 1493–1499.
15. Bresters, D., Cuypers, H.T., Reesink, H.W., Mauser-Bunschoten, E.P., van den Berg, H.M., Schaasberg, W.P., et al. (1994) Comparison of quantitative cDNA-PCR with the branched DNA hybridization assay for monitoring plasma hepatitis C virus RNA levels in haemophilia patients participating in a controlled interferon trial. *J. Med. Virol.* **43,** 262–268.

16. Hawkins, A., Davidson, F., and Simmonds, P. (1997) Comparison of plasma virus loads among individuals infected with hepatitis C virus (HCV) genotypes 1, 2, and 3 by quantiplex HCV RNA assay versions 1 and 2, Roche Monitor assay, and an in-house limiting dilution method. *J. Clin. Microbiol.* **35,** 187–192.
17. Sarrazin, C., Teuber, G., Kokka, R., Rabenau, H., and Zeuzem, S. (2000) Detection of residual hepatitis C virus RNA by transcription-mediated amplification in patients with complete virologic response according to polymerase chain reaction-based assays. *Hepatloogy* **32,** 818–823.
18. McHutchison, J.G., Blatt, L.M., Ponnudurai, R., Goodarzi, K., Russell, J., and Conrad, A. (1999) Ultracentrifugation and concentration of a large volume of serum for HCV RNA during treatment may predict sustained and relapse response in chronic HCV infection. *J. Med. Virol* **57,** 351–355.
19. Roth, W.K., Weber, M., and Seifried, E. (1999) Feasibility and efficacy of routine PCR screening of blood donations for hepatitis C virus, hepatitis B virus, and HIV-1 in a blood-bank setting. *Lancet* **353,** 359–363.
20. Sun, R., Schilling, W., Jayakar, H., Ku, J., Wang, J., Rosenstraus, M., et al. (1999) Simultaneous extraction of hepatitis C virus (HCV), hepatitis B virus, and HIV-1 from plasma and detection of HCV RNA by a reverse transcriptase-polymerase chain reaction assay designed for screening pooled units of donated blood. *Transfusion* **39,** 1111–1119.

12

Semi-Quantitative Detection of Hepatitis C Virus RNA by "Real-Time" RT-PCR

Joerg F. Schlaak

1. Introduction

The hepatitis C virus (HCV) is a major cause for liver-related morbidity and mortality worldwide. Quantification of HCV-RNA has become an important tool to tailor and monitor antiviral therapies for this disease *(1)*. Although a low viral load is associated with a higher efficacy of Interferonm (IFN)-α/ ribavirin combination therapies, patients with a high viral load respond more poorly to this treatment and require a longer duration of therapy. In addition, the HCV-RNA status in serum during treatment is used as a marker for the short-term success of therapy.

Several methods have been established to measure and quantify HCV-RNA in serum *(2–5)*. Until now, two main technologies have existed for assessing HCV-RNA levels. Quantitative reverse-transcriptase-polymerase chain reaction (RT-PCR) is the most sensitive method to determine HCV viral load, and the branched-chain DNA test has a wider range of detection.

Recently, two systems have been developed for quantitation of DNA and RNA that are based on "real-time" PCR. The TaqMan®system (Perkin-Elmer, Foster City, CA) measures PCR-product accumulation through a dual-labeled fluorogenic probe *(6)*, while the LightCycler®system (Boehringer Mannheim, Mannheim, Germany) uses a double-stranded DNA-binding dye for quantification of the PCR product *(7)*. Here, we describe the development of a sensitive, rapid, and cost-effective assay for the detection of HCV-RNA in serum that is based on "real-time" PCR using the LightCycler®system.

From: *Methods in Molecular Biology, vol. 193: RT-PCR Protocols*
Edited by: J. O'Connell © Humana Press Inc., Totowa, NJ

2. Materials

2.1. Sampling

1. Sterile cryo tubes (Nunc).
2. Red top tubes.

2.2. RNA Extraction

1. 1.5 mL-Eppendorf reaction tubes.
2. Filter pipet tips (Kisker).
3. QIAamp Viral RNA Mini extraction kit (Qiagen).
4. Ethanol, 99%.

2.3. Reverse Transcription

1. 5× reverse transcription buffer (Gibco-BRL).
2. 2.5 m*M* dNTP mix stock solution: Add 10 µL each of 100 m*M* dATP, dCTP, dGTP, and dTTP (Boehringer Mannheim) to 360 µL sterile distilled water.
3. Moloney murine leukemia virus (MMLV) reverse transcriptase (RT) (200 U/µL) (Gibco-BRL).
4. 100 m*M* dithiothreitol (DTT, Gibco-BRL).
5. 10 µ*M* of cDNA primer (SR1, *see* **Table 1**).
6. 40 U/µL RNasin ribonuclease inhibitor (Promega).

2.4. Preparation of HCV DNA Standard

1. TOPO TA cloning kit (Version E, Invitrogen).
2. QIAprep Spin Miniprep Kit (Qiagen).
3. 40 mg/mL X-Gal in DMF (dimethylformamide).
4. 37°C Shaking and non-shaking incubators.
5. Prepare Luria Bertani (LB)-medium (1.5%):
   a. 10 g tryptone, 5 g yeast-extract, 5 g NaCl in 1 L distilled water.
   b. Adjust pH to 7.5 and autoclave, add ampicillin (50 mg/L) prior to use.
6. Prepare agar plates:
   a. Add 7.5 g bacto-agar to 500 mL Luna Bertani (LB)-medium, autoclave, cool to 45°C, add 50 µg/mL ampicillin, and pour the plates.

2.5. Polymerase Chain Reaction

1. LightCycler real-time PCR machine, LightCycler capillaries, and DNA SYBR green kit (Roche Diagnostics).
2. 10 µ*M* of each PCR primer (SR1 and SF1, *see* **Table 1**).
3. Taqstart antibody (Clontech).

3. Methods

3.1. Sampling

1. Collect freshly drawn peripheral blood in red-top tubes, and allow blood to clot.

Table 1
Oligonucleotide Sequences used for cDNA-Synthesis and PCR

Name	Sequence (5' to3')	Position
SF 1	GCC ATG GCG TTA GTA TGA GT	−260 *bis* -241
SR 1*	TGC ACG GTC TAC GAG ACC T	−3 *bis* -21

* Primer used for cDNA-synthesis and PCR.

2. Separate serum in a centrifuge at room temperature.
3. Aliquot serum samples in 2-mL cryo tubes, snap-freeze in liquid nitrogen, and store at –80°C (*see* **Note 1**).

3.2. RNA Extraction

Several protocols are available for extraction of viral RNA from serum samples. In our laboratory, the most consistent results were obtained using the QIAamp Viral RNA Mini Kit®(*see* **Notes 2** and **3**). RNA extraction with this kit was performed according to the instructions and recommendations of the manufacturer.

1. Add 31 mL buffer (AVL) to 310 µg lyophilized carrier RNA.
2. Mix well and store at 4°C.
3. Pipet 560 µL of lysis buffer (buffer AVL/carrier RNA) and 140 µL serum into a 1.5-mL Eppendorf tube. Incubate at room temperature for 10 min.
4. Add 560 µL of ethanol (99%). Mix and apply 560 µL to the QIAamp spin column.
5. Centrifuge at 6000*g* for 1 min.
6. Place the QIAamp spin column into a new collection tube. Transfer the remaining lysis buffer/serum mix (approx 700 µL) onto the column.
7. Centrifuge at 6000*g* for 1 min.
8. Transfer 500 µL of washing buffer AW1 onto the column.
9. Centrifuge at 6000*g* for 1 min.
10. Transfer column to new collecting tube.
11. Add 500 µL washing buffer AW2.
12. Centrifuge at 20,000*g* for 3 min.
13. Transfer the column into a clean 1.5-mL Safe-Lock-Eppendorf reaction tube.
14. For elution, transfer 50 µL sterile distilled H_2O (prewarmed to 80°C) onto the column (*see* **Note 4**).
15. Centrifuge at 6000*g* for 1 min.
16. Store RNA at −80°C.

3.3. Reverse transcription (see *Notes 5* and *6)*

1. To denature secondary structures, incubate the RNA at 65°C for 10 min.
2. Then, transfer RNA immediately onto ice.
3. Add 10 µL of total RNA to 15 µL of reverse transcription mix containing:

a. 5 × RT-buffer (Gibco-BRL) 3.3 µL
b. DTT (Gibco-BRL) 1.7 µL
c. HCV Anti-sense primer SR1 (10 µM, *see* **Table 1**) 1 µL (*see* **Note 7**)
d. dNTP 40 m*M* 2.5 µL
e. MMLV RT (200 U/µL, Gibco-BRL) 1 µL
f. Sterile distilled H_2O 5.1 µL (*see* **Note 8**)
g. RNasin 0.4 µL

4. Incubate at 40°C for 60 min, and heat the reaction for 5 min at 94°C to inactivate the RT.
5. Store the cDNA at –20°C.

3.4. Preparation of HCV-DNA Standard (see *Note 9)*

3.4.1. Cloning of HCV Plasmid DNA

Cloning and transfection of HCV plasmid DNA using the TOPO TA cloning kit (Invitrogen) are performed according to the instructions of the manufacturer.

1. Generate fresh (*see* **Note 10**) PCR product from a serum sample from a highly viremic HCV patient with HCV primers SR1 and SF1 (*see* **Table 1**).
2. Incubate 0.5–2 µL of the PCR-product with 2 µL of sterile distilled water and 1 µL of TOPO vector for 5 min. at room temperature. Then, spin down briefly and transfer the mix onto ice.

3.4.2. Transformation (see ***Note 11)***

1. Add 2 µL of β-mercaptoethanol (0.5 *M*) to the mix. Mix gently.
2. Add 2 µL of the mix to freshly thawed, One-Shot(tm) competent cells of *E. coli*, mix and incubate on ice for 30 min.
3. Incubate for 30 s at 42°C, and do not shake.
4. Transfer on ice immediately and incubate for 2 min.
5. Add 250 µL of SOC-buffer.
6. Shake for 30 min at 37°C.
7. Transfer the mix onto ice.
8. Transfer 50 µL to Luria Bertani (LB)-ampicillin plates containing X-Gal and IPTG. Spread using an alcohol-flamed hockey-stick.
9. Incubate the plates overnight at 37°C.
10. Pick 10 white colonies and transfer them to 50-mL roller bottle containing LB-medium.
11. Incubate overnight at 37°C in a shaker.
12. Spin down at 700*g* for 20 min and remove supernatant. Continue with the plasmid extraction.

3.4.3. Plasmid Extraction Using Qiagen Miniprep Kit

Plasmid extraction with the Qiagen Miniprep kit (Qiagen) is performed according to the instructions of the manufacturer.

1. Resuspend the bacterial pellet in 250 µL of buffer P1 and transfer into a 1.5-mL Eppendorf tube.
2. Add 250 µL of buffer P2 and mix gently.
3. Add 350 µL of buffer N3 and mix gently.
4. Centrifuge for 10 min at 10,000*g*.
5. Transfer the supernatant to a column and centrifuge for 1 min at 10,000*g*.
6. Wash the column with 500 µL of buffer PB and centrifuge for 1 min at 10,000*g*.
7. Wash the column with 500 µL of buffer PE, centrifuge for 1 min at 10,000*g* and remove the supernatant.
8. Centrifuge the column again for 1 min at 10,000*g*.
9. Add 50 µL of sterile distilled water; after 1 min centrifuge the column for 1 min at 10,000*g*.
10. Store the isolated plasmid DNA at –20(C (*see* **Note 12**).
11. Measure the absorbance (Ultraviolet [UV], A_{260nm}) using a spectrophotometer and calculate the DNA concentration (*see* **Notes 13** and **14**).

3.5. Polymerase Chain Reaction

1. Incubate 2 µL of LightCycler DNA Master SYBR[a]®Green I with 0.16 µLTaqstart at room temperature (*see* **Note 15**).
2. Prepare the components as follows (*see* **Note 16**):

a. H_2O	7.4 µL
b. Primer SR1 10 *M*	2 µL
c. Primer SF1 10 *M*	2 µL
d. $MgCl_2$	1.6 µL

Add these to SYBR Green/Taqstart Mix.

5. Add 5 µL of cDNA to 15 µL of PCR-mix.
6. Pipet the mixture into a LightCycler glass capillary cuvet, cap this and transfer the cuvet into the LightCycler (*see* **Note 17**).

3.5.2. Programming the LightCycler

1. Click on LightCycler symbol on desktop screen.
2. Click on “Run.”
3. Click on “New Experiment” and name the file.
4. Set display mode to F1/1 and fluorometer gains (F1: 5; F2: 10; F3: 10).
5. Program and name the denaturation step (e.g., “HCV denaturation”).

Use the following parameters: Cycles: 1, Analysis mode: none.

Target temp	Incub. time	Temp. transition rate	Second target temp	Step size	Step delay	Acquisition mode
95	2:00	20.0	0	0	0	None

6. Add and name the amplification step (e.g., “HCV amplification”).

Use the following parameters: Cycles: 40, Analysis mode: quantification.

Target temp	Incub. time	Temp. transition rate	Second target temp	Step size	Step delay	Acquisition mode
95	0	20.0	0	0.0	0	None
60	5	20.0	0	0.0	0	None
72	15	20.0	0	0.0	0	Single

7. Add and name the melting step (e.g., "melting curve").

Use the following parameters: Cycles: 1, Analysis mode: melting curves.

Target temp	Incub. time	Temp. transition rate	Second target temp	Step size	Step delay	Acquisition mode
95	0	20.0	0	0.0	0	None
65	5	20.0	0	0.0	0	None
95	0	0.2	0	0.0	0	Cont

8. Add and name the cooling step (e.g., "cooling").

Use the following parameters: Cycles: 1, Analysis mode: none.

Target temp	Incub. time	Temp. transition rate	Second Target temp	Step size	Step delay	Acquisition mode
40	1:00	20.0	0	0.0	0	None

9. Enter the sample data. Click on "edit samples" and enter data.

Example:

Rotor #	Sample name	Type	Replicate of	Concentration
1	Std 1	Standard	0	3.00E+0
2	Std 2	Standard	0	3.00E+1
3	Std 3	Standard	0	3.00E+2
4	Std 4	Standard	0	3.00E+3
5	Std 5	Standard	0	3.00E+4
6	Patient A	Unknown	0	0.00E+0
7	Patient B	Unknown	0	0.00E+0

Temperature: 30, Number of Samples: 7

10. Click on "Done".
11. Click on "Run."
12. Name the file and save.
13. Click on "Done". The PCR process begins.
14. After the PCR is completed, click on "Select a Program" and select melting curve. Then, click on "melting curve".

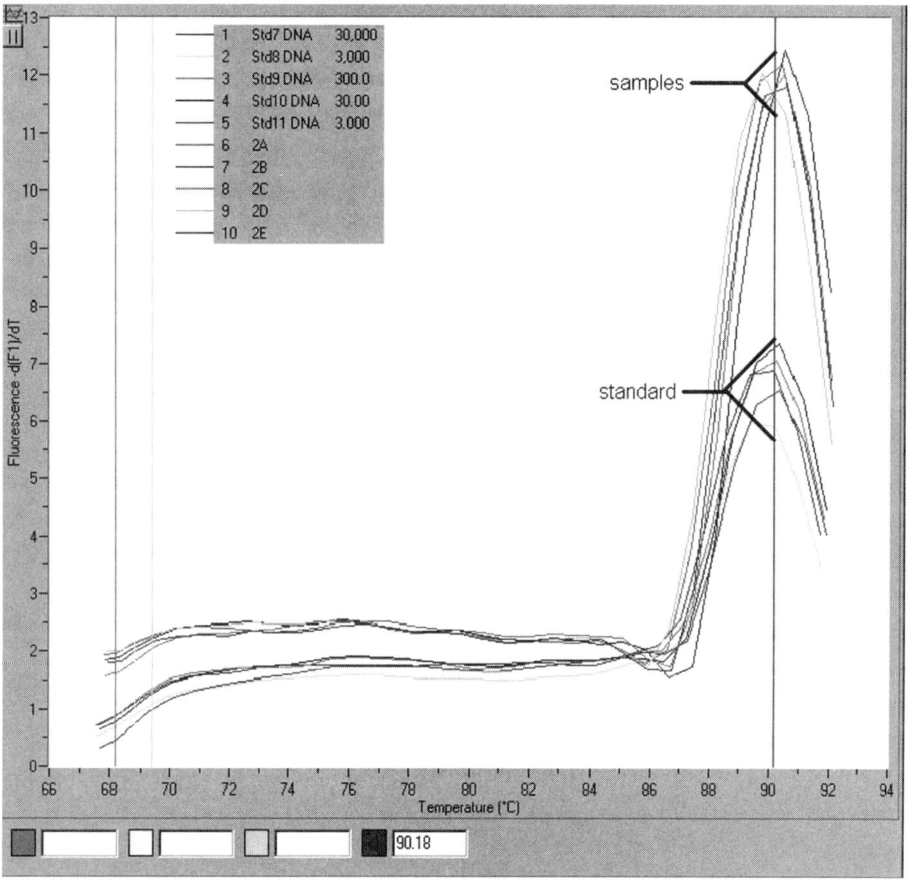

Fig. 1. Example of melting curves for the HCV-specific real-time RT-PCR products.

15. Adjust "°C to Average" parameter until reasonable melting curves appear (*see* **Note 18** and **Fig. 1**).
16. For analysis go to "data analysis" **(Fig. 2)**.

4. Notes

1. Avoid repeated freeze-thaw cycles and temperature changes to maintain viral RNA integrity.
2. Higher sensitivity can be obtained using more serum for RNA extraction (e.g., approx 35 HCV copies/mL when 2 mL serum is used; for this purpose use Qiagen Viral RNA Midi or Maxi kit).
3. In our studies, extraction of HCV-RNA using the Qiagen RNA/DNA Mini kit

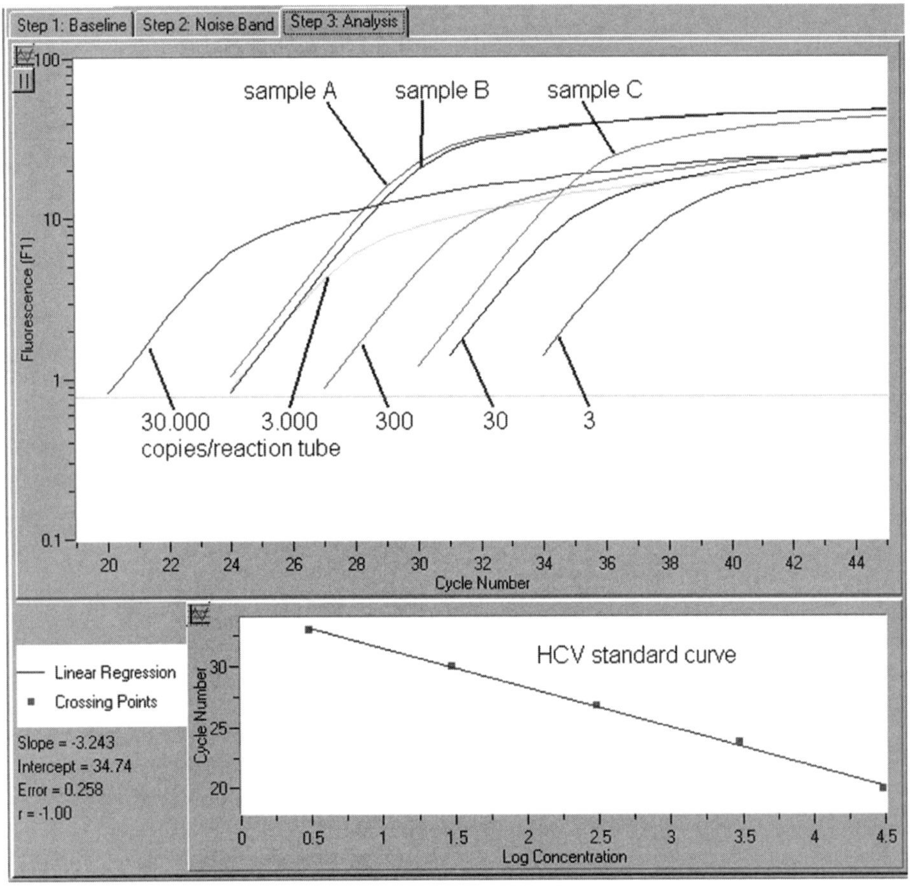

Fig. 2. Example of a HCV standard curve that enables quantification of the HCV samples.

(Qiagen) or High Pure Viral Nucleic Acid Kit (Roche) was less effective or more complicated, respectively.

4. A 5-min incubation of the QIAamp spin column filled with 50 µL of pre-warmed (80°C) sterile distilled water prior to centrifugation results in higher RNA yields.
5. Use only RNase-free materials; always wear gloves.
6. Reverse transcription in a separate reaction in a heating block leads to improved results compared to one-step RT-PCR, because less primer dimers occur, and a higher sensitivity is achieved. This is probably a result of the fact that enzymes of the reverse transcription interfere with the PCR reaction.
7. For reverse transcription of HCV-RNA, the use of specific primers for HCV leads to higher specificity and sensitivity.
8. May also use DEPC-treated water.

9. Initially, we had prepared a HCV-RNA standard according to the following procedure: For the in vitro transcription of the untransalated region (UTR), we have used a pHCV-UTR plasmid. This contains the complete HCV-UTR cDNA cloned into the pCRII-TOPO vector (Invitrogen) between the T7 and SP6 RNA polymerase-promoter regions. After linearization of the plasmid with *Eco*R V, the positive strand was synthesized by transcription using the SP6 RNA polymerase. After in vitro transcription, the plasmid was digested with DNase A and RNA was purified using the RNeasy®system (Qiagen). For routine purposes, this did not appear to be a useful approach since, at least in our studies, progressive degradation of RNA led to inconsistent results.
10. PCR products must be fresh, since the Taq-generated dA overhang upon which ligation into the cloning vector is based, may degrade over time.
11. Steps prior to transformation:
    a. Prepare a water bath or heating block at 42°C.
    b. Thaw SOC buffer.
    c. Warm up LB plates to room temperature for 30 min, add 40 µL X-gal (40 mg/mL), and spread over the surface using an alcohol-flamed hockey-stick.
    d. Thaw One-Shot competent *E. coli* cells (1 tube per transformation) slowly on ice.
    e. Place β-mercaptoethanol on ice.
12. Avoid vortexing of plasmid DNA.
13. As a control, determine the concentration of the HCV DNA standard using a commercial system (e.g., Roche Amplicor).
14. Using this system, 1 molecule (plasmid + insert) = 4.65 attogram.
15. The use of Taqstart antibodies increases the sensitivity and specificity, since formation of primer dimers is inhibited.
16. Prepare the PCR mix on ice.
17. Do not touch the stem of the capillaries, which must be kept clean to permit accurate fluorescence detection of the SYBR green-labeled PCR products.
18. Using this protocol, a melting temperature of 90°C should be obtained for the specific HCV amplicon. Fluorescence is continously collected during the slow temperature ramp. Melting curves are converted to melting peaks by plotting the negative derivative of the fluorescence with respect to temperature (–dF/dT) against temperature. The area under these melting peaks is related to the amount of product melting at that temperature.

References

1. EASL International Consensus Conference on Hepatitis C (1999) *J. Hepatol.* **30,** 956–961.
2. Goergen, B., Jakobs, S., Symmons, P., Hornes, E., Meyer zum Bueschenfelde, K.-H., and Gerken, G. (1994) Quantitation of HCV-replication using one-step competitive reverse transcription-polymerase chain reaction and a solid phase, colorimetric detection method. *J. Hepatol.* **21,** 678–682.
3. Mizutani, T., Ikeda, M., Saito, S., Sugiyama, K., Shimotohno, K., and Kato, N.

(1998) Single-step reverse transcription-polymerase chain reaction for the detection of hepatitis C virus RNA. *Microbiol. Immunol.* **42,** 549–553.
4. Kobayashi, M., Kumada, H., Arase, Y., Chayama, K., Kobayashi, M., Tsubota, A. et al. (1998) Usefulness of hepatitis C virus RNA counts by second generation HCV bDNA-probe in chronic hepatitis C based on the HCV genotype. *J. Gastroenterol.* **33,** 223–228.
5. Colucci, G., and Gutekunst, K. (1997) Development of a quantitative PCR assay for monitoring HCV viremia levels in patients with chronic hepatitis C. *J. Viral. Hepat.* **4,** 75–78.
6. Heid, C.A., Stevens, J., Livak, K.J., and Williams, P.M. (1996) Real-time quantitative PCR. *Genome Res.* **6,** 986–994.
7. Morrison, T.B., Weis, J.J., and Wittwer, C.T. (1998) Quantification of low-copy transcripts by continuous SYBR Green I monitoring during amplification. *Biotechniques* **24,** 954–962.

13

RT-PCR for the Assessment of Genetically Heterogenous Populations of the Hepatitis C Virus

Brian Mullan, Liam J. Fanning, Fergus Shanahan, and Daniel G. Sullivan

1. Introduction

Since the discovery of the Hepatitis C Virus (HCV) in 1989 *(1)*, known to that point as the infectious agent of Non-A Non-B hepatitis *(2)*, there has been an intensive focus on understanding the underlying biology of its disease process. Of significant interest has been the study of heterogenous populations of closely related, but genetically non-identical HCV virions, commonly termed quasispecies *(3,4)*, and their relationship to the pathogenesis of HCV infection. Studies of HCV quasispecies have predominantly focused on two viral genetic loci, Hypervariable Region 1 (HVR 1) and the Interferon Sensitivity Determining Region (ISDR) *(5)* (**Fig. 1**). HVR 1 encodes an approx 27 amino-acid sequence at the N-terminal of envelope glycoprotein 2 (E2), and sequence variation at this locus is used to characterize HCV quasispecies. HVR 1 acts as an immune escape epitope that, through continuous alteration of its antigenicity, is believed to hinder humoral immune-mediated clearance of the virus. The ISDR encodes an approx 40 amino-acid sequence in the C-terminal half of the nonstructural 5A (NS5A) protein. Protein kinase inhibition and transcriptional activation activities have been ascribed to NS5A, both of which are known to be modulated by sequence variation within the ISDR *(6,7)*. The current view is that although the role of the NS5A protein in HCV disease progression is important, pathologically relevant viral genomic sequence evolution may not be strictly confined to the ISDR alone. Sequence variation of other regions in

From: *Methods in Molecular Biology, vol. 193: RT-PCR Protocols*
Edited by: J. O'Connell © Humana Press Inc., Totowa, NJ

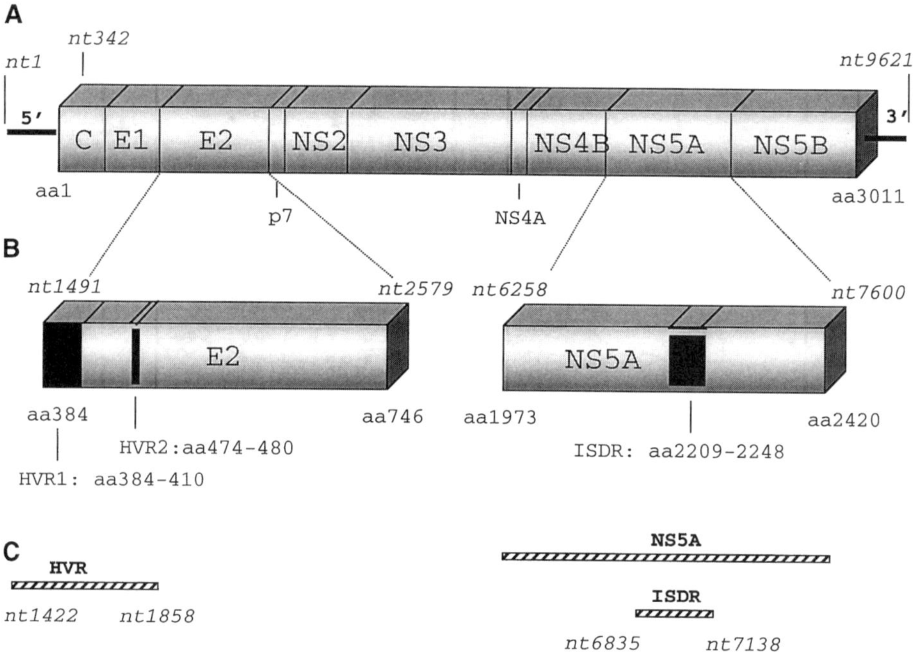

Fig. 1. Structure of the HCV genome and relative location of PCR products discussed in this chapter. **(A)** HCV genome comprising 5' and 3' untranslated regions (bold lines), flanking an open reading frame (box). Nucleotide (upper, italics), and amino acid (lower, plain) scales are shown. **(B)** Enlarged envelope 2 (E2) and nonstructural 5A (NS5A) genes showing (bold) HVR1 and 2, and ISDR, respectively. **(C)** Location of PCR products (hatched areas) in relation to HVR and ISDR loci. The NS5A PCR precisely covers the NS5A gene. Diagram based on HCV-H-strain nucleotide and amino-acid sequence ***(5)***.

NS5A, such as the V3 region ***(8)***, nuclear localization signals, and phosphorylation motifs may also play important roles ***(9)***.

In this chapter, we present detailed procedures for the PCR-amplification and cloning of HVR 1, ISDR, and NS5A RT-PCR amplicons from HCV-positive serum and liver tissue, which can be used to study viral quasispecies populations. These protocols will enable the determination of sequence information at these loci from material of clinical interest, which should prove useful to the understanding of HCV disease pathology. In addition, translational start-and-stop sites incorporated into the primers used to amplify NS5A enable these amplicons to be used in cellular protein-expression systems to assess the functionality of isolated NS5A variants ***(10–12)***. The assays described can also be applied to qualitative assessment of the HCV status of clinical material, and to detect viral replicative intermediates in potential in vitro HCV culture models. It should be noted that false-positives in PCR for negative-strand HCV RNA are common, and are principally due to false priming of cDNA during reverse transcription *(13)*.

This chapter is divided into individual sections that describe methods involved in viral RNA isolation, reverse transcription into cDNA templates, nested PCR, and cloning of these PCR products. The analysis of sequence data from quasispecies populations is beyond the scope of this chapter, and information on this topic can be found elsewhere *(14–16)*.

1.1. RNA Isolation

We routinely use the method described by Chomczynski and Sacchi *(17)*, with minor modifications. The basis of this method is the dissociation of viral particles from serum or liver proteins and lysis by chaotropic salt solution (guanidium isothiocyanate). RNA is separated from this homogenate by extraction with phenol and chloroform, and nucleic acid is then precipitated with isopropanol. This method yields RNA suitable for high-sensitivity PCR (detecting up to 10 copies of viral genome per PCR). We have also used spin-column-based formats from commercial suppliers (QiaAmp Viral RNA Mini Kit, Qiagen, Valencia, CA; High Pure Viral Nucleic Acid Kit, Roche Molecular Biochemicals, Indianapolis, IN) which provide equivalent results and are less laborious than acid-guanidium-phenol-chloroform purification *(18)*.

1.2. Reverse Transcription of RNA

Two principal types of reverse transcriptase are commercially available, and differ principally on the presence of Ribonuclease H (RNase H) enzyme activity, which can degrade single-stranded RNA as part of a double-stranded RNA:DNA duplex. Moloney murine leukemia virus (MMLV) RT contains no appreciable RNase H activity but cannot be used to synthesize cDNA templates at temperatures above 37–42°C. Avian myeloblastoma virus (AMV) RT does contain RNase H, but polymerizes at temperatures of 42–58°C, with decreasing processivity and enzyme half-life at higher temperatures. RNase H deletion/inactivation mutants of MMLV, such as Superscript (Gibco-BRL, Bethedsa, MD) are also available. Given the presence of secondary and tertiary RNA structures at the 5' and 3' ends of the HCV genome, higher temperatures will likely favor increased reverse transcription of such templates, and thus the use of AMV RT is preferential. Additionally, the intrinsic RNase H activity of AMV is sufficient to degrade most remaining RNA template, which can interfere with RT-PCR by competing with the cDNA as the PCR template *(19)*. We have tried other reverse transcriptases—e.g., Enhanced AMV (Sigma, St. Louis, MO), display THERMO-RT (Display Systems Biotech, Vista, CA)—

but have found no appreciable increase in the sensitivity of nested PCR assays as compared with AMV-generated cDNA templates. Typically, we use specific-primed cDNA synthesis using the anti-sense primer of the outer PCR primer set as the reverse transcription primer (unless undertaking PCR for negative strand HCV RNA, when the outer sense primer should be used), although random hexamers also work well. The reverse transcription involves melting of the RNA template and reverse transcription primer together at elevated temperatures followed by snap-chilling on ice. This serves to resolve RNA secondary structures and anneal the reverse transcription primer to its target site on the RNA genome. cDNA synthesis at 42°C is preceded by a short incubation at 23°C, which serves to lengthen the primer, ensuring that it does not dissociate at the higher temperatures used for polymerization. This is especially important when using random hexamers for priming. A final step at 95°C for 3–5 min inactivates the RT; the activity of Taq DNA polymerase can be reduced by an active RT ***(20)***.

1.3. Nested PCR

All amplification of cDNA templates described here is performed in two rounds (of 35 cycles each) of PCR—i.e., nested PCR. One-tube RT-PCR ***(21)*** and single-round PCR from cDNA has been described for HCV, but the use of nested PCR still predominates. The second round of amplification provides specificity and sensitivity that may be difficult to obtain with one round of amplification alone.

The primer sequences we use for nested PCR are shown in **Table 1**. Genotype specificity for the primer sets is against HCV 1b for all three PCRs. However, the HVR primer set has successfully amplified from genotype 1a. We include restriction-enzyme recognition sites at the 5' end of the primers to facilitate directional cloning of amplicons into vectors, which is more efficient than blunt-end cloning. These sites can be created by mutation of sequences on the viral genome (e.g., the HVR nested primers) or by "tagging" restriction sites onto the 5' end of the primer (e.g., the ISDR-nested primers). These tagged restriction sites require addition of extra bases to the 5' end to facilitate recognition and cleavage by the restriction enzyme ***(22)***. Restriction sites are chosen for their rare occurrence across the HCV genome as a whole, and their common occurrence in the multiple cloning sites of most cloning vectors (e.g., *Eco*R I, *BamH* I, *Xba* I, *Hind* III).

The detection limits of the HVR, ISDR, and NS5A RT-(nested) PCRs are approx 10, 50, and 200 viral genomes per PCR, respectively. These sensitivities represent PCR optimization on PE 2400/9600 thermal cyclers (Perkin-Elmer, Kenilworth, NJ) using 0.2-mL thin-wall MicroAmp tubes (Perkin-Elmer) and proofreading DNA polymerases (e.g., Pwo, Pfu). Amplifi-

Table 1
Sequences of Primers used for HCV RT-(Nested) PCR

Primer	[a]Position	[b]Sequence
HVR		
Reverse Transcription	1948–1964	5' ggcggccgcgtgttgtt 3'
Outer Sense	1272–1291	5' ggtcaccgcatggcttggga 3'
Outer Anti-sense	1842–1861	5' ggagtgaagcaatacactgg 3'
Inner Sense	1422–1441	5' gtggggatccgggctaaggt 3'
Inner Anti-sense	1836–1858	5' gtgaaggaattcactggaccaca 3'
ISDR		
Outer Sense	6722–6741	5' caggtacgctccggcgtgca 3'
Outer Anti-sense/RT	7275–7294	5' ggggccttggtaggtggcaa 3'
Inner Sense	6835–6865	5' aataggatcccagtgctcacttccatgctca 3'
Inner Anti-sense	7118–7138	5' aggtgaattcacggatatttccctctcatcc
[c]NS5A		
Outer Sense	6176–6195	5' cagcctcaccatcactcagc 3'
Outer Anti-sense/RT	7669–7688	5' gtggtgacgcagcagagagt 3'
Inner Sense	6243–6263	5' ttccaagctt**atg**tgttccggctcgtggctaaag 3'
Inner Anti-sense	7574–7591	5' attctaga**ctatta**cattgagcagcagacgacg 3'

*a*Nucleotide positions are based on HCV-J sequence ***(34)***.

*b*BamH I (ggatcc), *EcoR* I (gaattc), *Hind* III (aagctt), and *Xba* I (tctaga) restriction sites are underlined.

c For the inner sense and inner anti-sense primers, respectively, HCV homology is from bases 14 to 34, and 14 to 31; the remaining 5' primer sequences are not HCV-specific, and are included in the primers to incorporate restriction sites and start/stop codons. Translational start and stop codons are shown in bold print on the inner sense and inner anti-sense primers.

cation on different thermal cyclers, or with different tubes and non-proofreading polymerases (e.g., standard Taq DNA polymerase), may yield different detection limits, or require further optimization.

Given the powerful amplification capabilities of PCR, care must be taken in the design of work areas used for generating amplicons ***(23,24)***. One hundred copies of RNA template (10^{-15} g of nucleic acid) can be amplified to 10^{-7} g of DNA in 70 cycles of PCR. This degree of amplification engenders a significant risk of contamination of PCR by previously amplified products if both are handled in the same area. We extract RNA from serum in a Class III laminar flow-hood in a designated pre-amplification area (Area I). The reverse transcription (first and second rounds of PCR) are set up within this room, but at a different workstation (Area Ia). Products from the first round of PCR are added to the second-round PCR in a separate room (Area II). Nested PCR products are then resolved by agarose gel electrophoresis in a third room (Area III).

Each area should have its own dedicated equipment and solutions. Ideally, all areas should be situated in separate rooms. The use of barrier sealed tips for micropipetting (ART, Molecular BioProducts, San Diego, CA) is also recommended.

We routinely use proofreading enzymes for PCR amplification (i.e., enzymes with a 3' to 5' exonuclease activity that can edit mismatched bases during polymerization), such as Pwo (Roche Molecular Biochemicals, Indianapolis, IN) and Pfu (Stratagene, La Jolla, CA). These enzymes have up to a 10-fold higher rate of fidelity than that of non-proofreading enzymes ***(25,26)*** (e.g., standard Taq DNA polymerase). As these amplicons are used to analyze sequence data, high fidelity of amplification is an important requisite for generation of amplicons. Spurious incorporation of mutations by Taq DNA polymerase leading to false assessment of quasispecies diversity has already been shown ***(27, 28)***.

1.4. Ligation of Amplicons into Cloning Vectors

Cloning kits are commercially available that obviate the need for the ligation procedures described in this section. "TA" cloning kits, which utilize the deoxyadenosine (dA) residues added to the ends of PCR products by non-proofreading polymerases to directionally clone amplicons into vectors ***(29, 30)***, cannot be used with Pwo- or Pfu- generated amplicons, because proofreading polymerases do not add "dA" overhangs to PCR products. Kits such as the Zero Blunt Cloning Kit (Invitrogen, Carlsbad, CA) do not require preparation and digestion of PCR products and cloning vectors, and their quantitation prior to ligation. Amplicons are simply added (subsequent to confirmation of amplification by agarose gel electrophoresis) to prepared reactions, and left to incubate for 5 min (for TOPO-based kits) to overnight. It should be kept in mind that the target sites on these vectors for commonly used sequencing primers, e.g., T7, SP6, and M13, can be up to 100 basepairs (bp) from the site of insertion of the amplicon into the vector. Besides generating additional unwanted sequence data, the additional distance from sequence priming site to end of cloned insert may yield poor sequence read at the 3' end of the cloned amplicon if using sequencing systems based on high throughput for short-sequence reads, e.g., ABI 310 (Perkin-Elmer). The pBluescript vector (Stratagene) used here has generic SK and KS priming sites immediately contiguous with the site of amplicon insertion, which can be used for sequencing. Although the cloning procedures outlined may be slightly laborious, after obtaining the cloning vectors and bacterial strains that are readily available, the procedure is inexpensive, as opposed to the expensive nature of cloning kits.

2. Materials

1. RNA lysis buffer: 4.0 *M* guanidine thiocyanate (Sigma Chemical Co., St. Louis, MO), 25 m*M* sodium citrate, pH 7, 0.5% sarcosine; add 7.2 µL of β-mercapto-ethanol per mL.
2. AmpliWax PCR Gems; 0.2 mL MicroAmp reaction tubes (Perkin-Elmer, Foster City, CA).
3. AMV reverse transcriptase (Promega, Madison, WI).
4. Deoxynucleotide triphosphates, glycogen, Pwo DNA polymerase, ribonuclease inhibitor (Roche Molecular Biochemicals).
5. 0.05-µm dialysis filters (Millipore, Bedford, MA).
6. QiaPrep plasmid mini preparation kit; QiaQuick PCR purification kit (QiaGen, Valencia, CA).
7. Tryptone, yeast extract, agar (Difco, Detroit, MI).
8. Luria Bertani (LB) agar/X-gal/IPTG/amp: dissolve 10 g of tryptone, 5 g of yeast extract, 5 g of sodium chloride, and 15 g of agar per 1.0 L of distilled water. Autoclave. When cooled, add 100 µL of X-gal (2% w/v in dimethyl formamide), 100 µL of 100 m*M* IPTG, and 100 µg/mL of ampicllin per 20 mL LB agar plate. Pre-sterilize the IPTG and ampicillin solutions through 0.2-µm filters (Schleicher and Schuell, Keene, NH) before use. This is not necessary for the X-gal solution. Plates are stable at 4°C for up to 1 wk.
9. LB broth/amp: dissolve 10 g of tryptone, 5 g of yeast extract, 10 g of sodium chloride per 1.0 L of distilled water. Autoclave. Add ampicillin to 100 µg/mL prior to inoculation.

3. Methods

3.1. RNA Extraction (see *Note 1*)

3.1.1. Serum Viral RNA Extraction

1. Isolate RNA from 100 µL of serum. If a smaller volume is required, make the final volume of serum up to 100 µL with control serum, e.g., from a HCV-negative individual or from Sigma (normal human plasma).
2. Perform the extraction procedure, essentially as described by Chomczynski and Sacchi *(17)*, with minor variations. Add 100 µL of 2 *M* sodium acetate, pH 4, per mL of RNA lysis buffer, and add 400 µL of this solution to each 100 µL aliquot of serum to be extracted.
3. Place the extraction at 60°C for 10 min.
4. Add 0.5 mL of water-saturated phenol, pH 4.0, and 0.1 mL of a mixture of chloroform and isoamyl alcohol (ratio of 49:1). Mix the tubes after addition of each reagent, and mix the final suspension vigorously. Leave the tubes on ice for 15 min.
5. Separate the phases by centrifugation at 13,000*g* for 15 min.
6. Remove the upper aqueous phase and extract with an equal volume of chloroform:isoamyl alcohol (49:1).

7. Precipitate the RNA from the aqueous phase by addition of one volume of isopropanol.
8. Collect the precipitate by centrifugation at >13,000*g* for 25 min.
9. Remove all traces of the supernatant and then dissolve the pellet in 300 µL of RNA lysis buffer (with sodium acetate). Add 300 µL of isopropanol, vortex briefly, and centrifuge as in **step 8**.
10. Wash the pellet 3 × in 75% ethanol, which must be freshly prepared. Remove all traces of ethanol after the final wash.
11. Dry the RNA pellet for 5 min in a vacuum drier (Speed Vac), or at 65°C for 5–10 min in an open capped tube.
12. Redissolve the RNA in RNase-free water.

3.1.2. Liver Tissue RNA Extraction

1. Place a piece of liver obtained from a biopsy (approx 1.5-mm diameter, 2–4 mm in length) immediately in 500 µL of RNA lysis buffer, vortex, and store at −70°C.
2. To extract the RNA, remove 100 µL of the solution D/liver homogenate, and add to 400 µL of fresh RNA lysis buffer.
3. Add 50 µL of 2 M sodium acetate, pH 4, and perform the RNA extraction procedure as described in **Subheading 3.1.1.** for extraction of HCV RNA from serum (**steps 3–12**) (*see* **Notes 2** and **3**).

3.2. Reverse Transcription (Area Ia)

All reactions are prepared on ice.

1. Resuspend the dried RNA pellet in 9.0 µL diethyl pyrocarbonate (DEPC)-treated H_20. Add 4.0 µL of a 5.0 µ*M* stock of reverse transcription primer (= 20 pmols of primer) (*see* **Note 4**). Mix. Incubate at 75°C for 10 min, and then immediately snap-chill on ice. Centrifuge the tube briefly to bring down any condensate.
2. Reaction components for greater than two reverse transcriptions at one time are best prepared as a master mix, which is then added as a single aliquot to each reaction. For every 10 reactions add one extra to the mix to account for transfer losses (e.g., for 10 reverse transcriptions make a mix for 11).
   For each reverse transcription reaction, add 5X buffer (supplied with enzyme), ribonuclease inhibitor, and deoxynucleotide 5' triphosphates (dNTPs) to a master mix as described below:

Component	Volume (mL)	Final concentration
5X AMV buffer	4.0	1X
Ribonuclease inhibitor (40U/µL)	1.0	40 U/reaction
5 m*M* dNTP Mix	1.0	250 µ*M*
AMV RT (9–10 U/µL)	1.0	9–10 U/reaction

3. Add 7.0 µL of master mix to each reaction. Mix.
4. Incubate at 23°C for 10 min, followed by 42°C for 1 h and finally at 94°C for 3 min to inactivate the reverse transcriptase.

3.3. Nested PCR (see Note 5)

Both rounds of PCR are performed in a final vol of 50 µL per reaction. This volume can, however, be scaled up (to 100 µL) or down (to 25 µL) as needed. Negative controls (reverse transcription that had no viral RNA added, or water in place of template) should be amplified with each lot of PCRs. Ideally, one negative control should be included per PCR test sample. PCRs are performed using 0.2-mL, thin-wall MicroAmp Tubes (Perkin-Elmer). Cycling conditions shown are for PE 2400/9600 Thermal Cyclers (Perkin-Elmer). These cyclers have rapid ramping rates (time to change between temperatures). If other cyclers are used it may be necessary to lengthen times. We add 2.0 µL of the reverse transcription reaction to the first round of each PCR. The source of Mg^{2+} cations is $MgS0_4$ in the following procedures, as this is the Mg^{2+} formulation for use with *Pwo* DNA polymerase. For standard Taq DNA polymerase, use the 25 m*M* $MgCl_2$ provided with the enyzme. All PCR primer sequences used are shown in **Table 1**. Typical results of PCR amplification are shown in **Fig. 2**.

3.3.1. HVR PCR

3.3.1.1. First Round (Area IA)

This PCR is set up in two stages (lower and upper), with a separate master mix for each stage. A lower master mix is added first, a wax bead is then used to create a barrier between this and the upper master mix, which is subsequently added. All reactions are prepared on ice.

For each PCR, add dNTPs, PCR primers, DNA polymerase buffer, $MgSO_4$ (both supplied with enzyme), *Pwo* DNA polymerase, and water to a master mix as described in **step 1**:

1. Set up the master mix (lower layer) as follows:

Component	Volume (µL)	Final concentration
DEPC-treated H_2O	17.8	–
10 m*M* dNTP Mix	1.0	200 µ*M*
5 µ*M* outer primer mix	3.0	15 pmols each primer
25 m*M* $MgSO_4$	1.2	1.0 m*M* *
10X Polymerase Buffer	5.0	1X

* This includes the contribution of Mg^{2+} from the reverse transcription reaction (*see* **Note 6**).

2. Aliquot 28.0 µL for each reaction. Add 2.0 µL of the appropriate cDNA.
3. Add one wax bead (AmpliWax Gem). Incubate at 80°C for 5 min, then place the tubes on ice.
4. Set up the second master mix (upper layer):

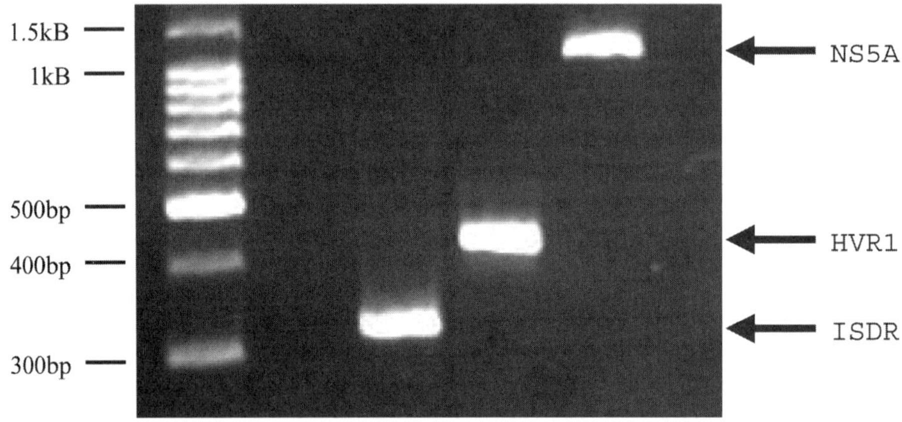

Fig. 2. Agarose gel electrophoresis of ISDR, HVR1, and NS5A amplicons. Amplicons were resolved on 3:1 NuSieve agarose:standard agarose gels. From left to right: 100-bp DNA mol-wt ladder, ISDR PCR (304 bp), HVR1 PCR (436 bp), and NS5A PCR (1,379 bp).

Component	Volume (μL)	Final concentration
DEPC-treated H_2O	17.5	–
10X polymerase buffer	2.0	1X
5U/μL DNA polymerase	0.5	2.5 U per reaction

5. Add 20.0 μL to each reaction.
6. On a thermal cycler program the following thermal profile:
   a. Step 1: 80°C, 3 min.
   b. Step 2: 94°C, 15 s.
   c. Step 3: 55°C, 30 s.
   d. Step 4: 72°C, 45 s.
7. There are 35 cycles (**steps 2–4**) in total:
   a. Cycles 1–10: Repeat **steps 2–4**.
   b. Cycles 11–35: For the remaining cycles add an incremental increase of 20 s per cycle to **step 4** (if this option is unavailable, replace with a 72°C, 4 min step).

3.3.1.2. Second (Nested) Round (Area Ia and II)

1. Prepare a master mix as follows:

Component	Volume (μL)	Final concentration
DEPC-treated H_2O	36.5	–
10 m*M* dNTP Mix	1.0	200 μ*M*
5 μ*M* nested primer mix	3.0	15.0 pmols each primer
25 m*M* $MgSO_4$	3.0	1.5 m*M*
10X Polymerase buffer	5.0	1X
5 U/μL DNA polymerase	0.5	2.5 U per reaction

2. Add 49.0 µL of master mix to each tube. Remove the tubes to Area II. Add 1.0 µL of each first round PCR to its corresponding second round tube. Mix.
3. On a thermal cycler program the following thermal profile (**steps 2–4** for 35 cycles):
   a. Step 1: 94°C, 1 min 30 s.
   b. Step 2: 94°C, 15 s.
   c. Step 3: 55°C, 30 s.
   d. Step 4: 72°C, 45 s.
4. Begin thermal cycling. Transfer the tubes from ice to the thermal cycler when the block temperature has reached 94°C. Cycling parameters (i.e., cycles 1–10, 20–35) are as per first-round PCR.

3.3.2. ISDR PCR

The principle of the reaction set up and cycling parameters are as for HVR 1 PCR.

3.3.2.1. First Round

1. Set up the master mix (lower layer) as follows:

Component	Volume (µL)	Final concentration
DEPC-treated H_2O	16.8	–
10 m*M* dNTP mix	1.0	200 µ*M*
5 µ*M* outer primer mix	3.0	15.0 pmols each primer
25 m*M* $MgSO_4$	2.2	2.0 m*M*
10X polymerase buffer	5.0	1X

2. Add 28.0 µL to each reaction. Add 2.0 µL of the appropriate cDNA.
3. Add one wax bead (AmpliWax Gem). Incubate at 80°C for 5 min. Place the tubes on ice.
4. Add 20.0 µL of upper-layer master mix to each reaction; the second master mix (upper-layer master mix) is identical to the upper layer master mix used for the first-round HVR PCR.
5. Thermal Cycling (**steps 2–4** for 35 cycles):
   a. Step 1: 80°C, 3 min.
   b. Step 2: 95°C, 15 s.
   c. Step 3: 65°C, 20 s.
   d. Step 4: 72°C, 45 s.

3.3.2.2. Second Round

1. Prepare a master mix as follows:

Component	Volume (µL)	Final concentration
DEPC-treated H_2O	36.5	–
10 m*M* dNTP Mix	1.0	200 µ*M*
5 µ*M* nested primer mix	3.0	15.0 pmols each primer
25 m*M* $MgSO_4$	3.0	1.5 m*M*
10X polymerase buffer	5.0	1X
5 U/µL DNA polymerase	0.5	2.5 U per reaction

2. Add 49.0 µL of master mix to each tube. Remove the tubes to Area II. Add 1.0 µL of each first-round PCR to its corresponding second-round tube. Mix.
3. On a thermal cycler, program the following thermal profile:
   a. Step 1: 94°C, 1 min 30 s.
   b. Step 2: 94°C, 15 s.
   c. Step 3: 65°C, 20 s.
   d. Step 4: 72°C, 45 s.
4. Begin thermal cycling. Transfer the tubes from ice to the thermal cycler when the block temperature has reached 94°C. Second round is for 30 cycles (**steps 2–4** only).

3.3.3. NS5A PCR

(These PCR products may require gel purification for some downstream applications.)

3.3.3.1. First Round

1. Set up the master mix (lower layer) as follows:

Component	Volume (µL)	Final concentration
DEPC-treated H_2O	18.8	–
10 m*M* dNTP Mix	1.0	200 µ*M*
5 µ*M* outer primer mix	3.0	15.0 pmols each primer
25 m*M* MgSO4	2.2	1.5 m*M*
10X polymerase buffer	3.0	1X

2. Add 28.0 µL to each reaction. Add 2.0 µL of the appropriate cDNA.
3. Add one wax bead (AmpliWax Gem). Incubate at 80°C for 5 min. Place the tubes on ice.
4. Add 20.0 µL upper-layer master mix to each reaction; the second master mix (upper-layer master mix) is identical to the upper layer master mix used for the first-round HVR PCR.
5. Place the tubes on a thermal cycler and cycle (**steps 2–4** for 35 cycles) as follows:
   a. Step 1: 80°C, 3 min.
   b. Step 2: 94°C, 20 s.
   c. Step 3: 58°C, 20 s.
   d. Step 4: 72°C, 3 min.
6. For cycles 20–35, add an increasing 20 s increment to **step 4** as per the primary HVR PCR.

3.3.3.2. Second Round

1. Set up a master mix (lower layer) as follows:

Component	Volume (µL)	Final concentration
DEPC-treated H_2O	16.0	–
10 m*M* dNTP mix	1.0	200 µ*M*

5 μ*M* nested primer mix	3.0	15.0 pmols each primer
25 m*M* $MgSO_4$	6.0	3.0 m*M*
10X polymerase buffer	3.0	1X

2. Add 29.0 μL to each reaction. Add 1.0 μL first-round PCR in Area II.
3. Add one wax bead (AmpliWax Gem). Incubate at 80°C for 5 min. Place the tubes on ice.
4. The second master mix (upper-layer master mix) is identical to the upper-layer master mix used for the first round HVR PCR.
5. Thermal cycling: as for first-round except for **step 2**: 55 °C, 20 s.

3.4. Restriction Digestion of Amplicons

1. Remove reaction components from PCRs using QiaQuick PCR Purification Kit (QiaGen). Use the protocol supplied by the manufacturer. Elute the purified amplicons in 50 μL of supplied 10 m*M* Tris-HCl, pH 8.5.
2. Digest the entire 50 μL PCR with 10 U of appropriate restriction enzymes at 37°C for 8 h (*see* **Table 2**). Heat inactivate the restriction enzymes at 80°C for 20 min.
3. Add 20 μg of glycogen and extract the digested amplicons with an equal volume of phenol:chloroform:isoamyl alcohol (25:24:1), pH 8.5, followed by chloroform:isoamyl alcohol (49:1); centrifuge at 13,000*g* in a table top centrifuge to separate the phases.
4. Precipitate the aqueous phase by adding a one-tenth volume of 3.0 *M* sodium acetate, pH 5.2, and 2 volumes of absolute ethanol. Incubate on ice for 15 min (*see* **Note 7**).
5. Centrifuge at 13,000*g* in a microfuge for 25 min and wash the pellet twice with 70% alcohol. Dry the pellet in a vacuum drier for 5 min.
6. The pBluescript II KS vector is digested and purified similarly (*see* **Note 7**). Quantify the amounts of both vector and amplicon by agarose gel electrophoresis (*see* **Note 8**).
7. Ligate the amplicon to the vector in a 3:1 (amplicon:vector) molar ratio (*see* **Table 3** for example).
8. Transform 2–5 ng of the ligated DNA into transformation-competent ***(30)*** *E. coli* XL1-Blue MRF' cells (Stratagene, La Jolla, CA). Plate the transformed cells on LB agar/X-gal/IPTG/amp plates and incubate at 37°C for 18 h.
9. 10 mL overnight cultures of LB broth/amp are inoculated with resulting white (vector with insert) colonies. High-quality plasmid DNA can be prepared using the QiaPrep Plasmid Mini Prep Kit (QiaGen), which yields DNA that is suitable for sequencing.

4. Notes

1. Solutions to be used in RNA extraction should be treated (shaking at 37°C) with 0.1% diethyl pyrocarbonate (DEPC) for 12–24 h prior to autoclaving. RNAse A is an ubiquitous and extremely stable enzyme, which is resistant to autoclaving alone. DEPC carboxymethylates amine groups, and thus inactivates RNAse by

Table 2
Sizes and Restriction Digesting Enzymes for PCR Products

PCR	Size (bp)(digested)	Restriction enzyme digestion for ligation
HVR	436 (425)	*EcoR* I and *BamH* I
ISDR	304 (294)	*EcoR* I and *BamH* I
NS5A	1379 (1365)	*Xba* I and *Hind* III

Table 3
Sample Calculation for Ligation

1. Size (bp) ISDR amplicon after digestion with *EcoR* I and *BamH* I 5 294 bp (M.W. of one bp 5 662)
   M.W. of digested ISDR amplicon 5 (294) × 662 5 1.94×10^5
   If purified digested amplicon DNA is estimated to be at 50 ng/µL:

   $$50\_10^{-9}/1{\cdot}94\_10^5 = 2{\cdot}57\_10^{-13} \text{ moles DNA/mL}$$

2. Size (bp) of pBluescript II KS+ vector after digestion with *EcoR* I and *BamH* I = 2,943 bp
   M.W. of digested vector = (2943) × 662 5 1.94×10^6
   If purified digested vector DNA is estimated to be at 200 ng/µL:

   $$200 \times 10^{-9}/1{\cdot}94\_10^6 = 1{\cdot}03\_10^{-13} \text{ moles DNA/mL}$$

3. For a 3:1 (insert : vector) molar ligation ratio:
   DNA

moles DNA	2.57×10^{-13}: 8.56×10^{-14}
mL DNA solution	1.0:0.83
ng DNA	50:166
Total ng	216

4. Add 1.0 µL of 50 ng/µL digested ISDR amplicon and 0.83 µL digested pBluescript vector, to 6.7 U T4 DNA ligase (New England Biolabs) (equal to 0.1 Weiss Unit) in 1.0X ligation buffer (supplied with enzyme) in a final volume of 10 µL. Incubate overnight at 16°C. Heat-inactivate the ligase at 65°C for 20 min.

covalent modification of the protein. Buffers with amine groups — notably Tris-HCl and EDTA — cannot be treated with DEPC. Autoclaving degrades the DEPC to ethanol and water, thus preventing it from subsequently covalently modifying amines in the RNA.

2. Incubation of the isopropanol precipitation at −70°C for 2 h to overnight enhances precipitation of nucleic acids from the solvent-extracted aqueous phase. Given the low amounts of genomic material typically encountered with HCV-positive sera, this practice is recommended.

3. RNA can be stored in the third 75% ethanol wash at −70°C without significant degradation. We have successfully amplified material stored for ⩾1.5 yr in this manner. Alternatively, the pellet can be dried and resuspended in 0.5% sodium dodecyl sulfate (SDS) or in 1–3 m*M* DTT with 40 U of ribonuclease inhibitor (ribonuclease inhibitor requires ⩾1.0 m*M* DTT for its activity) and stored at −70°C.
4. Although it is a common practice to use the anti-sense primer of the outer PCR primer pair to prime reverse transcription of RNA into cDNA templates, random hexamers can also be used. We recommend a final concentration of 100 nm in the reverse-transcription reaction if serum-derived RNA is used; 100 μ*M* if RNA has been extracted from cellular material such as biopsy tissue.
5. The two most common problems encountered with HCV RT-PCR are failure to amplify and contamination (product in negative-control PCR). The former is generally caused by degradation of RNA template prior to reverse transcription. The latter results principally from previously amplified products contaminating the pre-amplification area ***(23,24)***. Cloned amplicons also represent a source of contamination, and all plasmid handling should be done in the post-amplification area.
6. It is important to note that the reverse-transcription reaction contributes Mg^{2+} and dNTPs to the PCR, which can affect reaction performance. We generally use 2.0 μL of a 20.0 μL RT reaction (i.e., 10%) as cDNA template, and our PCRs are established to account for the Mg^{2+} and dNTP transfered with this amount of cDNA. If more cDNA is required to be added to the reaction, the following steps can be performed. The entire reverse transcription reaction can be drop dialyzed against sterile distilled H_2O using disc membranes ***(32,33)*** (Millipore). Pour 30 mL of sterile distilled water into a sterile Petri dish. Carefully place the dialysis membrane (shiny side up) onto the surface of the water, using membrane forceps (Sigma). Pipet the reverse transcription reaction (do not touch the membrane) onto the membrane. Allow dialysis to proceed for 1–2 h; then carefully remove the dialyzed cDNA. The post-dialysis volume is generally larger than the pre-dialysis volume. The filters retain nucleic acid, primers, and enzyme, yet allow the diffusion of all salts. The dialyzed reverse transcription can be lyophilized and resuspended in the outer, first-round PCR master mix. An additional 0.4 m*M* Mg^{2+} should be added to each such outer round PCR to account for Mg^{2+} lost from the reverse transcription reaction during dialysis. The anti-sense primer (except for HVR PCR) should be omitted from the outer-round PCR mix, as the 20 pmols of this primer used in the reverse transcription will be retained with the cDNA. These steps allow viral RNA from up to 100 μL serum to be added to one PCR.
7. The digested end-fragments of the amplicons will act as re-ligatable substrates during the ligation reaction. However, the 15-min ethanol/acetate precipation step will only precipitate a fraction of these small oligonucleotides. Thus, their removal from the digested amplicons is unnecessary. It is necessary, however, to remove the "stuffer" fragment from the digested pBluescript vector. This can be done by Sephacryl S300 (Amersham-Pharmacia, Uppsala, Sweden) spun-column chromatography, or by gel purification of the digested vector (e.g., by QiaEx II, Qiagen).

8. The following calculations can be used to quantitate DNA by agarose gel electrophoresis, and are based on a *Hind* III digest of lambda phage. A λ*Hind* III digest is prepared from a quantified lambda phage DNA stock (New England Biolabs, Beverly, MA). If 200 ng of this digest is run on a 1.0% agarose gel, fragments of 23,130, 9,416, 6,557, 4,361, 2,322, 2,027, and 564 bp are visible. Note: before electrophoresis, incubate the completed digest at 65°C for 10 min, then snap-cool on ice; this helps to separate the 23,130- and 4,361-bp fragments, which have a tendency to anneal. The undigested phage is 48,502 bp in size. Thus, the 23,130 bp fragment represents (23.130/48.502) × 100 = 47.68% of the 200 ng of DNA loaded. This is equal to (47.68/100) × 200 ng = 95.37 ng of DNA. Similarly, the quantity of DNA can be calculated for the other bands in the digest. The digested amplicon and vector to be quantified are run in twofold serial dilution (e.g., 1.0 and 2.0 µL of DNA solution), and the intensities of the bands are compared with those of the digest to give an approximate quantity. These calculations can also be applied to other commonly used DNA ladders, such as ϕX174 *HinF* I, λ*BstE* II.

References

1. Choo, Q. L., Kuo, G., Weiner, A. J., Overby, L. R., Bradley, D. W., and Houghton, M. (1989) Isolation of a cDNA clone derived from a blood-borne non-A, non-B viral hepatitis genome. *Science* **244**, 359–362.
2. Houghton, M. (1996) Hepatitis C viruses in Fields Virology (Fields, B.N., Knipe, D.M., Howley, P.M. et al. eds.) Lippincott-Raven, Philadelphia, PA, pp. 1035–1058.
3. Martell, M., Esteban, J. I., Quer, J., Genesca, J., Weiner, A., Esteban, R., et al. (1992) Hepatitis C virus (HCV) circulates as a population of different but closely related genomes: quasispecies nature of HCV genome distribution. *J. Virol.* **66,** 3225–3229.
4. Gomez, J., Martell, M., Quer, J., Cabot, B., and Esteban, J. I. (1999) Hepatitis C viral quasispecies. *J. Viral Hepat.* **6,** 3–16.
5. Major, M. E., and Feinstone, S. M. (1997) The molecular virology of hepatitis C. *Hepatology* **25,** 1527–1538.
6. Gale, M., Blakely, C.M., Kwieciszewski, B., Tan, S-L, Dossett, M., Tang, N. M., et al. (1998) Control of PKR protein kinase by hepatitis C virus nonstructural 5A protein: molecular mechanisms of kinase regulation. *Mol. Cell. Biol.* **18,** 5208–5218.
7. Fukuma, T., Enomoto, E., Marumo, F., and Sato, C. (1998) Mutations in the interferon-sensitivity determining region of hepatitis C virus and transcriptional activity of the nonstructural region 5A protein. *Hepatology* **28,** 1147–1153.
8. Inchauspe, G., Zebedee, S., Lee, D. H., Sugitani, M., Nasoff, M., and Prince, A. M. (1991) Genomic structure of the human prototype strain H of hepatitis C virus: comparison with American and Japanese isolates. *Proc. Natl. Acad. Sci. USA* **88,** 10292–10296.
9. Duverlie, G., Khorsi, H., Castelain, S., Jaillon, O., Izopet, J., Lunel, F., et al. (1998) Sequence analysis of the NS5A protein of European hepatitis C virus 1b isolates and relation to interferon sensitivity. *J. Gen. Virol.* **79,** 1373–1381.
10. Polyak, S. J., Paschal, D. M., McArdle, S., Gale Jr., M. J., Moradpour, D., and

Gretch, D. R. (1999) Characterization of the efffects of hepatitis C virus nonstructural 5A protein expression in human cell lines and on interferon-sensitive virus replication. *Hepatology* **29,** 1262–1271.

11. Khorsi, H., Castelain, S., Bonnet, M., Francois, C., Zawadzki, P., Capron, D., et al. (1998) Repression of the PKR protein kinase by the NS5A protein of HCV 1b isolates and response to IFN therapy in French patients with chronic hepatitis C. *Hepatology* **28,** Pt. 2, 409A.
12. Paterson, M., Laxton, C. D., Thomas, H. C., Ackrill, A. M., and Foster G. R. (1999) Hepatitis C virus NS5A protein inhibits interferon antiviral activity, but the effects do not correlate with clinical response. *Gastroenterology* **117,** 1187–1197.
13. Sangar, D. V., and Carroll, A .R. (1998) A tale of two strands: reverse transcription polymerase chain reaction detection of hepatitis C virus replication. *Hepatology* **28,** 1173–1175.
14. Morse, S. S., ed., *The Evolutionary Biology of Viruses,* Raven Press, NY, 1994.
15. Domingo, E., Holland, J. J., and Ahlquist, P. eds., *RNA Genetics Vol III: Variability of RNA Genomes,* CRC, Boca Raton, FL, 1988.
16. Holland, J. J., ed., (1992) Genetic Diveristy of RNA Viruses. *Curr. Top. Microbiol. Immunol.* **176.**
17. Chomczynski, P., and Sacchi, N. (1987) Single-step method of RNA isolation by acid guanidium-thiocyanate-phenol-chloroform extraction. *Anal. Biochem.* **162,** 156–159.
18. Goergen, B., Jakobs, S., Meyer zum Buschenfelde, K-H., and Gerken G. (1994) Evaluation of different serum RNA extraction protocols using quantitative HCV-PCR. *J. Hepatol.* **21,** 1144.
19. Sellner, L. N., Coelen, R. J., and Mackenzie, J.S. (1992) Reverse transcription inhibits Taq polymerase activity. *Nucleic Acids Res.* **20,** 1487–1490.
20. Hu, A-L. W., D'Alessio, J. M., Gerard, G. F., and Kullman, S. (1991) RT-catalyzed first-strand synthesis and the amplification of beta-actin mRNA using the polymerase chain reaction. *BRL-Focus* **13,** 20–29.
21. Whitby, K., and Garson, J. A. (1997) A single tube two compartment reverse transcription polymerase chain reaction system for ultrasensitive quantitative detection of HCV RNA. *J. Virol. Methods* **66,** 15–18.
22. Moreira, R., and Noren, C. (1995) Minimum duplex requirements for restriction enzyme cleavage near the terminii of linear DNA fragments. *Biotechniques* **19,** 56–59.
23. Kwok, S., and Higuchi, R. (1989). Avoiding false positives with RT-PCR. *Nature* **339,** 237–238.
24. Orrego, C. (1990) Organizing a laboratory for PCR work, in *PCR Protocols: A Guide to Methods and Applications,* Innis, M. A., Gelfand, D. H., Sninsky, J. J., White, T. J., eds., Academic Press, San Diego, CA, pp. 447–454.
25. Cline, J., Braman, J. C., and Hogrefe, H. H. (1996) PCR fidelity of Pfu DNA polymerase and other thermostable DNA polymerases. *Nucleic Acids Res.* **24,** 3546–3551.
26. Hengen, P. N. (1995) Fidelity of DNA polymerases for PCR. *Trends Biochem. Sci.* **20,** 324–325.

27. Bracho, M. A., Moya, A., and Barrio, E. (1998) Contribution of Taq polymerase-induced errors to the estimation of RNA virus diversity. *J. Gen. Virol.* **79,** 2921–2928.
28. Mullan, B., Kenny-Walsh, E., Collins, J. K., Shanahan, F., and Fanning, L. J. (2001) Inferred Hepatitis C virus quasispecies diversity is influenced by choice of DNA polymerase in Reverse Transcriptase-Polymerase Chain Reactions. *Anal. Biochem.* **289,** 137–146.
29. Clark, J.M. (1988) Novel non-templated nucleotide addition reactions catalyzed by procaryotic and eucaryotic DNA polymerases. *Nucleic Acids Res.* **16,** 9677–9686.
30. Holton, T. A., and Graham, M. W. (1991) A simple and efficient method for direct cloning of PCR products using dT-tailed vectors. *Nucleic Acids Res.* **19,** 1156.
31. Hanahan, D., Jessee, J., and Bloom, F. R. (1991) Plasmid Transformation of Escherichia coli and other bacteria. *Methods Enzymol.* **204,** 63–113.
32. Gorisch, H. (1988) Drop dialysis: time course of salt and protein exchange. *Anal. Biochem.* **173,** 393–398.
33. Marusyk, R., and Sergeant, A. (1980) A simple method for dialysis of small volume samples. *Anal. Biochem.* **105,** 403–404.
34. Kato, N., Hijikata, M., Ootsuyama, Y., Nakagawa, M., Ohkoshi, S., Sugimura, T., et al. (1990) Molecular cloning of the human hepatitis C virus genome from Japanese patients with non-A, non-B hepatitis. *Proc. Natl. Acad. Sci. USA* **87,** 9524–9528.

V

In Situ Localization of mRNA Expression

14

In Situ Immuno-PCR

A Newly Developed Method for Highly Sensitive Antigen Detection In Situ

Yi Cao

1. Introduction

Immunocytochemistry and *in situ* hybridization—the *in situ* staining approaches for detection of antigens, DNA, or RNA in intact cells and tissue sections— are common and powerful tools for biological and biomedical research and clinical diagnostics. By combining the exponential amplification power of the polymerase chain reaction (PCR) with *in situ* hybridization, *in situ* PCR and *in situ* reverse-transcription-polymerase chain reaction (RT-PCR) allows the detection of low copy numbers of nucleic acids in situ ***(1–3)***. However, with routine immunocytochemistry, detection of the minute numbers of target molecules accessible with *in situ* PCR or *in situ* RT-PCR is difficult. Immuno-PCR, which uses PCR amplification to increase the signal of immunoassays, exhibits high sensitivity and permits the detection of proteins at the level of a few hundred molecules ***(4–5)***. By combining the high sensitivity of *in situ* PCR amplification with the versatility and high specificity of immunocytochemistry, we have developed a new method known as *in situ* immuno-PCR to detect antigens at low levels in intact cells or tissue sections ***(6)***. The concept of *in situ* immuno-PCR is illustrated in **Fig. 1**. Our study demonstrated that *in situ* immuno-PCR is more sensitive for the detection of Hepatitis B-specific antigen (HBsAg) *in situ* compared with other signal-amplification systems (**Fig. 2**). *In situ* immuno-PCR may be the only technique available to detect minute quantities of biological macromolecules such as proteins, carbohydrates, and lipids in intact cells or tissue sections. This new method should find wide application in biological research and clinical diagnostics.

From: *Methods in Molecular Biology, vol. 193: RT-PCR Protocols*
Edited by: J. O'Connell © Humana Press Inc., Totowa, NJ

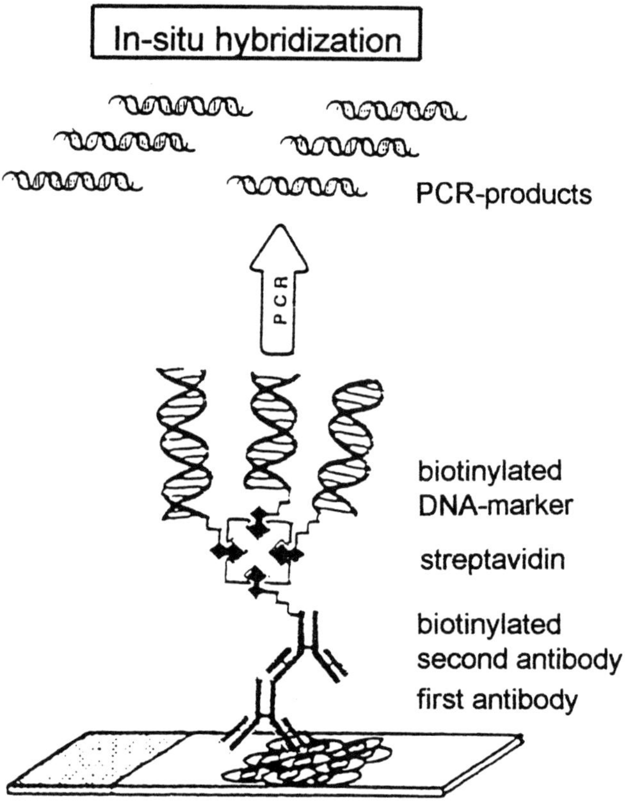

Fig. 1. Schematic outline of the *in situ* immuno-PCR procedure. Reprinted with permission from **ref. *(6)***.

2. Materials

1. Paraffin-embedded tissue sections on pretreated microscope slides (*see* **Note 1**).
2. Tris-buffered saline (TBS): 0.02 *M* Tris-HCl and 0.15*M* NaCl, pH 7.6.
3. Phosphate-buffered saline (PBS): 0.1*M* phosphate buffer and 0.15 *M* NaCl, pH 7.2.
4. Biotinylated goat anti-mouse Ig; streptavidin or avidin; avidin-biotin blocking kit (Vector, Burlingame, CA, USA).
5. Deoxyribonuclease (DNase) and ribonuclease (RNase) (Roche Molecular Biochemicals, Mannheim, Germany).
6. Normal goat serum.
7. Primary antibody.
8. Biotinylated DNA-marker (see Note 2).
9. PCR buffer: 10 m*M* Tris-HCl, 50 m*M* KCl, 4 m*M* $MgCl_2$, 0.01% gelatin, 0.05% Tween 20, 0.05% Nonidet P-40, pH 8.5.
10. Primer 1 and primer 2 (*see* **Note 2**).
11. PCR-kit (Roche Molecular Biochemicals, Mannheim, Germany).

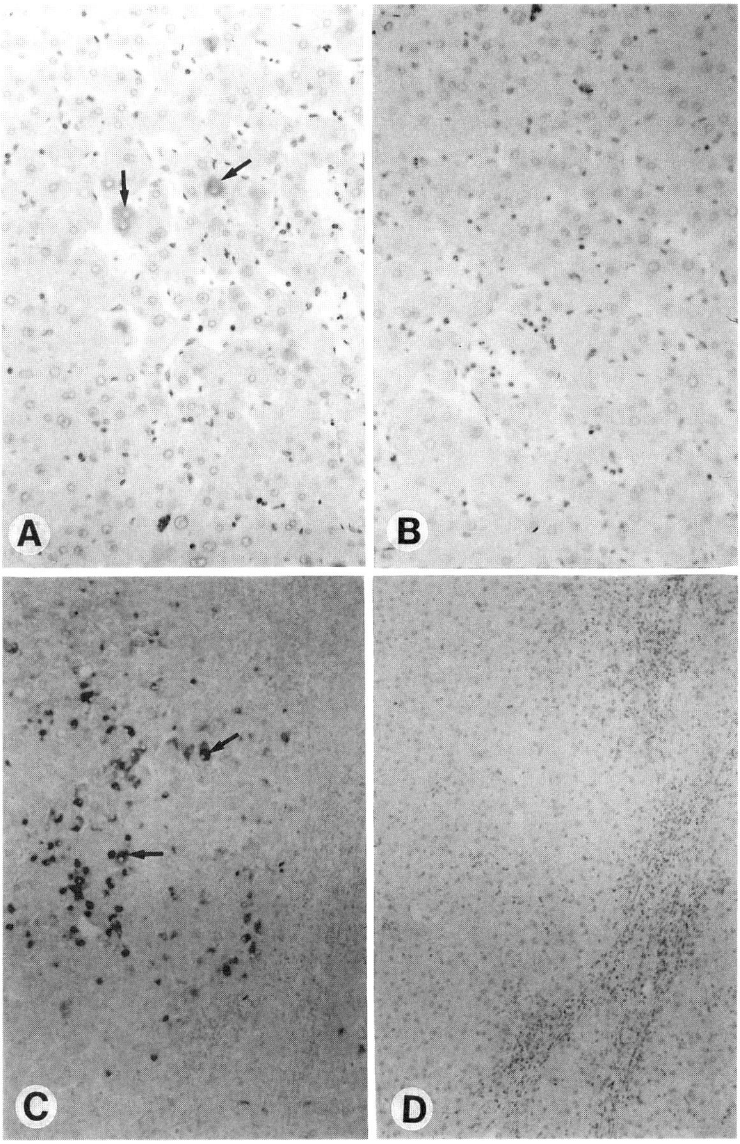

Fig. 2. Comparison of sensitivities of *in situ* immuno-PCR, the tyramide signal amplification (NEN Life Science Products, Boston, MA, USA), and the double alkaline phosphatase anti-alkaline phosphatase complex (APAAP; DAKO, Copenhagen, Denmark) method in the same samples. **(A,B)**: Sections of a liver cirrhosis: anti-HBsAg monoclonal antibody stains the cytoplasm of hepatocytes (arrows) with *in situ* immuno-PCR (A, scoring: +[< 25% positive cells]), but do not with the tyramide signal amplification **(B)**; **(C,D)**: Serial sections through a cirrhotic liver: *in situ* immuno-PCR staining shows positive reactivity in the cytoplasm of hepatocytes (arrows) with anti-HBsAg monoclonal antibody (**C**, scoring: ++ [25–50% positive cells]), but the double alkaline phosphatase anti-alkaline phosphatase complex method shows no signal **(D)**. Reprinted with permission from **ref.** ***(6)***.

12. 5X SSC: 50% formamide, 0.75 *M* NaCl, 0.075 *M* sodium citrate, pH 7.0.
13. Prehybridization buffer: 5X standard saline citrate (SSC) containing 0.1% bovine serum albumin (BSA), 0.1% ficoll, 0.1% polyvinyl pyrrolidone, 200 µg/mL salmon-sperm DNA, and 4 m*M* EDTA.
14. Dogoxigenin-labeled DNA probe (*see* **Note 2**).
15. Fab fragments of anti-dogoxigenin antibody conjugated with alkaline phosphatase (Roche Molecular Biochemicals).
16. The Fast Red Substrate System (DAKO, Copenhagen, Denmark).

3. Methods

3.1. Immunocytochemistry

1. Deparaffinize sections in xylene and rehydrate in 100%, 95%, 70%, and 40% ethanol; wash sections in water and TBS.
2. Block endogenous biotin using the avidin-biotin blocking kit by incubating the slides in avidin blocking solution for 15 min, then in biotin blocking solution for 15 min; wash sections in TBS (*see* **Note 3**).
3. Destroy intrinsic DNA and RNA using DNase (8 U/mL) / RNase (5 µg/mL) in TBS for 1 h at 37°C; then wash in TBS; heat the slides to 80°C for 3 min to inactivate DNase/RNase; wash in TBS.
4. Block nonspecific binding by incubating the sections in 10% normal goat serum for 30 min.
5. Incubate the sections with the primary antibody (mouse monoclonal) overnight at 4°C or 1 h at room temperature (*see* **Note 4**).
6. Wash the sections in TBS, then incubate with a biotinylated anti-mouse secondary antibody for 30 min.
7. Wash the sections in TBS, then incubate with streptavidin or avidin for 30 min.
8. Wash the sections in TBS, then incubate with the biotinylated DNA-marker for 60 min.

3.2. In Situ *PCR Amplification*

1. Incubate the sections with PCR buffer for 5 min.
2. Cover the sections with PCR reaction mixture (primer 1, 1 µ*M*; primer 2, 1 µ*M*; Taq DNA polymerase (2.5 U/50 µL), 250 µ*M* of each deoxynucleotide 5' triphosphate (dNTPs), 10 m*M* Tris-HCl, pH 8.5, 50 m*M* KCl, 4 m*M* $MgCl_2$.
3. Add an acid-cleaned cover slip to each section, and seal with rubber cement or nail polish.
4. Perform PCR amplification as follows: initial denaturation 95°C for 2 min; 30 cycles of denaturation at 95°C for 1 min, annealing at 55°C for 1 min, extension at 72°C for 1 min; final extension at 72°C for 5 min (*see* **Note 5**).

3.3. In Situ *Hybridization*

1. Pry off the cover slips using a fine blade washed in 100% alcohol, then post-fix sections with 100% alcohol for 20 min (*see* **Note 6**).

2. Heat the slides for 1 min on a 92°C heating block (*see* **Note 6**).
3. Wash the sections in 1X SSC with 0.05% Triton-X-100 for 5 min 3×.
4. Incubate the sections with prehybridization buffer for 1–3 h at 45°C.
5. Remove the prehybridization buffer and replace with hybridization mixture prepared by adding 1 L of digoxigenin-labeled DNA probe per 50 L prehybridization buffer.
6. Cover each section with an acid-cleaned cover slip and seal with rubber cement or nail polish.
7. Heat the slides for 5 min at 95°C to denature.
8. Incubate the slides overnight at 48°C (*see* **Note 7**).
9. Remove the cover slips with a fine blade. Wash the slides in 5X SSC with 50% formamide for 10 min at 45°C, in 2X SSC with 25% formamide 3X for 10 min at 42°C, twice for 30 min at 42°C, and in 0.2X SSC twice for 5 min at room temperature; wash in TBS.

3.4. Detection

1. Incubate the sections with 10% normal goat serum for 30 min.
2. Incubate the sections with Fab fragments of anti-digoxigenin antibody conjugated with alkaline phosphatase for 1 h at room temperature.
3. Wash the sections in PBS, then incubate with the Fast red substrate system under microscopy (*see* **Notes 8** and **9**).

4. Notes

1. Loss of tissue adherence is one of the most common problems. The pretreatment of slides with materials such as amino propylethoxysilane (APES), organosilane, or other adhesives is necessary to prevent sections falling off *(7)*.
2. Theoretically, an any segment of DNA can be used as DNA-marker for immuno-PCR and *in situ* immuno-PCR. Many DNA-markers were successfully used for immuno-PCR *(3,4,8–10)*. However, the additional requirements for DNA-markers of *in situ* immuno-PCR include (1) a sequence that is not homologous to sequences in the endogenous DNA and RNA; (2) no unspecific binding with the sections, and (3) lengths that did not exceed 1000 bp. When a DNA-marker is selected, corresponding primers (to amplify DNA-marker) and digoxigenin-labeled DNA probe (to detect PCR products) will be used. The DNA-marker can been biotinylated with Photoprobe biotin reagents (Vector). The biotinylated DNA-marker is estimated using the ABC kit (Vector) in dot-blot format. The digoxigenin-labeled DNA probe can be produced using PCR DIG Probe Synthesis kit (Roche Molecular Biochemicals). Labeling efficiency is estimated by a spot test with anti-digoxigenin-antibody conjugated to alkaline phosphatase (Roche Molecular Biochemicals).
3. In most cases, this step can be omitted. When an antibody-biotin-avidin bridge is used to link the DNA marker to target molecules in some tissues such as the liver and kidney, the pretreatment with the avidin-biotin blocking kit should be done to block endogenous biotin.

4. Antigens may be altered (destroyed, denatured, or masked) during tissue fixation and processing, which may diminish or abrogate their detection. The application of proteolytic or microwave pretreatment could be considered for antigen retrieval.
5. The temperature profile of *in situ* PCR depends on the selected DNA-marker and the primers.
6. Post-fixation and heating of slides are very important procedures for immobilization of the PCR products *in situ.*
7. The temperature of *in situ* hybridization may be varied according to the melting temperature of the probe.
8. The color development should be done under microscopy in order to obtain the strongest staining and the weakest background.
9. Negative controls for *in situ* immuno-PCR should consist of (a) a comparable dilution of mouse IgG instead of the primary antibody; (b) omission of primers and DNA polymerase; (c) competition with a 20-fold excess of unlabeled DNA probe in the presence of digoxigenin-labeled DNA probe; and (d) treatment of sections with deoxyribonuclease before hybridization.

References

1. Haass, A.T., Retzel, E.F., and Staskus, K.A. (1990) Amplification and detection of lentiviral DNA inside cells. *Proc. Natl. Acad. Sci. USA* **87,** 4971–4975.
2. Patterson, B. K., Till, M., Otto, P., Goolsby, C., Furtado, M., McBride, L. J., et al. (1993) Detection of HIV-1 DNA and messenger RNA in individual cells by PCR-driven in situ hybridization and flow cytometry. *Science* **260,** 976–979.
3. Komminoth, P., and Werner, M. (1997) Target and signal amplification: approaches to increase the sensitivity of in situ hybridization. *Histochem. Cell Biol.* **108,** 325–333.
4. Sano, T., Smith, C.L., and Cantor, C.R. (1992) Immuno-PCR: very sensitive antigen detection by means of specific antibody-DNA conjugates. *Science* **258,** 120–122.
5. Ruzicka, V., März, W., Russ, A., and Gross, W. (1993) Immuno-PCR with a commercially available avidin system. *Science* **260,** 698–699.
6. Cao, Y., Kopplow, K., and Liu, G.-Y. (2000) *In situ* immuno-PCR to detect antigens. *Lancet* **356,** 1002–1003.
7. Nuovo, G.J. (2000) In situ localization of PCR-Amplified DNA and cDNA in *Methods in Molecular Biology: In Situ Hybridization Protocols,* Darby, I.A., ed., Humana Press, Totowa, NJ, pp 217–238.
8. Case, M., Major, G.M., Bassendine, M.F., and Burt, A.D. (1997) The universality of immuno-PCR for ultrasensitive antigen detection. *Biochem. Soc. Trans.* **25,** 374S.
9. Sanna, P.P., Weiss, F., Samson, M. E., Bloom, F. E., and Pich, E. M. (1995) Rapid induction of tumour necrosis factor α in the cerebrospinal fluid after intracerebroventricular injection of lipopolysaccharide revealed by a sensitive capture immuno-PCR assay. *Proc. Natl. Acad. Sci. USA* **92,** 272–275.
10. Chang, T.C. and Huang, S.H. (1997) A modified immuno-polymerase chain reaction for the detection of β-glucuronidase from *Escherichia coli. J. Immunol. Methods* **208,** 35–42.

15

RT-PCR from Laser-Capture Microdissected Samples

Tatjana Crnogorac-Jurcevic, Torsten O. Nielsen, and Nick R. Lemoine

1. Introduction

Laser-capture microdissection (LCM) is a powerful method for efficient procurement of defined cell populations from whole tissue sections. It enables the study of cell biology and molecular pathology at the level of DNA, RNA, and protein *(1,2,3)*. The principle of LCM, as applied by the Pixcell system (Arcturus Engineering Inc., Mountain View, CA) comprises a laser beam of variable size that activates a thermoplastic polymer film on a cap placed in direct contact with cell(s) chosen for microdissection. The melted ethylene vinyl acetate film flows into and replaces dehydrated interstitial and cellular compartments, "capturing" targeted cells onto the cap (**Fig. 1**). The cap is then placed directly in contact with the appropriate isolation buffer in a standard microfuge tube.

The quantity of material recovered after LCM is usually limited and typically requires amplification techniques for analysis. Reverse-transcriptase-polymerase chain reaction (RT-PCR) is one of the methods used to study gene expression in LCM material, but obtaining RNA of adequate quality for quantitative analysis is usually challenging. The protocol described here details the steps involved in slide preparation, fixation and staining, LCM, RNA isolation, and RT-PCR.

2. Materials

1. Cryostat apparatus (*see* **Note 1**); baked, clean plain-glass microscope slides (*see* **Note 2**) (Blue Star Micro slides, Chance Propper Ltd), Tissue-Tek O.C.T. embedding compound (Miles, Inc., Elkhart, IN), and plastic boxes for storage of the slides.

From: *Methods in Molecular Biology, vol. 193: RT-PCR Protocols*
Edited by: J. O'Connell © Humana Press Inc., Totowa, NJ

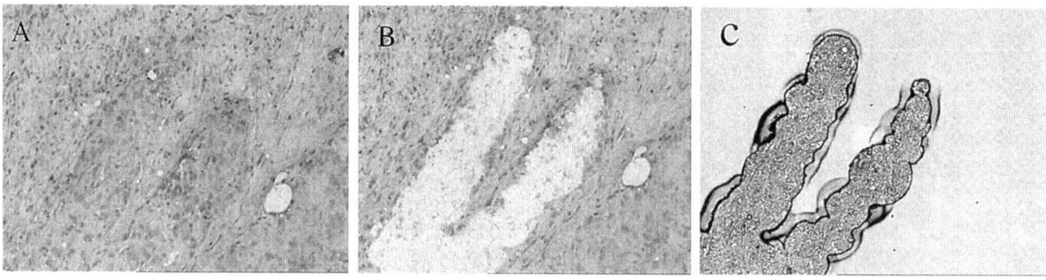

Fig. 1. Laser-capture microdissection: (**A**) human melanoma sample before microdissection, (**B**) after microdissection, and (**C**) the microdissected cells captured onto the cap.

2. Metal slide holders and Coplin jars for staining; diethyl pyrocarbonate- (DEPC)-treated water; 70%, 95%, and absolute ethanol; Mayer's Hematoxylin; Scott's tap water substitute concentrate (Sigma); eosin Y; xylene (*see* **Note 3**).
3. LCM apparatus, (PixCell or PixCell II) and CapSure transfer caps and pads (Arcturus Engineering Inc., Mountain View, CA).
4. For RNA isolation: GITC buffer (4 *M* guanidinium thiocyanate, 0.02 *M* sodium citrate, 0.5% sarcosyl); β-mercaptoethanol; 2 *M* sodium acetate; phenol saturated with DEPC-treated water; chloroform:isoamyl alcohol; 2 mg/mL glycogen; isopropanol (these solutions are included in the Stratagene Micro RNA Isolation kit); 40 U/μL RNasin RNase inhibitor (Promega); 10 U/μL DNase I enzyme and 10× DNase reaction buffer comprising 100 m*M* Tris-HCl pH 8.4, 500 m*M* KCl, 15 m*M* $MgCl_2$ and 0.01% gelatin (GenHunter Corporation).
5. For cDNA preparation: 500 μg/mL stock of oligo-$(dT)_{12\text{-}18}$ primer (alternatively, 50–250 ng of random primers or 2 pmol of a gene- specific primer can be used per cDNA reaction); SuperScript II RNaseH-reverse transcriptase (RT), 5× first-strand buffer (250 m*M* Tris-HCl, pH 8.3, 375 m*M* KCl, 15 m*M* $MgCl_2$) and 0.1 *M* dithiothreitol (DTT) (all from Gibco-BRL), 10 m*M* deoxynucleotide 5' triphosphate (dNTP) mix (10 m*M* each dATP, dGTP, dCTP, and dTTP).
6. For RT-PCR reaction: synthesized cDNA, PCR primers, 10× PCR buffer, 50 m*M* $MgCl_2$, 10 m*M* dNTP mix and a thermo-stable DNA polymerase.

3. Methods

3.1. Preparation of Slides

1. Cut the surgical tissue specimens into small pieces, immediately embed in Tissue Tek O.C.T mounting medium on a cork disk, snap-freeze in liquid nitrogen, and store at -80°C until use (*see* **Note 4**).
2. Cut the frozen-tissue block into serial 5–12-μm-thick sections (*see* **Note 5**). Up to two sections are positioned at the center of a cleaned and baked plain-glass microscope slide (*see* **Note 6**).
3. After cryosectioning, keep the slides with tissue sections previously prepared on dry ice in airtight, plastic slide boxes (made RNase-free with RNase Away, Gibco-BRL) for transport and store at –80°C until microdissection.

3.2. Staining and LCM

(*See also* the Arcturus website at: http://www.arctur.com). Immediately after removal of tissue sections from cold storage to room temperature, fix and stain as follows (*see* **Note 7**):

1. Fix the slides with 70% EtOH for 1 min.
2. Rinse briefly with DEPC-treated H_2O.
3. Stain with Mayer's hematoxylin for 30 s
4. Rinse briefly with DEPC-treated H_2O.
5. Put into Scott's solution bath until stain is differentiated or "blued" (usually 30–60 s).
6. Wash with 70% EtOH for 30 s.
7. Wash with 95% EtOH for 30 s.
8. Stain with eosin Y for 20 s.
9. Dehydrate in 95% EtOH (first bath, 30 s).
10. 95% EtOH (second bath, 30 s).
11. 100% EtOH (third bath, 30 s).
12. Immerse in xylene for 2 × 5 min.
13. Air dry the slides for at least 5 min (*see* **Note 8**).
14. Prepare fresh aliquots of 200 μL GITC buffer and 1.6 μL of β-mercaptoethanol in 0.5 mL RNase-free microcentrifuge tubes.
15. LCM. When the first slide to be dissected has been mounted on the inverted microscope, load the cover slip optical adaptor instead of a microdissection transfer cap; this allows you to make a histologic review of the slide (*see* **Note 9**).
16. After reviewing the section, next load a CapSure microdissection cap and swing into position over the tissue specimen. Perform LCM of the chosen cells or tissue area as effectively as possible on all areas that can be accessed, and reposition the glass slide to bring other areas of tissue into the working field. The same cap can be used on further serial sections until its working area is effectively filled (*see* **Note 10**).
17. Now, remove the cap, and touch to a CapSure pad to remove non-specifically transferred material (*see* **Note 11**), and snap into place in the aperture of a 0.5-mL microfuge tube with the guanidine isothiocyanate (GITC) lysis buffer prepared earlier.
18. Invert the tube for 2 min, spin to bring its contents to the bottom, and keep on ice. Further caps used to dissect the same sample can be pooled into one lysis buffer tube. Following microdissection, store the tubes at –80°C until RNA isolation.

3.3. RNA Isolation and DNase Treatment (see Note 12)

RNA isolation is based on the Stratagene RNA Micro-Isolation kit and Arcturus protocol (http://www.arctur.com).

1. Thaw GITC lysis buffer solution with LCM material and transfer it to a 1.5-mL tube.
2. Add 1/10 vol (20 μL) of 2 *M* sodium acetate, 1× vol (220 μL) of phenol (bottom layer), and 0.3× vol (60 μL) of chloroform:isoamyl alcohol.

3. Vortex and chill the tube on ice for 15 min.
4. Centrifuge for 30 min at 4°C at 13,000*g*.
5. Transfer the upper aqueous phase to a fresh tube and add 2 μL (1/100 vol) of glycogen (2 mg/mL solution). Glycogen carrier facilitates the precipitation of small amounts and low concentrations of RNA, and produces a visible pellet.
6. Add 200 μL of cold isopropanol and precipitate the RNA at -80°C for 30 min (or overnight).
7. Centrifuge for 30 min at 4°C at 13,000*g*.
8. Carefully discard the supernatant, wash the pellet with 400 μL of cold 70% ethanol, and spin for an additional 15 min at 13,000*g*.
9. Remove all the supernatant and air-dry the pellet. The dried pellet can be stored at –80°C.
10. DNase treatment (MessageClean kit from GenHunter can be used as an alternative): Resuspend the nucleic acid pellet in 16 μL of DEPC-treated water; add: 2 μL of 10× DNase reaction buffer, 1 μL (40 U/μL) of RNasin RNase inhibitor (Promega), and 1 μL (10 U/μL) of DNase I (GenHunter). Incubate at 37°C for 30 min.

The samples are then re-extracted with phenol-chloroform as described in **Subheading 3.3.2., step 9**, scaled down for a 20 μL-vol. The washed and dried pellet is stored at –80°C (*see* **Note 13**).

3.4. RT-PCR

1. Resuspend the RNA pellet in 23 μL of DEPC-treated water (*see* **Note 14**).
2. Add 2 μL of 500 μg/mL oligo(dT)$_{12–18}$ primer.
3. Heat to 70°C for 10 min, quickly chill on ice, and add:
   a. 8 μL 5× first-strand buffer (Gibco-BRL),
   b. 4 μL 0.1 *M* DTT (Gibco-BRL),
   c. 2 μL of a 10 m*M* dNTP mix (10 m*M* each)
4. Split the specimen into two: one half represents the test reaction (+RT), and the other one is a control of DNase I treatment (–RT).
5. Warm both tubes at 42°C for 2 min.
6. Add 1 μL (200 U) of SuperScript II (Gibco-BRL) to the test specimen (+RT), and 1 μL of H_2O to the negative control (-RT).
7. Incubate the tubes at 42°C for 50 min.
8. Terminate the reactions by heating to 70°C for 15 min. Store the cDNA product at –20°C.

3.5. PCR

1. 2–5 μL of the synthesized cDNA is used as template, and an appropriate vol of:
   a. 10 × PCR buffer.
   b. 10 m*M* PCR primers.
   c. 50 m*M* $MgCl_2$.
   d. 10 m*M* dNTP mix.
   e. Thermo-stable DNA polymerase.

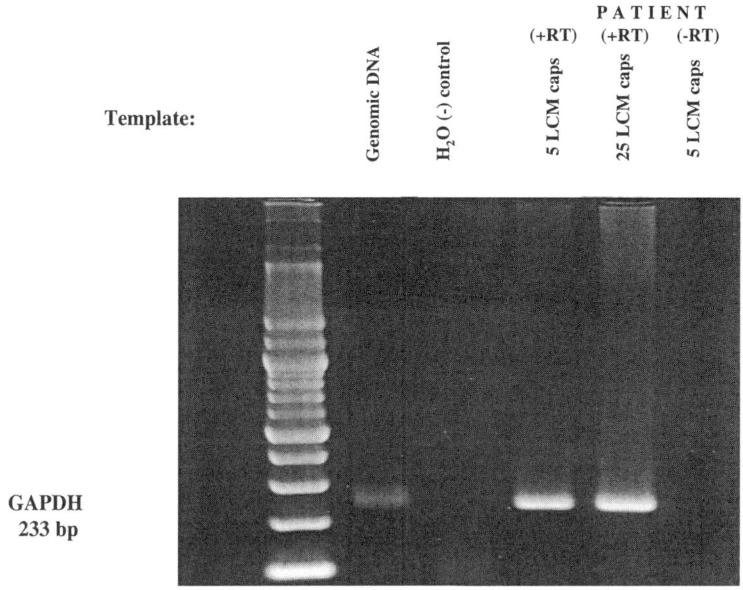

Fig. 2. RT-PCR of microdissected melanoma samples.

The concentrations of PCR components and cycling should be optimized using appropriate positive control test reactions prior to the use of any microdissected cDNA template. Because of its scarcity, LCM-cDNA should only be used under fully optimized PCR conditions. 10–25% of the LCM-derived cDNA provides sufficient template for PCR in most cases (**Fig. 2**) (*see* **Note 15**).

4. Notes

1. The cryostat is cleaned thoroughly with EtOH and loaded with a fresh blade for each specimen. Use your own set of brushes, designated only for RNA work.
2. Put plain glass microscopic slides in metal racks, wash for 2 h under running water with dishwashing liquid, rinse intensively in MilliQ water, dry briefly in the oven, wrap in aluminium foil and then bake for 4 h at 200ºC.
3. All solvent stocks should be dedicated for RNA work and handled as such. Hematoxylin and eosin (H&E) should be filtered, and dilutions of EtOH and Scott's solution made in DEPC-treated water. Coplin jars are washed and baked as described for the slides. All solutions are exchanged prior to a new LCM session.
4. Good-quality RNA can only be obtained from freshly snap-frozen tissues ***(4)***. If possible, RNA quality from a sample should be checked prior to cryosectioning and LCM (*see* **Note 13**). We have noticed a great variability in quality of tissue blocks, partly dependent on the time lapse between surgical resection and sample freezing.

5. 7 µm-Thick sections seem optimal for both histology and microdissection. Sections can be cut thicker (10–12 µm), allowing more material to be obtained from fewer microdissected slides. The trade-off is more complicated histologic interpretation, difficulties in accurate dissection, and potential problems in achieving transfer to the CapSure microdissection cap. Simple surface epithelia (e.g., endocervix) can be dissected from thicker sections than complicated microacinar structures (e.g., pancreas).
6. Two sections can be placed on each slide, but only the middle 3/4 of the working space on the slide should be used to permit contact of the slide with the microscope stage vacuum ports and full access of the laser to all parts of the tissue sections.
7. All tissue sections to be microdissected in one session should be fixed and stained simultaneously in small batches of 5 or 10 slides, depending on the tissue and number of cells to be dissected, so that the entire procedure can be finished within a few hours (working as quickly as possible).
8. Using two sequential xylene steps at the end of the staining procedure improves subsequent transfer efficiency. Slides can be safely left in xylene longer if necessary (e.g., the final slides in a large batch). As soon as one air-dried slide is mounted onto the microscope, the next slide can be brought out of xylene to begin air-drying; this will ensure no delays waiting for the xylene to evaporate off the slides.
9. Because a cover slip cannot be used during the LCM procedure, the reduction in refractive index and scattering of the light passing through the tissue makes the sections appear darker in general, and video images can be difficult to interpret. Therefore, it is useful to arrange the slides in order of cutting and perform H&E staining on one section at the beginning and then each 15–20 sections (or alternatively from the beginning, middle, and end of the cut block). These slides, which are cover-slipped after H&E staining and stored at room temperature, they provide a histological reference and permit planning of the microdissection.
10. One frequent problem with LCM is poor transfer of laser-dissected tissue from the slide to the cap. Causes include poor contact of cap to tissue (especially near folds in the tissue section, or when nearby areas of the cap have already been used), incomplete dehydration and drying, sections that are too thick, and inadequate laser power. The simplest remedies are lifting and resetting the cap, selecting an unused area of the cap, using an entirely new cap, and increasing the laser-pulse power. However, if laser-pulse power is increased much over 100 mW, (default settings for laser power and amplitude are 50 mW and 30 ms, respectively) this may lead to poor efficiency of nucleic acid isolation. So, optimization is needed for each tissue type within the individual LCM setting.
11. Another common problem is nonspecific transfer of undissected tissue from the glass slide to the cap; the same relatively loose contact between ethanol-fixed tissue and glass that facilitates LCM unfortunately also facilitates nonspecific transfer. This can occur during any lifting of the cap, and is most common around the cap edges, which are often outside the field of view. Large amounts of

unwanted tissue stuck to the edges of the cap can defeat the purpose of microdissection, so it is vital that the caps be touched down onto CapSure pads (the glue strip on unused PostIt notes works just as well and at a fraction of the price) prior to placement in lysis buffer.

12. It is possible to analyze microdissected frozen tissue sections by RT-PCR without prior RNA isolation. The procedure is based on cell lysis by three cycles of freezing and thawing in a buffer containing dithiothreitol (DTT) and RNase inhibitor ***(5)***.
13. A control of RNA quality after LCM can be assessed by nonselective dissection of the remaining material on slides using "machine-gun" mode. The settings applied are: laser beam of 60-µm diameter, laser power and amplitude increased to 100 mW and 50 ms, respectively, and delay between pulses of 500 ms. Isolate total RNA as described previously, and resuspend it in 3 µL of DEPC-treated water. Add 1 µL of loading dye and 1 µL of 1/100 dilution of SYBRGold dye (Molecular Probes) directly to the sample. SYBRGold is a cyanine dye that exhibits approx 1000-fold fluorescence enhancement upon binding to nucleic acids. Set a 1% agarose minigel in 50 mL of 1× TAE (made with DEPC-treated water), and run the prepared RNA sample in an RNA-dedicated gel tank at 50 V for approx 20 min in 1× TAE/DEPC-H_2O buffer. Alternatively, dilute the stock of SYBRGold 10,000× (10 µL in 100 mL of water or TE buffer), and post-stain the gel for half an hour on a shaker, well protected from light. In both cases, about 25 ng of RNA can be seen (three caps filled with approx 5000 cells usually give enough material) and RNA quality can be assessed based on the 28S and 18S ratio (ideally, 2:1). The gel can be visualized and photographed with Polaroid 667 black-and-white film using a SYBRGold gel stain photographic filter (S-7569) and ultraviolet (UV) transilluminator, or alternatively, a non-UV Dark Reader transilluminator (GRI).
14. To quantitate the amount of RNA after LCM, the RiboGreen RNA Quantitation Kit (Molecular Probes) can be used. It is based on an ultrasensitive fluorescent nucleic acid stain (excitation 485 nm, emission 530 nm). The linear range for quantitation with the RiboGreen reagent extends from 1 ng/mL to 1 mg/mL RNA, using two dye concentrations. For the high- or low-range assay, RiboGreen reagent is diluted 200 or 2000-fold, respectively, and added to a standard RNA at known concentrations (ribosomal RNA is included as a standard in the kit). Fluorescence emission intensity is then plotted vs RNA concentrations to obtain a standard curve. The RNA concentration of LCM samples after measurement can then be determined. On a fluorescence microplate reader, as little as 100 pg of RNA in 100 µL of low-range assay volume, can be detected. As the dye binds to DNA as well as RNA, DNase I treatment is necessary prior to measurements of RNA concentration.
15. The amount of tissue required and the number of sections that need to be cut will vary depending on the abundance of the desired cell populations within the tissue, the abundance of the mRNA transcript to be analyzed, and the patience and efficiency of the microdissection operator. Five to ten serial sections will yield

enough material for successful amplification in most cases. During the reverse transcription step, the RNA sample is split into two to provide a negative control that is not exposed to RT. The absence of a PCR product in this negative control proves that any RT-PCR product in the test specimens is derived from RNA, and not from DNA that was incompletely digested during the DNase I step. If the chosen PCR primers recognize sequences from different exons spanning an intron, this step may not be necessary.

References

1. Emmert-Buck, M.R., Bonner, R.F., Smith, P.D., Chuaqui, R.F., Zhuang, Z., Goldstein, S.R., et al. (1996) Laser capture microdissection. *Science* **274,** 998–1001.
2. Simone, N.L., Bonner, R.F., Gillespie, J.W., Emmert-Buck, M.R., and Liotta, L.L. (1998) Laser-capture microdissection: opening the microscopic frontier to molecular analysis. *Trends Genet.* **14,** 272–276.
3. Sirivatanauksorn, Y, Drury, R., Crnogorac-Jurcevic, T., Sirivatanauksorn, V., and Lemoine, N.R. (1999) Laser-assisted microdissection: applications in molecular pathology. *J. Pathol.* **189,** 150–154.
4. Hiller, T., Snell, L., and Watson, P. (1996) Microdissection RT- PCR analysis of gene expression in pathologically defined frozen tissues. *BioTechniques* **21,** 38–44.
5. To, M.D., Done, S.J., Redston, M., and Andrulis, I.L. (1998) Analysis of mRNA from microdissected frozen tissue sections without RNA isolation. *Am. J. Pathol.* **153,** 47–51.

16

Mycobacterium paratuberculosis Detected by Nested PCR in Intestinal Granulomas Isolated by LCM in Cases of Crohn's Disease

Paul Ryan, Simon Aarons, Michael W. Bennett, Gary Lee, Gerald C. O'Sullivan, Joe O'Connell, and Fergus Shanahan

1. Introduction

It has long been questioned whether *Mycobacterium paratuberculosis* plays a role in the aetiology of Crohn's disease *(1,2)*. Several methods of detection (serology, culture, and PCR) have been employed to prove or disprove a link, but without broad consensus to date *(3,4)*. With granulomatous disease, it is reasonable to expect a (partially) causative infectious agent to be present in the granuloma, analogous to cases of M. tuberculosis *(5)*. In this chapter we outline a method using Laser-Capture Microdissection (LCM) to isolate granulomata in cases of Crohn's disease, and explain how to test for the presence of *M. paratuberculosis* using nested PCR. Formalin-fixed, paraffin-embedded tissue samples are examined using the LCM microscope. Granulomata are isolated and processed along with corresponding full-thickness samples. The DNA is harvested from all samples, and PCR is used to test the ability to amplify normal human genes—amplifying a 133-base-pair (bp) segment of the human *adenomatous polyposis coli (APC)* gene. To test for the presence of *M. paratuberculosis*, nested PCR amplifying a 193-bp and then a 155-bp internal fragment of the *M. paratuberculosis* IS900 region is used. This specific insertion sequence (IS)—IS900—is a multicopy DNA insertion element that is unique to the genome of *M. paratuberculosis (6)*. Products are visualized by gel electrophoresis and staining with SYBR Green 1 nucleic acid gel stain. Tissue from

From: *Methods in Molecular Biology, vol. 193: RT-PCR Protocols*
Edited by: J. O'Connell © Humana Press Inc., Totowa, NJ

a positive control, culture-positive specimen—taken from an animal infected with *M. paratuberculosis*—together with purified *M. paratuberculosis* bacterial isolates—are used as positive controls. Similarly treated samples of several types of non-Crohn's granulomatous disorders are used as disease controls. The specificity of amplified PCR products can then be confirmed by restriction mapping or by direct sequencing.

2. Materials

1. Analytical-grade xylene, ethanol, sterile distilled water.
2. PixCell II® laser-capture microdissector, and CapSure® laser-capture caps (Arcturus Engineering, Mountain view, CA).
3. Dneasy® Tissue kits for DNA extraction (QIAGEN, Crawley, UK).
4. The forward and reverse PCR primers used are as follows: human APC gene: 5'-GGA CTA CAG GCC ATT GCA GAA-3' and 5'-GGC TAC ATC TCC AAA AGT CAA-3'; *M. paratuberculosis* outer PCR primers: 5'-GCC CGG ATG CGC CAC GAC TT-3' and 5'-GCG CGG CAC GGC TCT TGT TGT A-3'; *M. paratuberculosis* nested PCR primers: 5'-GCG CCA CGA CTT GCA GCC TCT G-3' and 5'-CGC GTT CCA GCG CCG AAA GTA TT-3'. The DNASTAR Lasergene Primerselect program (DNASTAR Inc., Madison, WI) was used to design the *M. paratuberculosis* primers, which are specific for the *M. paratuberculosis* IS900 region. All primers were synthesized by Sigma-Genosys Ltd. (Pampisford, UK).
5. An Eppendorf Mastercycler gradient® thermocycler (Eppendorf, Hamburg, Germany) was used to perform PCR, using the HotStarTaq® PCR mastermix kit (QIAGEN).
6. Molecular-biology-grade agarose.
7. 50 × TAE solution.
8. SYBR Green 1 nucleic acid stain (Amresco Inc., OH).
9. Specimens. All samples were derived from formalin-fixed, paraffin-embedded tissue. Positive control tissue was intestine from an *M. paratuberculosis*-infected bull. This was obtained from a local authority veterinary laboratory where the bull had tested positive by culture for *M. paratuberculosis*. Disease-negative control samples were obtained from archival samples of various types of non-inflammatory bowel disorders. Test samples were drawn from archival samples of Crohn's disease.

3. Methods

3.1. Slide Preparation

3.1.1. Sectioning of Tissue

1. From the paraffin block, cut 5-μm-thickness sections. Use consecutive slices with the first slide stained with hematoxylin and eosin (H&E) to determine the position of the granulomata and other tissues.
2. Mount the sections without fixative on uncoated, uncharged microscope slides.

3.1.2. Deparaffinization of Sections

Deparaffinize the slides as follows:

1. Prepare four slide baths, one of each containing xylene, 100% ethanol, 95% ethanol, and 75% ethanol.
2. Immerse the first slide in the xylene bath for 10 min, lift it out for 5 s and repeat immersion for another 10 min.
3. Immerse the slide in 100% ethanol for 30 s, lift it out for 5 s and repeat immersion for another 30 s.
4. Repeat this process in 95% EtOH, 75% EtOH, then 95% EtOH again and 100% EtOH, each time for 2 × 30 s.
5. Immerse the slide in xylene for 2 × 10 min as before.
6. Allow the slide to air-dry until chalky white.

3.2. Laser-Capture Microdissection

The slides are now ready to be dissected as per instructions for use of the LCM microscope. Details regarding the use of LCM are provided on the Arcturus Engineering website, www.arctur.com. This system uses a laser to melt cellular material onto a specially designed plastic cap, which sits over the slide as prepared above in **Subheading 3.1.2.**

1. Use the H&E-stained slide as a map to guide to the area of interest—in this case, the granulomata.
2. The laser-guiding light is aimed at the target granulomata in the tissue section. Perform microdissection by applying a pulse of the laser.
3. Lift off the cap. When the cap is lifted off the slide, the chosen cells, which are now stuck to the plastic, lift off with the cap (*see* **Fig. 1**) (*see* **Note 1**).
4. Place the cap onto the top of appropriately sized sterile microfuge tubes to prevent contamination.
5. As a control for the presence of the target gene in the tissue, corresponding full-thickness samples are also collected by scraping a slide prepared as above in **Subheading 3.1.2.** and collecting the cells into a separate microfuge tube.

3.3. DNA Isolation

1. Isolate DNA from all samples using the DNeasy® Tissue kit from QIAGEN as per the manufacturer's instructions. This results in purified DNA in a final vol of 100 μL per sample. The microfuge tubes used to hold the sample caps and slide full-thickness scrapes are used in the initial stages of DNA isolation—i.e.. cell lysis.

3.4. PCR (see *Notes 2* and *3)*

3.4.1. Control PCR for DNA Yield

Confirmation of the effective isolation of DNA is achieved by PCR amplification of a normal human gene, namely a 133-basepair (bp) fragment of the *adenomatous polyposis coli (APC)* gene.

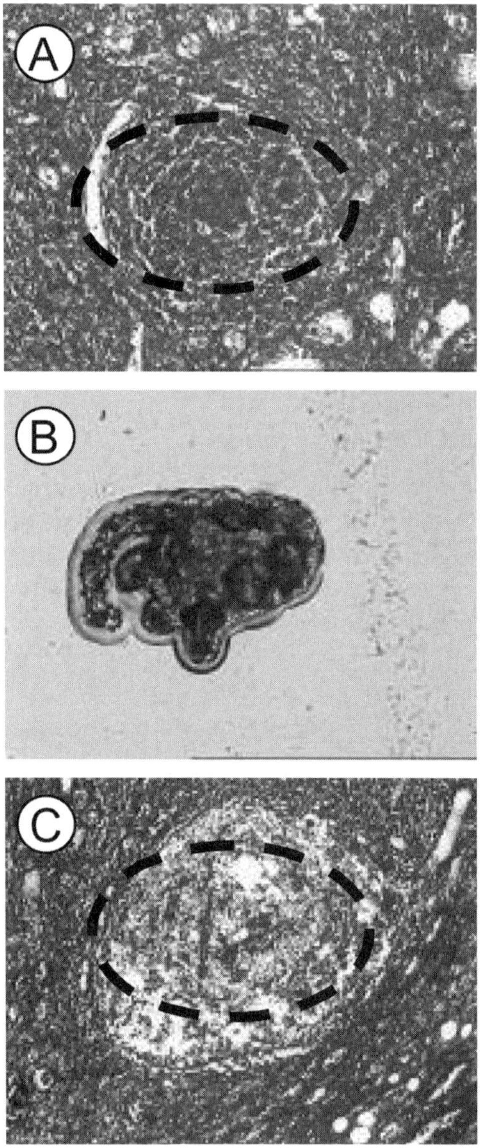

Fig. 1. Laser-capture microdissection. (**A**) The broken line marks a granuloma in the target human colon tissue section before microdissection. The laser beam of the microdissector was guided to this area of the tissue. H&E staining of a reference section assisted identification of the granuloma. (**B**) After the laser pulse, cells from the granuloma are captured onto a plastic cap, as shown. (**C**) The broken line indicates where cells have been microdissected out of the granuloma by the laser beam. In this example, some cellular residue remains in the microdissected area.

1. Prepare a PCR reaction mix containing: 1 µL of primer mix, 15 µL of HotStarTaq master mix, 9 µL of sterile distilled H_2O and 5 µL of isolated DNA.
2. Perform thermal cycling as follows: an initial denaturation step at 95°C for 15 min, followed by 40 cycles of: 94°C for 1 min, 55°C for 1 min, and 72°C for 1 min. A final extension step is performed for 2 min at 72°C (*see* **Note 4**).
3. Analyze the PCR products by running on a 1× Tris-Acetate-EDTA (TAE), 2% agarose gel. Stain using SYBR Green nucleic acid stain as per manufacturer's instructions, and visualize at 302 nm with a transilluminator.

3.4.2. M. paratuberculosis *Primary PCR*

The PCR primers used amplify a sequence specific to *M. paratuberculosis* at the 5' end of the IS900 sequence of which there are several copies per bacterium. The resulting PCR product is 193-bp in length.

1. Prepare a PCR reaction mix containing: 1 µL of primer mix, 15 µL of HotStarTaq master mix, 9 µL of sterile distilled H_2O and 5 µL of isolated DNA.
2. Perform thermal cycling as follows: an initial denaturation step at 95°C for 15 min, followed by 40 cycles of: 94°C for 1 min, 68°C for 1 min, and 72°C for 30 s. A final extension step is performed for 2 min at 72°C.
3. Analyze the PCR products by running on a 1× TAE, 2% agarose gel. Stain using SYBR Green nucleic acid stain as per manufacturer's instructions and visualize at 302 nm with a transilluminator (*see* **Fig. 2**).

3.4.3. M. paratuberculosis Nested PCR

Oligonucleotides that amplify a sequence internal to the primary PCR product were selected. These nested primers amplify a 155-bp sequence.

1. Prepare a PCR reaction mix containing: 1 µL of primer mix, 15 µL of HotStarTaq Master Mix, 13 µL of sterile distilled H_2O and 1 µL of PCR product from the primary PCR.
2. Perform thermal cycling as follows: an initial denaturation step at 95°C for 15 min, followed by 40 cycles of: 94°C for 1 min, 68°C for 1 min, and 72°C for 30 s. A final extension step is performed for 2 min at 72°C.
3. Analyze the PCR products by running on a 1× TAE, 2% agarose gel. Stain using SYBR Green nucleic acid stain as per manufacturer's instructions and visualize at 302 nm with a transilluminator (*see* **Fig. 2**).

3.5. Confirmation of PCR Product Specificity

The PCR product can be purified, and its specificity can be confirmed by direct nucleic acid sequencing using the PCR oligonucleotides as primers. The primary *M. paratuberculosis* PCR product spans nucleotides 575–767 (193-bp) of the IS900A sequence on the EMBL GenBank DNA sequence database, accession number X16293. The nested primers span nucleotides 583–737 (155 bp) of the *M. paratuberculosis* IS900A sequence (*see* **Note 5**).

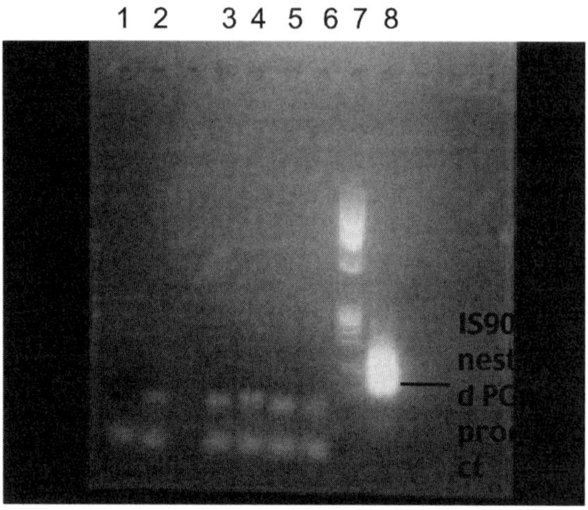

Fig. 2. PCR detection of the *Mycobacterium paratuberculosis* IS900 gene in laser-microdissected intestinal granulomata. Lane 1: negative control (nested PCR); lane 2: negative control (primary PCR); lanes 3–6: test samples: DNA from microdissected intestinal granulomata from Crohn's patients (all negative); lane 7: size markers; lane 8: positive control sample: DNA from a microdissected intestinal granuloma from a Mycobacterium paratuberculosis-infected animal.

3.6. Conclusion

The uncertainty surrounding the role of *M. paratuberculosis* in the etiology of Crohn's disease may be attributed to a number of factors. The tissue samples tested varied between studies. Some studies used biopsies or full-thickness samples of bowel, and this may have led to labeling of contaminating luminal *Mycobacteria* as pathogenic. Another confounding factor is the variation in methodologies. Some studies tried to amplify sequences > 300 bp from formalin-fixed, paraffin-embedded sections, which may not be reliable. The LCM method described here should overcome these difficulties through the specific isolation of genomic DNA from granulomata, where pathogenic Mycobacteria have previously been detected; also, the use of appropriately designed, nested PCR primers should increase the sensitivity and specificity of detection. Using this method, we have successfully detected *M. paratuberculosis* in some granulomata, but only in a minority of cases to date.

4. Notes

1. After the cap is placed onto the slide and cells are dissected, do not try to move the same cap onto another area of interest, as it will not rest flush with the slide.

This is because dissected cells will lie on top of other cells and prevent contact with the cap.

2. Formalin fixation results in widespread crosslinkage between nucleic acids and proteins. As a result, DNA extracted from formalin-fixed tissues is fragmented into sequences of variable size *(7)*. Thus, all primers were chosen to amplify sequences of <250 bp, as it has been shown that it is increasingly difficult and unreliable to amplify longer sequences in formalin-fixed, paraffin-embedded tissue *(8)*.
3. In order to minimize the risk of contamination of the PCRs, mixing of PCR reagents was done in a separate laboratory area from where the template DNA was added. Before addition of the template DNA, the PCR reaction mix was first placed in ultraviolet (UV) light of wavelength 254 n*M* and intensity of 30,000 μJ/cm^2 for 2×2 min to denature any contaminating, double-stranded DNA. Filtered pipet tips were used throughout the entire process to prevent contamination by aerosolized DNA.
4. When dealing with small cell numbers as in the case of dissected granulomata, sometimes an additional round of 25 cycles is necessary to reveal detectable product.
5. Restriction mapping can also be used to confirm PCR product specificity; the amplified IS900 sequence contains unique restriction sites for *Sma*I, *Xma*I, *Pvu*I, and *Pvu*II, which facilitate convenient restriction map analysis.

References

1. Dalziel, T.K. (1913) Chronic intestinal enteritis. *B.Med.J.* **2,** 1068–1070.
2. Burnham, W.R., Lennard-Jones, J.E., Stanford, J.L., and Bird, R.G. (1978) Mycobacteria as a possible cause of inflammatory bowel disease. *Lancet* **30,** 693–696.
3. Sanderson, J.D., Moss, M.T., Tizard, M.L., and Hermon-Taylor, J. (1992) *Mycobacterium paratuberculosis* DNA in Crohn's disease tissue. *Gut* **33,** 890–896.
4. Rowbotham, D.S., Mapstone, N.P., Trejdosiewicz, L.K., Howdle, P.D., and Quirke P. (1995) *Mycobacterium paratuberculosis* DNA not detected in Crohn's disease tissue by fluorescent polymerase chain reaction. *Gut* **37,** 660–667.
5. Fidler, H.M., Thurrell, W., Johnson, N.M., Rook, G.A., and McFadden, J.J. (1994) Specific detection of *Mycobacterium paratuberculosis* DNA associated with granulomatous tissue in Crohn's disease. *Gut* **35,** 506–510.
6. Green, E.P., Tizard, M.L., Moss, M.T., Thompson, J., Winterbourne, D.J., McFadden, J.J., et al. (1989) Sequence and characteristics of IS900, an insertion element identified in a human Crohn's disease isolate of *Mycobacterium paratuberculosis. Nucleic Acids Res.* **17,** 9063–9073.
7. Goelz S.E., Hamilton, S.R., and Vogelstein, B. (1985) Purification of DNA from formaldehyde fixed and paraffin embedded human tissue. *Biochem. Biophys. Res. Commun.* **130,** 118–126.
8. Hung, J., Kishimoto, Y., Sugio, K., Virmani, A., McIntire, D.D., Minna, J.D., et al. (1995) Allele-specific chromosome 3p deletions occur at an early stage in the pathogenesis of lung carcinoma. *JAMA* **273,** 1908.

17

RT-PCR-Based Approaches to Generate Probes for mRNA Detection by *In Situ* Hybridization

Joe O'Connell

1. Introduction

The ability to detect mRNA in tissue sections provides a powerful tool for localizing gene expression in complex tissues comprising mixed cell populations. Detection of gene expression at the mRNA level is particularly important to confirm the identity of cells expressing soluble or secreted proteins. Such proteins may be immunohistochemically detected in association with cells other than those expressing them, and can yield a diffuse pattern of staining. The specific mRNA, however, will only be detected in cells that actually express the protein.

The process of detecting mRNA by *in situ* hybridization is fraught with technical difficulties. Since mRNA is notoriously unstable, because of the ubiquity and stability of the ribonucleases present in cells and tissues, preservation of mRNA integrity during tissue handling is of paramount importance. After taking adequate precautions to ensure that the integrity of the mRNA is maintained, the next most important factor in the process is the quality of the probe used to detect the mRNA. Homology searches of gene databases will confirm that the selected target sequence within the mRNA is relatively unique, exhibiting minimal homology to other known genes or related members of a gene family. If the target is a member of a closely homologous gene family, gene alignment of the family members will enable an area with the least homology to other family members to be targeted by the probe.

As to the probe itself, there are a number of options. Synthetic oligonucleotide probes are certainly the easiest to manufacture, and because of their relatively short sequence (usually 20–30 bp), they can be designed to have almost no significant homology to other genes, even closely related ones. However,

From: *Methods in Molecular Biology, vol. 193: RT-PCR Protocols*
Edited by: J. O'Connell © Humana Press Inc., Totowa, NJ

because of their short length, oligonucleotide probes do not form very stable hybrids with their target sequences. This usually means that relatively low stringency hybridization conditions are required to encourage hybridization, and this in turn can lead to a high level of nonspecific background hybridization. Longer probes—of approx 300 bp in length—form very stable hybrids with their target sequences, enabling much more robust and stringent hybridization conditions to be applied.

The disadvantage of long probes relative to oligonucleotide probes is that they are somewhat more difficult to synthesize. Custom oligonucleotide synthesis is widely available commercially, is relatively inexpensive, and enables a variety of labels to be incorporated into the synthesis. Traditionally, long probes required a plasmid clone of a fragment of the target cDNA to be available. If the plasmid contained a promoter sequence for an RNA polymerase, such as the T3 or T7 RNA polymerase, an RNA probe corresponding to the cloned insert could be generated by in vitro transcription, with the inclusion of an appropriately labeled ribonucleotide triphosphate.

The advent of reverse-transcriptase-polymerase chain reaction (RT-PCR) means that long probes for the detection of mRNA can now be made without the need to clone the target sequence.This chapter presents two simple RT-PCR-based techniques to generate long probes. The first enables the synthesis of a labeled RNA probe, or "riboprobe." Riboprobes are one of the most widely used and successful types of probe for hybridization to mRNA, since RNA-RNA hybrids exhibit very high stability. Essentially, the technique involves performing RT-PCR amplification of the target sequence from a suitable source of RNA. For the PCR, an anti-sense primer is used, which has the T7-promoter sequence tagged onto its 5' end. The resultant PCR product will then include the target sequence flanked by the T7-promoter sequence, so that an RNA probe can be synthesized by in vitro transcription of the PCR product using the T7 RNA polymerase (*see* **Fig. 1**).

The second technique uses RT-PCR itself to generate the probe. The target sequence is first obtained by standard RT-PCR. The PCR product is then used as a template for asymmetric PCR. This involves using a single primer (the anti-sense primer) in a PCR reaction, containing an appropriately labeled deoxynucleotide 5' triphosphate (dNTP). The asymmetric PCR will result in the synthesis of a single-stranded, labeled DNA fragment corresponding to the anti-sense of the target sequence, and therefore a suitable probe for detection of the mRNA. Because there is no exponential amplification during asymmetric PCR, a very high number of cycles is used to generate a good yield of product. Single-stranded DNA probes are much less subject to degradation

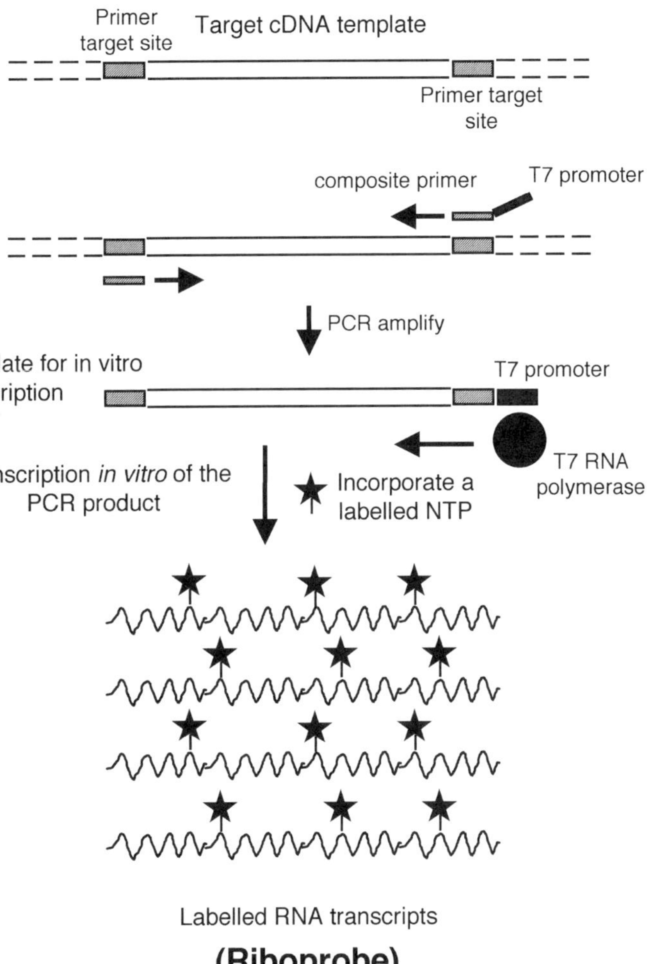

Fig. 1. RT-PCR in the synthesis of RNA probes (riboprobes) for *in situ* hybridization. RT-PCR is used to generate template DNA for in vitro transcription to generate the riboprobe. The RT-PCR employs primers to amplify a unique region of the target cDNA. By tagging the sequence of the T7 RNA polymerase promoter (pT7) onto the 5' end of the antisense primer, the resultant RT-PCR product will include the pT7 sequence. Following purification, this RT-PCR product serves as a template for in vitro transcription using T7 RNA polymerase. By including an appropriately labeled rNTP (e.g., digoxigenin-11-UTP, or biotin-16-UTP) in the transcription reaction, a labeled RNA probe is synthesized corresponding to the anti-sense sequence of the target mRNA. This probe can be used directly for *in situ* hybridization to detect the target mRNA.

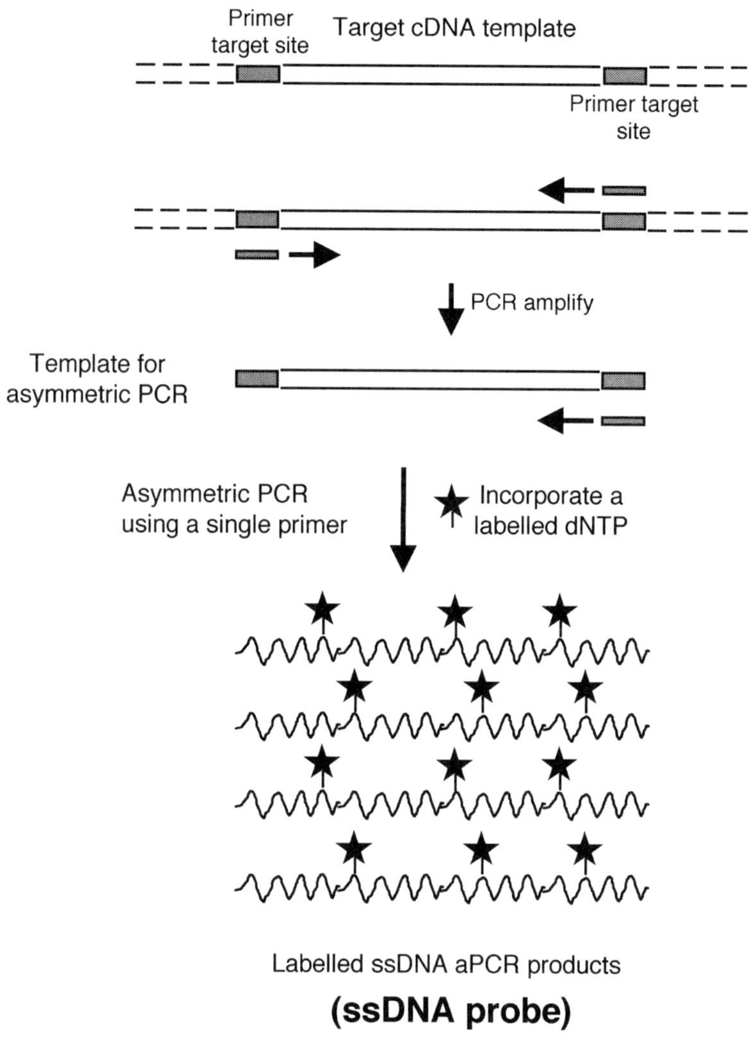

Fig. 2. Asymmetric PCR (aPCR) in the synthesis of ssDNA probes for *in situ* hybridization. RT-PCR is used to generate template DNA for subsequent aPCR to generate the ssDNA probe. The RT-PCR employs primers to amplify a unique region of the target cDNA. Following purification, this RT-PCR product serves as a template for aPCR using only the anti-sense primer; the sense primer is omitted. Since amplification by aPCR is not exponential, a large number of cycles are used (60–70) to synthesize a good yield of ssDNA product. By including an appropriately labeled dNTP (e.g., digoxigenin-11-dUTP, or biotin-16-dUTP) in the aPCR, a labeled ssDNA probe is synthesized corresponding to the anti-sense sequence of the target mRNA. This probe can be used directly for *in situ* hybridization to detect the target mRNA.

during hybridization than RNA probes, and thus can be used at very low concentration. A single asymmetric PCR can generate enough highly labeled probe for an enormous number of *in situ* hybridizations (*see* **Fig. 2**).

Finally, I will present a typical technique for performing *in situ* hybridization on formalin-fixed, paraffin-embedded tissue sections. This is just one of a considerable variety of techniques that have been employed for the detection of mRNA in tissues.

2. Materials

1. A program for designing primer pairs, such as the DNASTAR Lasergene Primerselect program (DNASTAR Inc., Madison, WI).
2. Access to a recent version of the European Molecular Biology Laboratory (EMBL) DNA sequence database (e.g., on CD-ROM from DNASTAR) to perform a homology search of the selected primers. This ensures that the primers have no significant homology to any other known gene. A homology search is also performed to test the sequence-specificity of the amplicon sequence, which will ultimately be used as a probe for *in situ* hybridization.
3. Access to automated DNA synthesis for primer synthesis.
4. Cellular source of the target mRNA. In our example, we wished to generate a probe to detect mRNA for Fas ligand (FasL/CD95L) in tissues. To obtain cellular RNA containing mRNA for FasL, we used the colon carcinoma cell line SW620 which expresses functional FasL ***(1)***. SW620 cells were obtained from the European Collection of Cell Cultures (ECACC), Salisbury, UK. Cells were cultured in Dulbecco's Modified Eagle's Medium (DMEM) supplemented with 10% fetal calf serum (FCS) and antibiotics (Gibco-BRL, Grand Island, NY). Cells were maintained at 37°C in a humidified atmosphere containing 5% CO_2.
5. RNA isolation. Guanidine thiocyanate cell lysis buffer: 4.0 *M* guanidine thiocyanate (Sigma Chemical Co., St. Louis, MO), 25 m*M* sodium citrate, pH 7, 0.5% sarcosine. For extraction, use molecular-biology-grade, water-saturated phenol (Sigma); ethanol (Sigma) is used for precipitation of the RNA, and 70% ethanol for washing the dried pellet. RNA pellets are redissolved in water pretreated with diethyl pyrocarbonate (DEPC; Sigma) to inactivate any contaminating ribonucleases. DEPC, which is highly toxic, is added at 0.1 % to a bottle of distilled water in a fume-hood, and the lid is replaced. The water is left to incubate with the DEPC for several hours, with occasional vigorous shaking to disperse the DEPC throughout the water. Finally, the DEPC-treated water is autoclaved. Heat breaks down DEPC to ethanol and carbon dioxide, so autoclaved DEPC-water is not toxic. (Several kits are also commercially available for quick RNA isolation (e.g., RNeasy® kit from Qiagen, Crawley, UK).
6. cDNA synthesis: AMV reverse transcriptase (RT) and its buffer, and the ribonuclease inhibitor RNasin, are obtained from Promega (Madison, WI); random hexanucleotide primers (Boehringer Mannheim, GmbH, Mannheim, Germany) are used at 125 n*M* in the cDNA synthesis.

7. PCR: dNTPs (Promega); Taq DNA polymerase and buffer (Promega); UlTma DNA polymerase (Perkin-Elmer, Norwalk, CT); a kit for purifying PCR products (e.g., QIAquick® PCR purification kit from Qiagen); digoxigenin-11-dUTP or biotin-16-dUTP (Boehringer Mannheim, GmbH).
8. In vitro transcription: rNTPs, digoxigenin-11-UTP, or biotin-16-UTP, T7 RNA polymerase, and 10X T7 RNA polymerase transcription buffer are obtained from Boehringer Mannheim GmbH.
9. Electrophoresis: molecular-biology-grade agarose (Promega); ethidium bromide (Sigma); *Hae*III-digested ϕX174 DNA size markers (Promega).
10. *In situ* hybridization: aminopropylethoxysilane (APES); glycine; Triton X-100; paraformaldehyde; acetic anhydride; triethanolamine; deionized formamide; dextran sulphate; 25× standard saline citrate (SSC) solution; phosphate buffered saline (PBS) tablets; DTT; yeast tRNA; herring sperm DNA; NBT; BCIP (Sigma); proteinase K; alkaline-phosphatase conjugated sheep anti-digoxigenin antibody (Boehringer Mannheim).

3. Methods

3.1. RT-PCR to Generate Template for Probe Synthesis

1. Select a PCR primer pair to amplify the cDNA of interest (FasL in this instance). Our PCR primer pair was designed using the DNASTAR Lasergene Primerselect program (DNASTAR Inc.). The primer pair was chosen to span introns in the genomic sequence (*see* **Note 1**). Each primer was screened by a homology search of the EMBL DNA sequence database to ensure that it had no significant homology to any other known gene. The selected FasL PCR primer pair had the following sense and anti-sense sequences respectively: 5'-GGATTGGGCCTGGG GATGTTTCA-3' and 5'-TTGTGGCTCAGGGGCAGGTTGTTG-3'. These primers amplify a fragment of FasL cDNA (codons 96–210) to yield a 344-bp PCR product. The composite pT7-FasL anti-sense primer has the following sequence: 5'-<u>CTAATACGACTCACTATAGGG</u>TTGT GGCTCAGGGGCAG GTTGTTG-3'. The underlined sequence is the sequence of the T7-promoter; the remainder of the sequence is the "regular" FasL anti-sense primer sequence (*see* **Note 2**). The nucleotide sequence of the FasL amplicon (ultimately constituting the sequence of the probe) showed no significant homology to any other sequence in the EMBL DNA sequence database.
2. Isolate total RNA from the treated cells by first lysing in 0.5 mL of guanidine thiocyanate lysis buffer. Add 50 µL of 2 *M* sodium acetate, pH 4, 0.5 mL of water-saturated phenol, and 0.1 mL of a mixture of chloroform and isoamyl alcohol (ratio of 49:1). Mix the tubes after addition of each reagent, and mix the final suspension vigorously. Leave the tube on ice for 15 min, then centrifuge at 10,000*g* for 20 min in a microfuge. Remove the RNA-containing aqueous phase to a new tube, and add 2–2.5 volumes of absolute ethanol. Leave on ice for 15 min to precipitate the RNA. Collect the precipitate by centrifugation at 10,000*g*

for 30 min in a microfuge. Wash the pellet in 70% ethanol, dry, and redissolve the RNA in 10–20 µL of DEPC-treated water. Several kits are also commercially available for quick RNA isolation (e.g., RNeasy® kit from Qiagen).

3. Synthesize cDNA using the AMV RT (Promega, Madison, WI) and random hexanucleotide primers (Boehringer Mannheim, GmbH). Perform the cDNA synthesis in a final reaction volume of 30 µL containing: 1 U of Avian myeloblastoma virus (AMV) RT, 40 U of RNasin, 0.5 m*M* of each of the four dNTPs, 125 n*M* of random hexanucleotide primers, and approx 1 µg of total RNA. Incubate for 90 min at 42°C. Dilute the cDNA by adding 70 µL of T.E. (10 m*M* Tris-HCl, pH 8.0, 1 m*M* ethylenediaminetetraacetic acid [EDTA]), and store at –20°C.
4. Amplify the target (FasL) cDNA by RT-PCR. For the first method of probe synthesis (in vitro transcription method), use the FasL sense and pT7-FasL anti-sense composite primers; this generates a FasL amplicon which incorporates the pT7 sequence at its 3' end. For the second method of probe synthesis (asymmetric PCR method), use the "regular" FasL sense and anti-sense primers. Use PCR primers at a final concentration of 0.1 µ*M* each, dNTPs at 50 µ*M* and $MgCl_2$ at 1.5 m*M*. Perform the PCR to generate the FasL amplicon using a proof-reading heat-stable DNA polymerase, such as the UlTma DNA polymerase (Perkin-Elmer), in order to minimize misincorporation that may lead to sequence changes in the resultant probe. Use 1.0 U of UlTma DNA polymerase per 50 µL reaction, and 1 µL of cDNA as template. Use the following program of thermal cycling: denaturation at 96°C for 15 s; annealing at 55°C for 30 s and extension at 72°C for 3 min. Perform "hot start" by heating the reaction to 80°C before adding the polymerase. Perform 35–40 cycles. Analyze PCR products by electrophoresis through 2% agarose gels and view under ultraviolet (UV) light following ethidium bromide staining. *Hae*III-digested φX174 DNA size markers can be used. PCR product specificity should ideally be confirmed, either by restriction mapping or DNA sequence analysis
5. Purify the FasL RT-PCR product. A number of kits are commercially available for purifying PCR products (e.g., QIAquick® PCR purification kit from Qiagen). Its important to purify the full-length FasL amplicon (344 bp) to eliminate primers and primer artifacts that could generate truncated, nonspecific probes in the subsequent probe synthesis steps.
6. Analyze the purified RT-PCR product by agarose gel electrophoresis through a 2% agarose gel.

3.2. Probe Synthesis

3.2.1. Method 1: In Vitro Transcription to Generate a Riboprobe

The purified FasL RT-PCR amplicon generated with the FasL-pT7 anti-sense primer now serves as a template for riboprobe generation. The riboprobe is synthesized using in vitro transcription catalyzed by T7 RNA polymerase in the presence of a labeled rNTP.

1. Using the purified FasL amplicon (generated by RT-PCR using the pT7-FasL anti-sense primer) from **Subheading 3.1.5.** as template, set up an in vitro transcription reaction as follows:

template DNA (100 ng)	× μL
10X T7 RNA polymerase transcription buffer	2 μL
RNase-free water	20 − (6.5 + ×) μL
rNTP 10 m*M* stock*	2 μL (1 m*M*)
RNasin ribonuclease inhibitor	0.5 μL (0.5 U)
T7 RNA polymerase	2 μL (40 U)
Final reaction volume	20 μL

* Note: for the unlabeled control probe, a mix of the four rNTPs (10 m*M* each) is used; for probe labeling, a labeling mix of the rNTPs is prepared containing 10 m*M* ATP, GTP, CTP, and appropriate concentrations (10 m*M* total) of labeled and unlabeled UTP (e.g. 2.5 m*M* digoxigenin-11-UTP, 7.5 m*M* UTP). The ratio of labeled to unlabeled UTP will determine the amount of label incorporated into the probe (*see* **Note 3**).

2. Incubate the in vitro transcription reaction at 37°C for 2 h.
3. Analyze the synthesized RNA by agarose gel electrophoresis through a 2% agarose gel (*see* **Note 4**).
4. The in vitro transcribed RNA can be used directly for *in situ* hybridization (*see* **Note 5**).

3.2.2. Method 2: Asymmetric PCR to Generate a Single-Stranded DNA Probe

The purified FasL RT-PCR amplicon generated with the "regular" FasL sense and anti-sense primers now serves as a template for probe generation by asymmetric PCR (aPCR). The single-stranded DNA probe is synthesized using aPCR in the presence of a labeled dNTP.

1. Set up a PCR reaction as described in **Subheading 3.1.4.**, except that only the anti-sense FasL primer is added to the mix; the sense primer is omitted. Use 1 μL of the purified FasL amplicon from **Subheading 3.1.5.** as a template for the aPCR. Also, include an appropriately labeled dNTP, such as digoxigenin-11-dUTP or biotin-16-dUTP. The ratio of labeled dUTP to unlabeled dTTP will determine the relative number of digoxigenin or biotin labels introduced into the probe. For example, a labeled-dUTP:unlabeled-dTTP of 1:3 will result in approximately every third incorporated dUTP/dTTP containing a label, or approximately one label per every 12 nucleotides in the sequence. For use as a competitive control in subsequent *in situ* hybridization, also synthesize an unlabeled probe by performing aPCR with unlabeled dNTPs.
2. Perform hot-start as usual, and use the same thermal cycling program as used in the FasL RT-PCR, except that 60 cycles are performed. Since amplification is not exponential, an increased number of cycles is necessary to obtain a good yield of probe.

3. Check the probe synthesis (labeled and unlabeled probes) by agarose gel electrophoresis on a 2% agarose gel (*see* **Note 4**). The concentration of each probe can be estimated from the gel (*see* **Note 6**). The probes are ready for direct use in *in situ* hybridization (*see* **Note 7**).

3.3. Localization of mRNA Expression by In Situ *Hybridization*

A wide range of different techniques are employed for *in situ* hybridization to detect mRNA, with a diversity of variations for each protocol. This chapter presents a typical protocol that we have used successfully for mRNA detection in paraffin-embedded human tissue sections (*see* **Note 8**).

1. For *in situ* hybridization, use paraffin-embedded human tissue sections (4-µm thick), mounted on APES-treated slides (*see* **Note 9**).
2. Perform the following prehybridization treatments: wash 2 × 5 min each in PBS, PBS; 0.1 *M* glycine, PBS; 0.3% triton X-100, and PBS again. Perform digestion for 30 min at 37°C with proteinase K (10 µg/mL in 100 m*M* Tris-HCl, pH 8.0, 50 m*M* EDTA, pH 8.0).
3. Fix for 5 min at 4°C in 4% paraformaldehyde in PBS, and then acetylate for 2 × 5 min in fresh 0.25% acetic anhydride, 0.1 *M* triethanolamine (pH 8.0).
4. Incubated the sections at 37°C for 10 min in a prehybridization buffer consisting of 50% deionized formamide in 4× SSC.
5. Perform hybridization at 42°C overnight in hybridization buffer containing the following: 50% formamide, 10% dextran sulphate, 1× Denhardt's reagent, 4X SSC, 10 m*M* dithiothreitol (DTT), 500 µg/mL yeast tRNA, and 1 mg/mL heat-denatured herring sperm DNA, containing 1 ng/µL digoxigenin-labeled riboprobe.
6. After hybridization, wash the tissue sections with increasing stringency to 0.1X SSC at 37°C.
7. Detect the hybridized probe immunologically using alkaline-phosphatase conjugated sheep anti-digoxigenin antibody (Boehringer Mannheim), and visualize with NBT/BCIP (purple/black precipitating product).
8. Control slides involve direct, internal competitive inhibition of hybridization. This is performed exactly as described in **steps 1–7** above except that a 10-fold excess of "cold" unlabeled riboprobe is added to the digoxigenin-labeled riboprobe prior to hybridization. This should result in direct, competitive displacement of positive hybridization, leading to a marked reduction in signal intensity, thus confirming the specificity of hybridization (*see* **Fig. 3**).

4. Notes

1. The use of intron-spanning primers generates an RT-PCR product that is mRNA-specific. This ensures that the resultant probe will bind specifically to the target mRNA, since binding of the probe to the genomic DNA sequence will be interrupted because of the presence of introns. Because the probe can only partially bind to the genomic DNA sequence, stringent hybridization conditions should prevent background because of binding to genomic DNA.

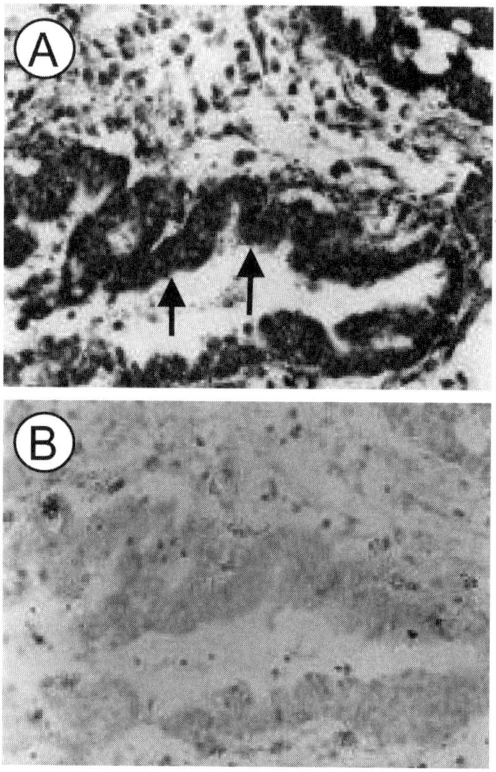

Fig. 3. Detection of Fas ligand (FasL) expression in a human gastric adenocarcinoma using *in situ* hybridization *(3)*. FasL mRNA was detected in a formalin-fixed, paraffin-embedded gastric carcinoma section by *in situ* hybridization, using a digoxigenin-labeled FasL-specific riboprobe. The riboprobe was synthesized using the approach described in **Subheading 3.2.1.** **(A)** Positive purple hybridization signals (arrows) were obtained within neoplastic gastric epithelial cells in the tumor. **(B)** Control hybridization was performed on a serial section from the same tumor with the addition of a 10-fold excess of unlabeled probe together with the labeled probe. The unlabeled probe caused direct, competitive displacement of the labeled probe, confirming the specificity of hybridization; staining is negligible in the corresponding area of neoplastic epithelial cells that were positive in (**A**).

2. It has been shown that the yield of RNA synthesized from the T7-promoter (pT7) can be increased significantly (by approx 1 order of magnitude) by adding an additional dC nucleotide onto the 5' end of the canonical pT7 sequence *(2)*. In the pT7-FasL composite primer used here, this modified pT7 sequence was used in order to obtain maximal transcriptional activity for riboprobe synthesis.

3. As an alternative to digoxigenin-11-UTP, biotin-16-UTP can also be incorporated into the probe, enabling avidin-based detection of the biotinylated riboprobe in the *in situ* hybridization.
4. Note that unlike dsDNA, because of secondary structures, the mobility of RNA or ssDNA during agarose gel electrophoresis does not depend on size. Thus, the probe may appear to have a different mobility than predicted from the dsDNA size markers. Also, the labeled probe usually exhibits a different mobility than the unlabeled control probe.
5. At the end of the in vitro transcription, DNase digestion of the template DNA is not necessary. The yield of transcribed RNA is usually in about 20-fold excess over the amount of DNA template; thus, the presence of a relatively low level of DNA will not significantly interfere with probe binding during *in situ* hybridization. Also, the dsDNA template will not serve as a good substrate for probe binding. Purification of the riboprobe is generally unnecessary, since the various components of the in vitro transcription will be diluted to negligible levels in the *in situ* hybridization.
6. Despite differences in electrophoretic mobility between dsDNA and RNA or ssDNA, the concentration of the RNA or ssDNA probes can be estimated by reference to the band intensity of the dsDNA size markers. For example, if 100 ng of *Hae* III-digested ϕX174 DNA-size markers are run on the gel, the 600-bp band will contain 11 ng of nucleic acid.
7. Because the aPCR technique generates a high concentration of probe, with a high rate of label incorporation, the probe will be diluted considerably for use in *in situ* hybridization. For this reason, we find that purification of the probe is generally unnecessary, because the various components of the aPCR will be diluted to negligible levels in the *in situ* hybridization.
8. Although mRNA can be successfully detected in paraffin-embedded human tissue sections, the integrity of the mRNA within the specimen can very depending on the handling of the tissue prior to, during, and after fixation. Frozen tissue specimens may provide more consistent preservation of mRNA integrity, and are commonly used for *in situ* hybridization. However, paraffin-embedded, formalin-fixed tissues tend to result in better tissue morphology throughout the *in situ* hybridization protocol. One parameter that we have found to influence the success of mRNA detection in paraffin-embedded tissue is the time interval between cutting the tissue sections and performing the *in situ* hybridization protocol; preferably, *in situ* hybridization should be performed as soon as possible after cutting the section (within a day or so).
9. Since the *in situ* hybridization protocol involves numerous washing steps and incubations, it is essential to use microslides that allow strong adherence of the tissue section; otherwise, the tissue section can become detached and lost during the protocol. APES-treated slides provide for very stong adherence of the tissue section.

Acknowledgments

This work was supported by grants from the Health Research Board of Ireland (HRB) and a "New Blood" fellowship award from the Wellcome Trust-HRB to Joe O'Connell.

References

1. O'Connell, J., O'Sullivan, G.C., Collins, J.K., and Shanahan, F. (1996) The Fas counterattack: Fas-mediated T cell killing by colon cancer cells expressing Fas ligand. *J. Exp. Med.* **184,** 1075–1082.
2. Baklanov, M.M., Golikova, L.N., and Malygin, E.G. (1996). Effect on DNA transcription of nucleotide sequences upstream to T7 promoter. *Nucleic Acids Res.* **24,** 3659–3660.
3. Bennett, M.W., O'Connell, J., O'Sullivan, G.C., Roche, D., Brady, C., Kelly, J., Collins, J.K. and Shanahan, F. (1999). Fas ligand is expressed by human gastric adenocarcinomas: a potential mechanism of immune escape in stomach cancer. *Gut* **44,** 156–162.

VI

Differential mRNA Expression

18

Amplified RNA for Gene Array Hybridizations

Valentina I. Shustova and Stephen J. Meltzer

1. Introduction

The human genome contains perhaps more than 50,000 genes. A typical cell in the human body is believed to express from 15,000 to 25,000 proteins. The majority of these proteins are expressed in all the cells, and are known as housekeeping genes. These are proteins involved in vital cell functions, including production of energy, oxidation, biosynthesis, regulation of cell cycle, and cell growth. The rest of the proteins are expressed differentially in various cell types. For example, some genes are expressed only in liver cells, muscle cells, or neurons, and not in any other cell types. These differentially expressed genes define the phenotype and function of a particular cell type. The identification of genes solely expressed in one cell type may shed light on mechanisms of cellular function and gene interaction.

Notably, some genes are only expressed during a short time in a cell's life cycle, or are expressed at a particular stage of growth and development. The identification and study of transiently expressed genes will provide invaluable information about the mechanisms of body development and aging. Studies are now proving that a number of human diseases are associated with "turning on" or "turning off" particular genes. Of particular interest are those genes that control cell proliferation and the cell cycle. The discovery and study of these genes will greatly advance our understanding of mechanisms of neoplastic transformation and the development of malignancy. Approaches to manipulate the function of the "diseased" genes might have great therapeutic benefit for cancer patients, as well as for other "gene-controlled" diseases.

The ability to study differentially expressed genes has been greatly enhanced recently with the development of DNA microarray technology and bio-

From: *Methods in Molecular Biology, vol. 193: RT-PCR Protocols*
Edited by: J. O'Connell © Humana Press Inc., Totowa, NJ

informatics analysis *(1,2)*. This method allows the rapid, simultaneous identification and study of multiple, differentially expressed genes in a single sample of cells or tissue. It employs glass slides covered with a number of known and unknown probes that hybridize with cDNAs generated from RNAs from the cells or tissue under investigation *(3–5)*. Labeling of cDNA with different colors allows for detection of overexpressed and underexpressed genes in samples compared to control cDNA *(6)*. The use of DNA microarrays on glass appears to be more accurate than similar techniques employing nylon membranes, where radioactively labeled libraries are hybridized with probes on different membranes. This may result in a higher margin of error because of membrane-to-membrane variability *(7,8)*.

Subheading 2. describes protocols successfully used in our laboratory for identifying differentially expressed genes during cancerous transformation in premalignant Barrett's esophagus.

2. Materials

2.1. RNA Extraction

1. Tissue specimens. We use small quantities of tissue, either from a biopsy or a small piece of surgically resected tissue *(9)*.
2. Homogenizer: Ultra-Turrax T25 homogenizer (Janke and Kunkel, IKA-Labortechnik, Germany).
3. TRIZOL Reagent (Gibco-BRL, Grand Island, NY).

2.2. RNA Amplification

1. High-pressure liquid chromatography (HPLC)-purified T7-oligo(dT) primer: 5'-TCT AGTCGACGGCCAGTGAATTGTAATACGACTCACTATAGGGCG(T)27-3'.
2. Super Script Preamplification system (Gibco-BRL).
3. cDNA Synthesis System (Gibco-BRL).
4. T4 DNA polymerase (Gibco-BRL).
5. AmpliScribe T7 Transcription Kit (Epicentre Technologies).
6. Microcon-100 columns (Millipore, Bedford, MA).
7. dN6 primer, 8 μg/μl (Roche Molecular Biochemicals, Mannheim, Germany).
8. Deoxynucleotide triphosphate (dNTPs), 100 m*M* (Promega, Madison, WI).
9. Recombinant RNasin RNase inhibitor (Gibco-BRL).

2.3. Microarray Hybridization

1. Cyanine 5-dUTP and cyanine 3-dUTP (Perkin-Elmer).
2. Microcon YM-30 columns (Millipore).
3. Denhardt's blocking solution (Sigma Chemical Co., St Louis, MO).
4. Poly(dA), 8 mg/mL (Pharmacia).
5. Yeast tRNA (4 mg/mL; Sigma).
6. Human Cot I DNA (10 mg/mL; Gibco-BRL).

7. Prepare the following hybridization wash solutions:

	2X SSC+0.1% SDS	1X SSC	0.2X SSC	0.05X SSS
dH2O	179 mL	190 mL	198 mL	200 mL
20X SSC	20 mL	10 mL	2 mL	0.5 mL
20% SDS	1 mL	—	—	—

SSC: Standard saline citrate. SDS: Sodium dodecyl sulfate.

3. Methods

3.1. RNA Extraction

1. Add 2–5 mL of TRIZOL Reagent for 20–100 mg of tissue, in a 15-mL glass (Corex) tube. Homogenize the tissue at full speed in the Ultra Turrax homogenizer. Leave at room temperature for 10 min.
2. Add 0.2 mL of chloroform per 1 mL of TRIZOL reagent. Cap the sample tube and shake it vigorously for 15 s.
3. Incubate at room temperature for 3 min.
4. Centrifuge the sample at 12,000*g* for 15 min at 4°C.
5. Carefully transfer the upper aqueous phase to a microcentrifuge tube.
6. Precipitate the RNA by adding 0.5 mL of isopropyl alcohol per 1 mL of TRIZOL reagent used for **step 2**.
7. Incubate for 10 min at room temperature.
8. Centrifuge at 12,000*g* for 10 min at 4°C.
9. Remove the supernatant.
10. Wash the RNA pellet once with 75% ethanol (1 mL). Mix by vortexing.
11. Centrifuge at 7500*g* for 5 min at 4°C.
12. Carefully remove the supernatant using a pipet and vacuum-dry the pellet.
13. Dissolve the RNA in 25–100 µL of RNase-free water.
14. Incubate for 10 min at 60°C and vortex for 10–15 s to aid dissolution of the RNA.
15. Measure the O.D. ratio at A260/A280 with an ultraviolet (UV)-visible spectrophotometer.
16. Check the quality of the RNA by running on a 1.2% agarose gel.

3.2. RNA Amplification

3.2.1. First and Second-Strand cDNA Synthesis

1. Use 10 µL of total RNA (1–4 µg).
2. Add 1 µL of 0.5 mg/mL T7-oligo(dT) primer.
3. Incubate the samples at 70°C for 10 min. Chill them on ice immediately. Spin down each sample by brief centrifugation.
4. Add:
   a. 4 µL of 5X first-strand cDNA reaction buffer (from the cDNA synthesis kit).
   b. 2 µL of 0.1 *M* dithiothreitol (DTT).
   c. 1 µL of 10 m*M* dNTPs.

d. 1μL of RNasin.
e. 1 μL of Super Script II RT.

5. Incubate the samples at 42°C for 1 h.
6. Add :
   a. 30 mL of second-strand synthesis buffer.
   b. 3 mL of 10 mM dNTPs.
   c. 4 mL of DNA polymerase I.
   d. 1 mL of E. coli Rnase H.
   e. 1 mL of E. coli DNA ligase.
   f. 92 mL of RNase-free water.

   Mix all components gently and incubate at 16°C for 2 h.
7. Add 2 μL of T4 DNA polymerase. Incubate at 16°C for 10 min (*see* **Note 1**).
8. Extract the double-stranded DNA by adding 150 μL of chloroform and 150 μL of phenol (1:1). Shake 20×. Spin at 3000*g* for 4 min at 4°C.
9. To remove unincorporated nucleotides and salts, transfer the aqueous layer to a Microcon-100 column prerinsed with 500 μL of RNase-free water and bring the volume in the column up to 500 μL with RNase-free water. Spin at 3000*g* for 8 min at room temperature. Wash the sample 3× by running 500 μL of RNase-free water through the Microcon-100 column.
10. Place the column containing the sample in a new tube upside down, and spin at 3000*g* for 8 min to recover the desalted sample.
11. Dry the sample to 8 μL using a speed-vacuum drier.

3.2.2. T7 RNA Polymerase Amplification

T7 RNA polymerase is used to generate anti-sense RNA (aRNA) from the double-stranded cDNA, to which a T7-promoter sequence has been incorporated via the T7-oligo(dT) primer used in the reverse transcription step (**Subheading 3.2.1.**).

1. To 8 μL of double-stranded cDNA add:
   a. 2 μL of 10 × Ampliscribe T7 buffer..
   b. 2 μL of each 10 m*M* ATP, CTP, GTP, UTP.
   c. 2 μL of 0.1 *M* DTT.
   d. 2 μL of T7 RNA polymerase.
2. Incubate at 42°C for 3 h.
3. Add 1 μL of RNase-free DNase (from the T7 transcription kit), and incubate the sample at 37°C for 15 min.
4. Desalt the sample in a Microcon-100 column as described in **Subheading 3.2.1., steps 9–10**, and wash the sample 3× with 500 μL of RNase-free water.
5. Measure the optical density (OD) ratio at A_{260}/A_{280} with an ultraviolet (UV)-visible spectrophotometer.
6. Check the quality of the RNA by running on a 1.2% agarose gel (*see* **Fig. 1**) (*see* **Note 2**).

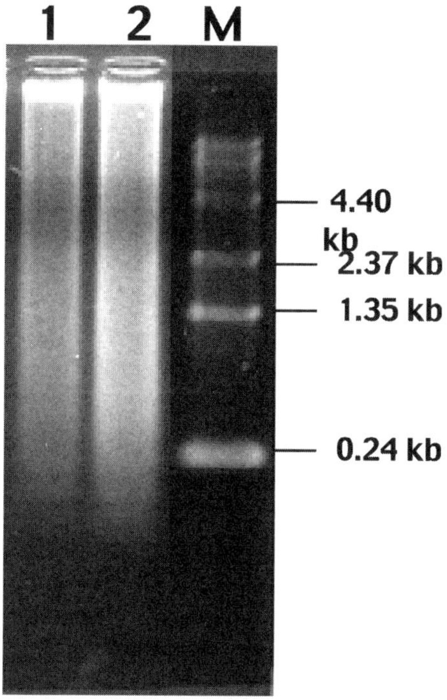

Fig. 1. A 1% agarose gel showing aRNA after ethidium bromide staining. 2 µg of aRNA were loaded per lane; the aRNA appears as a smear. (**1**) aRNA from normal lung tissue; (**2**) aRNA from lung tumor tissue; (**M**) Size markers; 0.24 – 9.5 kb RNA ladder. The marker sizes are indicated at the right of the panel.

3.2.3. Re-Amplifying RNA from aRNA

When the amount of starting total RNA is low (e.g., 10–100 ng), the resultant amount of aRNA may be increased by performing a re-amplification step. This amplification involves converting the first-round aRNA into cDNA, and then performing a second T7-mediated transcription of aRNA from the amplified cDNA (*see* **Note 3**).

1. Use 0.2–1 µg of aRNA. The volume of the aRNA must to be 10 µL.
2. Add 1 µL of 1 mg/mL random hexamers.
3. Heat the samples at 70°C for 10 min, then place on ice. Spin briefly in a microcentrifuge.
4. Add:
   a. 4 µL of 5× first-strand cDNA synthesis buffer.
   b. 2 µL of 0.1 *M* DTT.
   c. 1 µL of 10 m*M* dNTPs.
   d. 1 µL of RNasin.
   e. 1 µL of Superscript II RT.

5. Incubate the sample at room temperature for 5 min.
6. Transfer the tube to 37°C for 1 h.
7. Add 1 µL of RNase H and mix gently.
8. Incubate the samples at 37°C for 20 min.
9. Denature the hybrids at 95°C for 2 min.
10. Transfer the tubes on ice for 2–5 min, then spin briefly, and on ice add 1 µL of 0.5 mg/mL T7-oligo(dT) primer.
11. Heat the samples to 70°C for 5 min, spin briefly, and incubate at 42°C for 10 min to anneal the T7-oligo(dT) primer. Transfer the tube onto ice, then spin briefly.
12. Add:
    a. 30 µL of second-strand cDNA synthesis buffer.
    b. 3 µL of 10 m*M* dNTPs.
    c. 4 µL of DNA polymerase I.
    d. 1 µL of RNase H.
    e. 90 µL of RNase-free water.
13. Incubate the samples at 16°C for 2 h.
14. Add 2 µL of T4 DNA polymerase and mix gently (*see* **Note 1**).
15. Continue to incubate the samples at 16°C for 10 min.
16. Add 150 µL of chloroform and 150 µL of phenol. Shake 20 ×. Spin at 3000*g* for 4 min at 4°C.
17. Remove the upper aqueous layer and desalt using a Microcon-100 column as described in **Subheading 3.2.1.**, **steps 9–10**; wash the sample 3× with 500 µL of RNase-free water.
18. Measure the OD ratio at A_{260}/A_{280} with a UV-visible spectrophotometer.
19. Check the quality of the RNA by running on a 1.2% agarose gel.

3.3. Microarray Hybridization

3.3.1. Reverse Transcription Setup

1. In a sterile microfuge tube, prepare the following mix:
    a. 8 µL of first-strand cDNA synthesis buffer.
    b. 2 µL of dN6 primer (8 µg/µL).
    c. 4 µL of 10X low-dT-dNTP mix (5 m*M* dATP, dCTP, and GTP; 2 m*M* dTTP).
    d. 4 µL of Cy-dUTP (1 m*M* Cy3-dUTP or Cy5-dUTP).
    e. 4 µL of 0.1 *M* DTT.
    f. 1 µL of Rnasin.
    g. 2–2.5 µg of amplified RNA in 17 µL of DEPC-H_2O (*see* **Note 4**).
2. Mix well and heat at 65°C for 5 min, then cool down to 42°C.
3. Add 2 µL of SuperScript II RT. Incubate for 30 min at 42°C and add another 1 µL of SuperScript II RT for 30 min at 42°C.
4. Add 5 µL of 500 m*M* ethylenediaminetetraacetic acid (EDTA) and heat at 65°C for 1 min.
5. Add 10 µL of 1 *M* NaOH and incubate at 65°C for 15 min to hydrolyze the RNA.
6. Add 25 µL of 1 *M* Tris-HCl, pH 7.4, to neutralize the pH.

3.3.2. Probe Clean-Up

1. Add 500 µL of Tris-EDTA (TE) (10 m*M* Tris-HCl, pH 7.4, 1 m*M* EDTA) to the probe sample and transfer to a Microcon 30 concentrator (*see* **Note 5**).
2. At this time, turn on a heating block and set at 100°C.
3. Add 400 µL of TE to the sample tube and transfer the sample to the Microcon 30 column (*see* **Note 6**).
4. Spin the samples at 6000–7000*g* at room temperature. Check the volume. Continue spinning as necessary until the volume is reduced to 20–30 µL.
5. To combine the probes, carefully invert one column of probe (Cy3) into a fresh tube and spin at 12,000*g* for 1 min. Transfer the recovered probe to the Microcon column containing the other probe (Cy5).
6. Add 400 µL of TE into the column and spin down to 20–30 µL.
7. Wash the probe a final time (to remove unincorporated fluor) by adding 450 µL of TE and reduce volume to 16 µL. Measure the final volume carefully, using a pipetman.
8. Invert the column into a fresh tube and spin for 1 min at 13,000*g* to recover the probe. If recovered probe volume is <16 µL, increase volume up to 16 µL with TE; if recovered probe volume is > 16 µL, concentrate to 16 µL using a speed-vacuum drier ***(10–12)***.

3.3.3. Hybridization Setup

1. Add to the tube with probe:
   a. 1 µL of 50X Denhardt's blocking solution.
   b. 1 µL of poly(dA).
   c. 1 µL of yeast tRNA.
   d. 1 µL of human Cot I DNA.
   e. 2.6 µL of 20X SSC.
2. Heat for 2 min at 100°C and add 0.6 µL of 10% SDS.
3. Spin for 10 min at 13,000*g* to pellet any particulate matter and to facilitate cooling.
4. Allow the probe to cool at room temperature for at least 15 min.
5 Add 15 µL of sterile deionized H_2O to the reservoir pool of the hybridization chamber.
6. Place a slide in the hybridization cassette (Telecham).
7. Pipet the probe onto the cover slip and quickly (but carefully) place the cover slip on top of the slide in such a manner as to reduce the formation of air bubbles under the cover slip surface. Air bubbles that do form usually dissipate during hybridization ***(13,14)***.
8. Incubate the slide in the hybridization cassette overnight (10–16 h) at 65°C.

3.3.4. Washing

1. Remove the slide from the hybridization chamber, place in a slide rack and submerge in the staining dish filled with 2X standard saline citrate (SSC) and 0.1% sodium dodecyl sulfate (SDS). Plunge the slide gently until the cover slip falls

away. Tilt the slide at an angle while plunging it to ensure that the cover slip falls away from the array without scratching it.
2. Wash for 1 min in 1X SSC, plunging gently.
3. Wash for 1 min in 0.2X SSC, plunging gently.
4. Wash for 1 min in 0.05X SSC, plunging gently.
5. Quickly spin-dry the slide (in a slide rack) in a centrifuge set at 700*g* for 3 min.
6. Scan as soon as possible. The hybridized microarray is scanned using a microarray scanner, such as the Gene Pix 4000 (Axon, CA). Slides are scanned at two wavelengths that correspond to each probe's fluorescence. Images are obtained and read using the Gene Pix 3.0 software that reads, averages, and records the intensities of each pixel within the spots (there are 50–100 pixels per spot). All the readings are normalized relative to the background around each spot. Collected image data that are converted into the numeric format by Gene Pix 3.0 software will then be imported into Cluster software for final analysis and interpretation (*see* **Fig. 2**).

4. Notes

1. Transcription templates should be linear double-stranded DNA with blunt or 5'-protruding ends. Templates containing 3'-protruding ends can produce spurious transcripts due to non-specific initiation. 3'-protruding ends can be readily converted to blunt ends with T4 DNA polymerase.
2. We do not routinely analyze the aRNA by gel electrophoresis because it uses up a significant amount of the aRNA (approx 2 μg per gel lane). For measurement of aRNA concentration, use a spectrophotometer at wavelengths of 260 and 280 nm. The quality of the aRNA can also be analyzed by RT-PCR using β-actin, COX-2, G-protein, or other primers.
3. It is important to consider that after each round of aRNA amplification, the resultant aRNA is slightly shorter in size. This is because the aRNA obtained from the initial round of amplification is primed with random hexamers to generate an amplified cDNA for use as a template for the second T7 RNA transcription. Unlike the oligo(dT) primer used in first-strand cDNA synthesis, which primes from the 3' poly-A-tailed termini of the mRNA, random hexamers prime randomly from internal positions in the aRNA. Thus, the second cDNA synthesis will not be full-length. For example, one round typically yields aRNA within a size range of approx 0.5–3 kb; two rounds yields a size range of approx 0.4–2.4 kb; three rounds—0.2–1.5 kb, and so on. This limits the number of amplifications that can be performed.
4. We consider 2–2.5 μg of amplified RNA to be the minimum amount of RNA required for this step. Use more if available.
5. Prior to use, spin 500 μL of TE through the Microcon column to remove glycerol from the membrane.
6. Avoid contact between the pipet tip and the Microcon membrane, because the membrane is fragile.

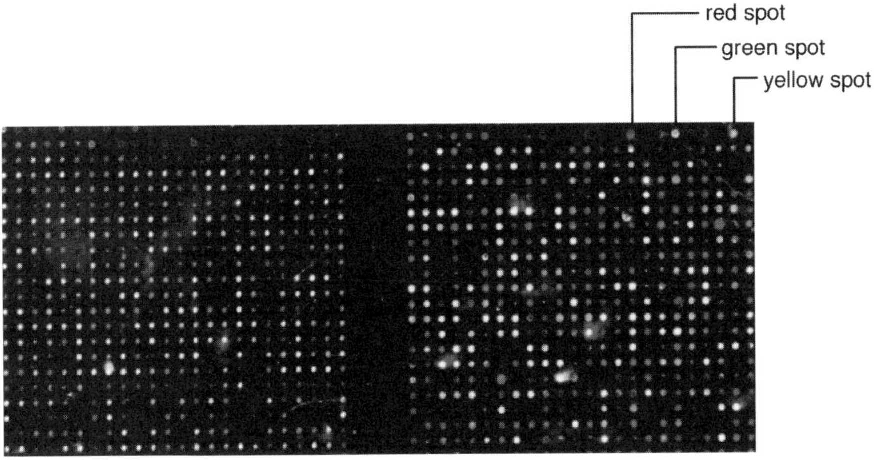

Fig. 2. The array was hybridized with probes made from normal colon (labeled with Cy 3) and colon tumor (labeled with Cy 5) tissues. Color overlays are composite images of Cy 3 and Cy 5 fluorescence signals. Green represents Cy 3 > Cy 5; red represents Cy 5 > Cy 3; and yellow represents Cy 3 = Cy 5. Spots that hybridized more strongly to Cy 5-labeled probes appear red and therefore denote genes that are upregulated in the tumor. Spots that hybridized more strongly to Cy 3-labeled targets appear green; these genes are downregulated in the tumor. Spots that hybridized equally well to both labels appear yellow and therefore represent genes whose expression is unaltered in the transition from normal to tumorous tissue. In the figure, examples of green, yellow, and red spots are indicated. Figure gray-scale here.

References

1. Perou, C. M., Sorlie, T., Eisen, M. B., van de Rijn, M., Jeffrey, S. S., Rees, C. A., et al. (2000) Molecular portraits of human breast tumors. *Nature* **406,** 747–752.
2. Alizadeh, A. A., Eisen, M. B., Davis, R. E., Ma, C., Lossos, I. S., Rosenwald, A., et al. (2000) Distinct types of diffuse large B-cell lymphoma identified by gene expression profiling. *Nature* **403,** 503-511.
3. Schena, M. (1996) Genome analysis with gene expression microarrays. *Bioessays* **18,** 427–431.
4. Cheung, V. G., Morley, M., Aguilar, F., Massimi, A., Kucherlapati, R. and Childs, G. (1999) Making and reading microarrays. *Nat. Genet.* **21** 15–19.
5. Schena, M., Shalon, D., Davis, R. W., and Brown, P. O. (1995) Quantitative monitoring of gene expression patterns with a complementary DNA microarray. *Science* **270,** 467–470.
6. Chee, M., Yang, R., Hubbell, E., Berno, A., Huang, X. C., Stern, D., et al. (1996) Accessing genetic information with high-density DNA arrays. *Science* **274,** 610–614.
7. Duggan, D. J., Bittner, M., Chen, Y., Meltzer, P., and Trent, J. M. (1999) Expression profiling using cDNA microarrays. *Nat. Genet.* **21,** 10–14.

8. Schena, M., Heller, R. A., Theriault, T. P., Konrad, K., Lachenmeier, E., and Davis, R. W. (1998) Microarrays: biotechnology's discovery platform for functional genomics. *Trends Biotechnol.* **16,** 301–306.
9. Luo, L., Salunga, R. C., Guo, H., Bittner, A., Joy, K. C., Galindo, J. E., et al. (1999) Gene expression profiles of laser-captured adjacent neuronal. *Nat. Med.* **5,** 117–122.
10. Heller, R. A., Schena, M., Chai, A., Shalon, D., Bedilion, T., Gilmore, J., et al. (1997) Discovery and analysis of inflammatory disease-related genes using cDNA microarrays. *Proc. Natl. Acad. Sci. USA* **94,** 2150–2155.
11. DeRisi, J., Penland, L., Brown, P. O., Bittner, M. L., Meltzer, P. S., Ray, M.,et al. (1996) Use of a cDNA microarray to analyse gene expression patterns in human cancer. *Nat. Genet.* **14,** 457–460.
12. Graves, D. J., Su, H. J., McKenzie, S. E., Surrey, S., and Fortina, P. (1998) System for preparing microhybridization arrays on glass slides. *Anal. Chem.* **70,** 5085–5092.
13. Kozal, M. J., Shah, N., Shen, N., Yang, R., Fucini, R., Merigan, T. C., et al. (1996) Extensive polymorphisms observed in HIV-1 clade B protease gene using high-density oligonucleotide arrays. *Nat. Med.* **2,** 753–759.
14. Sapolsky, R. J., and Lipshutz, R. J. (1996) Mapping genomic library clones using oligonucleotide arrays. *Genomics* **33,** 445–456.

19

Semi-Quantitative Determination of Differential Gene Expression in Primary Tumors and Matched Metastases by RT-PCR

Comparison with Other Methods

Benno Mann and Christoph Hanski

1. Introduction

Analysis of differential antigen expression in tumor and metastases vs normal tissue is frequently performed in oncological research. If suitable antibodies are available, its specificity is tested in Western blot, and the distribution of the antigen in the tissue is followed by immunohistochemistry. The information on the relative amount of the antigen in each tissue is obtained by quantitative evaluation of Western blots or enzyme-linked immunosorbent assay (ELISA) data. However, if no antibody is available for antigen of which the coding sequence is known, Northern blotting is the procedure of choice to quantify differences in gene expression at the transcriptional level. This is the second-best solution, since the mRNA level does not always correctly reflect the amount of the antigen at the protein level. Furthermore, the contamination with mRNAs from undesired cells and RNA losses caused by endogenous RNases may distort the completed picture. In fact, when tissue is investigated, total-RNA or whole-cell-lysates are always inadvertently contaminated with material from the nondesired cells.

To assure maximal cellular homogeneity of the preparation, when gastrointestinal tumors are investigated, the normal-mucosa specimens should be taken far away from the primary carcinoma (usually at a distance of 10 cm), because the transitional mucosa can already show altered gene expression. Only the mucosal layer, separated from submucosa by dissection, should be used for analysis. The tumor specimen should be taken where the carcinoma borders

From: *Methods in Molecular Biology, vol. 193: RT-PCR Protocols*
Edited by: J. O'Connell © Humana Press Inc., Totowa, NJ

the transitional mucosa and not from the center, where necrosis is often present. Finally, it is important to use only metastatic tissue itself, and not the surrounding healthy tissue. The tissue preparation must be carried out quickly to prevent cell lysis and the liberation of endogenous RNases. These precautions at the macroscopic level allow the assumption that the main component of the obtained mixture originates from the target tissue, and that the original amount of RNA is not significantly reduced by digestion.

In most cases, when human tissue is investigated, only small amounts of patients' tissue are available: several micrograms when specimens are obtained through biopsies, and usually less than 1 g when specimens are obtained during surgery. Consequently, the amount of available total RNA is frequently too small for Northern-blot analysis, and reverse-transcriptase-polymerase chain reaction (RT-PCR) is an alternative method yielding information on relative mRNA amounts. Quantitative information from the RT-PCR, however, is limited by the previously-mentioned weak points inherent to the mRNA isolation procedure. A further source of errors are the intricacies of the reverse transcription and the PCR (*see* **Notes 3–8**). Because of these uncertainties the integral quantitative data obtained by RT-PCR must be verified by *in situ* hybridization. *in situ* hybridization with specific anti-sense cRNA probes complementary to the analyzed transcript can be performed on paraffin-embedded or frozen sections. In this case, the mRNA expression can be visually assigned to a particular cell population. In contrast to RT-PCR, *in situ* hybridization permits analysis of the variability in gene expression within a particular tissue locale—e.g. expression only at the infiltrating margin of a carcinoma. However, Northern-blot or RT-PCR analysis using total RNA derived from approx 1 g of tissue represents an integral value comparable to 2000 paraffin-embedded 5-μm-thick sections. Thus, the integral techniques such as Northern blot or RT-PCR are not substitutes but rather necessary complements to *in situ* techniques.

The following protocol is an example of how to approach the problem of differential gene-expression analysis in patient tissue samples using the RT-PCR technique. The objective in our example was to compare the expression of FasL in the normal colonic mucosa, colorectal carcinomas, and their liver metastases. The additional question was how far the activated tumor infiltrating lympocytes (TIL) —which are known to express FasL—contribute to the integral results. Therefore, in addition to FasL, the expression of the T-cell marker, CD25 (the interleukin-2 receptor), was semi-quantitatively analyzed using RT-PCR in the same tissue samples. The results, the limitations of quantitative gene-expression analysis using RT-PCR, and the possible caveats are demonstrated and emphasized through comparison with the results also obtained by three other techniques: *in situ* hybridization, Western blot, and immunohistochemistry.

2. Materials

2.1. Patient Tissue Samples

1. Either extract RNA and protein immediately from the tissue after surgical resection, or flash-freeze in liquid nitrogen for later analysis. In our example, tissue samples from primary colorectal carcinomas, their liver metastases, and specimens of normal colonic mucosa (taken at a distance of 10 cm from the tumor) were obtained at the time of surgery from 13 patients with UICC stage IV colorectal carcinoma (CRC). Total RNA was immediately isolated from samples of 10 patients, and protein lysates from samples of six patients. Tissue samples from normal liver tissue were obtained from four patients. Samples of normal, tumor, and metastatic tissues from 12 patients (all UICC stage IV) were fixed in 3.7% formalin and embedded in paraffin.
2. RNA was isolated using the RNAClean RNA extraction kit (Hybaid-AGS, Heidelberg, Germany).
3. Prepare a solution of phosphate-buffered saline (PBS) as follows: 0.1 *M* phosphate buffer and 0.15 *M* NaCl, pH 7.2.
4. RNase free DNase I, 10 U/µL and RNase-inhibitor, 40 U/µL (Roche Molecular Biochemicals, Mannheim, Germany).
5. For cDNA synthesis: random hexanucleotide primers, 3 µg/µL; Reverse transcriptase, 200U/µL (Superscript) and 5× reverse transcriptase buffer; deoxynucleotide 5' triphosphate (dNTP) mix (2.5 m*M* each) (all from Gibco-BRL, Karlsruhe).
6. PCR primer sequences, their positions on the corresponding cDNA sequences, and their PCR product sizes: FasL: forward 5' GGA TTG GGC CTG GGG ATG TTT CA 3' (bp 474–496), reverse 5' TTG TGG CTC AGG GGC AGG TTG TTG 3' (bp 817–794) (344-bp product); CD25: forward 5' TTC CAG GTG AAG AGA AGC CTC AGG 3' (bp 521–544), reverse 5' TCT GTT CCC GGC TTC TTA CCA AGA 3' (bp 807–784) (287 product); pyruvate dehydrogenase (PDH): forward 5' GGT ATG GAT GAG GAG CTG GA 3' (located in exon 1), reverse 5' CTT CCA CAG CCC TCG ACT AA 3' (located in exon 2) (103-bp product from mRNA, 185-bp from genomic DNA). Primer pairs for CD25 were selected and tested for specificity in the HUSAR® program (genius@embnet.dkfz-heidelberg.de); primer pairs for FasL ***(1)*** and PDH ***(4)*** were previously described. All primers were obtained from TIB MOLBIOL Syntheselabor (Berlin, Germany).
7. Taq DNA polymerase and 10X PCR buffer containing 100 m*M* Tris-HCl (pH 8.8), 500 m*M* KCl, 0.8% Nonidet P40, 15 m*M* $MgCl_2$ (MBI Fermentas, Vilnius, Lithuania).
8. Thermal cycler (Omnigene, Hybaid, MWG-Biotech, Germany).
9. ChemMate™ slides (Dako Diagnostika GMBH, Hamburg, Germany).
10. Vector for cloning PCR products to generate probes for *in situ* hybridization: pTAQ (R&D Systems, Wiesbaden, Germany).
11. For probe labeling for *in situ* hybridization, use [35S]-UTP (1300 Ci/mmol, NEN, Bad Homburg, Germany).

12. Amersham nuclear emulsion for autoradiography after *in situ* hybridization (Amersham, Bra unschweig, Germany).
13. Western blot lysis buffer: 1% Triton X-100, 20 m*M* Tris-HCl, pH 7.6, 140 m*M* NaCl, 1.5 m*M* $MgCl_2$, 1 m*M* EGTA, 10% glycerol, 10 µg/mL DNase, 1 m*M* $NaVO_3$, 50 m*M* NaF, 1 m*M* PMSF, 1 m*M* benzamidine and 10 µg /mL of each of the protease inhibitors aprotinin, leupeptin and pepstatin.
14. Western blot blocking solution: 5% dry milk, 1% bovine serum albumin (BSA) in 1X PBS.
15. For Western blotting, anti-human FasL murine monoclonal antibody (Mab) (raised to residues 116–277 of human FasL, Transduction Laboratories, Lexington, KY) was used; a rabbit anti-murine IgG antibody conjugated to peroxidase (Dako) was used as secondary antibody; detection was performed using the ECL-detection system (Amersham).

3. Methods

3.1. RNA Isolation

We isolate total RNA from tissue specimens using the RNAClean kit (Hybaid-AGS, Heidelberg, Germany). This procedure is much less time-consuming and yields a higher amount of RNA from the tissue, than the isolation via CsCl gradients, as described previously *(3)* (*see* **Note 1**). The tissue specimen is freshly placed in a Falcon tube in the operating theatre and cooled on ice. The total RNA is isolated as rapidly as possible from the fresh tissues as follows:

1. Wash the tissue once with PBS solution.
2. Add 2 mL of RNAClean solution per 100 mg wet wt of tissue.
3. Homogenize the tissue in the RNAClean solution.
4. Transfer aliquots of 1 mL of the homogenate to 1.5-mL tubes.
5. Add 100 µL of chloroform per tube.
6. Vortex thoroughly.
7. Cool on ice for 5 min.
8. Centrifuge for 15 min at 12,000*g* at 4°C.
9. Transfer the clear upper phase (approx 500 µL) in a new 1.5-mL tube.
10. Add an equal volume of isopropanol.
11. Vortex thoroughly.
12. Cool on ice for 15 min.
13. Centrifuge for 15 min at 12,000*g* at 4°C.
14. Remove the supernatant, and save the pellet at the bottom of the tube.
15. Wash the pellet with 70% ethanol.
16. Centrifuge for 8 min at 7500*g* at 4°C.
17. Repeat the washing with ethanol and centrifuge again.
18. Dry the pellet at 37°C until it becomes transparent (approx 10 min).
19. Dissolve the pellet in 20–50 µL DEPC H_2O for 5 min at 56°C.
20. Determine the RNA concentration spectrophotometrically (*see* **Note 2**).

3.2. cDNA Synthesis

1. Perform DNase digestion of the total RNA prior to cDNA synthesis to rule out contamination with genomic DNA. Set up the DNase digest as follows:

a. 20 mg of total RNA	X µL
b. RNase free DNase I, 10 U/µL (Roche Molecular Biochemicals)	20 µL
c. 1 *M* Tris-HCl (pH 6.5) to yield a final conc. of 50 m*M*	2.5 µL
d. 100 m*M* $MgCl_2$ to yield a final conc. of 10 m*M*	5 µL
e. 100 m*M* DTT (Gibco-BRL) to yield a final conc. of 10 m*M*	5 µL
f. RNase-inhibitor, 40 U/µL (Boehringer Manheim)	0.5 µL
g. Add DEPC-H_2O to a final volume of:	50 µL

2. Digest for 2 h at 37°C , then denature the DNase at 100°C for 5 min.
3. Next, the cDNA synthesis reaction is carried out. First, mix the RNA with the primers for reverse transcription as follows:

a. 2.5 µg DNase-digested RNA from the reaction in **step 1**	6.25 µL
b. Random hexanucleotide primers, 3 µg/µL (Gibco BRL, Karlsruhe)	2.5 µL
c. Add DEPC-H_2O to a final volume of:	22 µL

4. Denature for the RNA-primer mix for 10 min at 70°C, then cool on ice to allow primer annealing.
5. Prepare and add 28 µµL of the following master-mix:

a. 5× reverse transcriptase buffer (Gibco-BRL)	10 µL
b. dNTP mix (2.5 m*M* each) (Gibco-BRL), final conc. of 0.5 m*M*	10 µL
c. 100 m*M* DTT to yield a final conc. of 10 m*M*	5 µL
d. RNase inhibitor, 40 U/µL (Boehringer, Manheim, Germany)	0.5 µL
e. Reverse transcriptase, 200U/µL (Superscript, Gibco-BRL)	2.5 µL
f. Add DEPC-H_2O to a final volume of:	28 µL

6. Incubate the cDNA synthesis reaction (final volume of 50 µL) for 1 h at 37°C.
7. Terminate the reaction by boiling for 10 min and cool the samples on ice.

3.3. PCR

1. In addition to detection of the target FasL mRNA, perform RT-PCR for CD25 mRNA as a marker for the presence of activated lymphocytes in the tissue samples. To control for the quantity of cDNA, perform RT-PCR for the housekeeping gene pyruvate dehydrogenase (PDH), which is expressed equally in normal colonic mucosa and colorectal carcinomas *(2)*. The PDH PCR can also be used to test for contamination of the RNA with genomic DNA, since the intron-spanning PDH primers were selected to yield a 103-bp product from mRNA and a 185-bp product from genomic DNA. Although the primers for FasL and CD25 also span introns in their genomic sequences to ensure mRNA-specific amplification, the yield of mRNA-specific products can be affected by competition with any contaminating DNA.

2. Set up the PCR reaction mixture as follows:

a. Newly synthesized cDNA	X µL
b. 10X PCR buffer (MBI Fermentas)	2.5 µL
c. dNTP mixture (containing 10 m*M* of each dNTP), final conc. 0.3 m*M*	0.75 µL
d. 10 pmol of each primer	1 µL
e. 0.5 U of Taq-DNA polymerase (MBI Fermentas)	0.5 µL
f. Add H_2O to a final volume of:	25 µL

   For the FasL and PDH PCRs, use 2.5 µL of the cDNA template; for the CD25 PCR, use 1 µL of the cDNA template in a doubled reaction volume (50 µL final vol); for the FasL PCR, increase the amount of Taq DNA polymerase to 1.25 U. Keep all components on ice until the tube is transferred into the PCR machine.
3. Overlay the mixture with 50 µL of mineral oil to prevent evaporation unless the PCR machine has a heated lid.
4. Perform thermal cycling as follows for the PDH PCR:
   Initial denaturation at 95°C for 2 min, then 35 cycles of: denaturation at 95°C for 45 s, annealing at 58°C for 45 s, extension at 72°C for 1 min; a final elongation step at 72°C for 5 min.
   The only differences in the temperature program are that for FasL, 40 cycles are performed, and annealing is at 65°C for 1 min; for CD25, denaturation and annealing are both for 1 min, and extension is for 2 min For the CD25 PCR, perform "hot-start" by holding the tube at 80°C after the initial denaturation step before adding the Taq DNA polymerase. To determine the level of CD25 mRNA in the samples, remove an aliquot of PCR products at 30, 35, 40, and 45 cycles.
5. Separate the PCR products by electrophoresis on a 1.5% agarose gel, and visualize by ethidium bromide staining. Estimate the intensity of the specific bands as none (–), weakly positive (–), positive (+) or strongly positive (++) (*see* **Notes 3–5**).

3.4. Additional Detection Methods

Since the *in situ* hybridization, Western blot, and immunohistochemistry methods are not the primary focus of this chapter, and were used only to deliver auxillary information, we will refer to specialized manuals for more detailed protocols.

3.4.1. In Situ *Hybridization*

1. Preparation of sections. For *in situ* hybridization, cut 5-µM-thick tissue sections of paraffin-embedded tissue and adhere onto ChemMateTM slides (Dako Diagnostika GMBH).
2. Probe synthesis. Generate the probe for FasL by amplification of FasL cDNA from a FasL-expressing cell line, e.g., from the anaplastic large-cell lymphoma cell line Karpas 2995, using primers specific for the FasL sequence *(5)*. Clone the fragment obtained with these primers (820 bp, comprising nucleotides 67–886)

into the vector pTAQ (R&D Systems). Confirm the sequence of the cloned insert on a DNA sequencer (e.g., 373 A, Applied Biosystems, Foster City, USA). Linearize the plasmid, and generate the sense and anti-sense cRNA probes by run-off transcription using the T7 or T3 RNA polymerases and [^{35}S]-UTP (1300 Ci/mmol, NEN, Bad Homburg, Germany). Perform limited alkaline hydrolysis for 50 min to generate short cRNA fragments of 100–150 bp in length. For *in situ* hybridization, expose dewaxed and rehydrated paraffin sections to 0.2 N HCl and 0.5 mg/mL of pronase (Boehringer-Mannheim), acetylate with 0.25% acetic anhydride in 0.1 *M* triethanolamine pH 8.0, and dehydrate through graded ethanol solutions. Hybridize the slides with 4×10^5 cpm of the labeled probe (specific radioactivity of the probe should be approx 8×10^6 cpm/μg RNA).

3. Hybridization. Perform a prehybridization step, then hybridize the probe. Following hybridization, remove nonspecifically bound probe by RNase digestion and washing under stringent conditions. These steps are described in detail elsewhere *(5)*.
4. Autoradiography. Perform autoradiography by dipping the dehydrated slides into Amersham nuclear emulsion (Amersham) and exposing for 100 d.
5. Evaluation. After counterstaining with hematoxylin and eosin (H&E), score the staining of cells for intensity as: – (identical to the background), + (2× stronger than the background) and ++ (more than 2 × stronger than the background).

3.4.2. Western Blot

1. Prepare lysates from whole-tissue samples by pulverizing approx 1 g wet wt of snap-frozen tissue under liquid nitrogen and incubating in 10 mL of western blot lysis buffer.
2. Centrifuge at 13,000*g* for 30 min.
3. Determine the protein concentration, and use 5 μg of total protein for sodium dodecyl sulfate polyacrylamide gel electrophoresis (SDS-PAGE) (12.5% polyacrylamide gel).
4. Electroblot the gel onto PVDF membrane.
5. Saturate the membranes for 1 h in Western blot blocking solution.
6. Incubate the membranes overnight at 4°C with 2.5 μg/mL of anti-human FasL murine MAb (Transduction Laboratories) in blocking solution.
7. Incubate the membranes for 1 h with a rabbit anti-murine IgG antibody conjugated to peroxidase (Dako).
8. Detection the bound antibody with the ECL-detection system (Amersham).

3.4.3. Immunohistochemistry

1. Preparation of sections. Deparaffinize 5-μM-thick, paraffin-embedded tissue sections in xylene, and rehydrate through graded alcohols into PBS. Perform epitope retrieval by immersing the slides in 10 m*M* citrate buffer, pH 6.0, and boiling for 10 min in a pressure cooker, followed by cooling in PBS. Block endogeneous peroxidase activity by incubation for 15 min with a solution of 0.6% H_2O_2 in methanol.

2. Immunohistochemical staining. To prevent unspecific binding, incubate the sections in 20% heat-inactivated fetal calf serum (FCS) in PBS for 30 min. Incubate in 0.6 μg/mL of anti-FasL antibody overnight at 4°C. Incubate with a secondary antibody conjugated to peroxidase, and develop the color by incubating in a solution of 3,3'-diaminobenzidine tetrahydrochloride.
3. Evaluation. After counterstaining with H&E, score the staining for intensity as: – (none), + (weak), ++ (moderate), and +++ (strong).

3.5. Analysis fo Results

3.5.1. RT-PCR

RT-PCR was applied to RNA isolated from normal colonic (N), carcinomatous (T), and metastatic (M) tissue of 10 patients. **Figure 1** shows the results from five randomly selected cases out of 10 investigated patients.

1. PDH RT-PCR showed the specific amplimer band at 103 bp of comparable intensity in all analyzed samples, without additional bands. This indicates that no contamination with genomic DNA occurred, and that the efficiency of cDNA synthesis was similar in all samples.
2. FasL-mRNA was detectable in 5 out of 10 normal tissues, in 3 out of 10 primary carcinomas, and in 9 out of 10 liver metastases (*see* examples in **Fig. 1**). In three of four investigated samples of normal liver tissue, FasL-mRNA was weakly detectable (not shown).
3. mRNA for the T-cell activation marker CD25 was detected in 3 out of 10 N samples, in all T samples, and in 5 out of 10 metastases after the standard RT-PCR procedure performed with 40 cycles of PCR (*see* examples in **Fig. 1**). In order to semi-quantify the level of CD25-mRNA present in mRNA from each type of sample, CD25-RT-PCR was performed at different cycle numbers: 30, 35, 40, and 45 (**Fig. 2**). The percentage of samples that were positive for CD25 mRNA was high at high cycle numbers (at 45 cycles, for example, 80–100% of all samples were positive). However, at moderate cycle numbers, within the linear range of the CD25-RT-PCR, (e.g., 35), fewer samples were positive; only about 10% of the metastatic samples were positive (*see* **Fig. 2**). Thus, the FasL mRNA detected in the liver metastases was unlikely to be derived from CD25-positive T-cells.

3.5.2. In Situ *hybridization*

In most samples, FasL-mRNA was detectable by *in situ* hybridization only after 100 or more days of exposure. The usually weak signals after this long exposure time indicated that the expression of FasL-mRNA is very low. In the normal colonic mucosa, FasL-mRNA expression was detectable only in one case at the tip of the crypt. By contrast, in 4 of 12 primary carcinomas, FasL-mRNA expression was detectable in the majority of cells. Of 12 investigated liver metastases, 6 showed expression of FasL-mRNA. In 5 of 12 patients, all

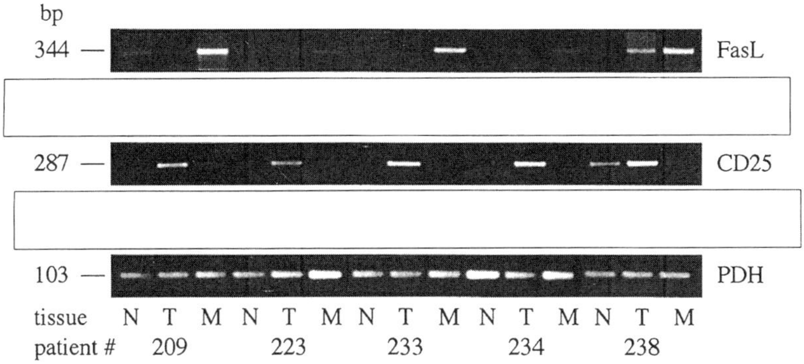

Fig. 1. Detection of FasL, CD25, and PDH mRNA by RT-PCR in patient-matched normal colonic mucosa (N), primary colon tumors (T), and their liver metastases (M). Results for five out of the ten patients analyzed are shown. Specific sequences for each gene were amplified from cDNA, separated by agarose gel electrophoresis, and visualized by ethidium bromide staining. RT-PCR for PDH was performed to verify that the efficiencies of RT-PCR amplification were similar in each sample. Reprinted with permission from **ref.** *(7)*.

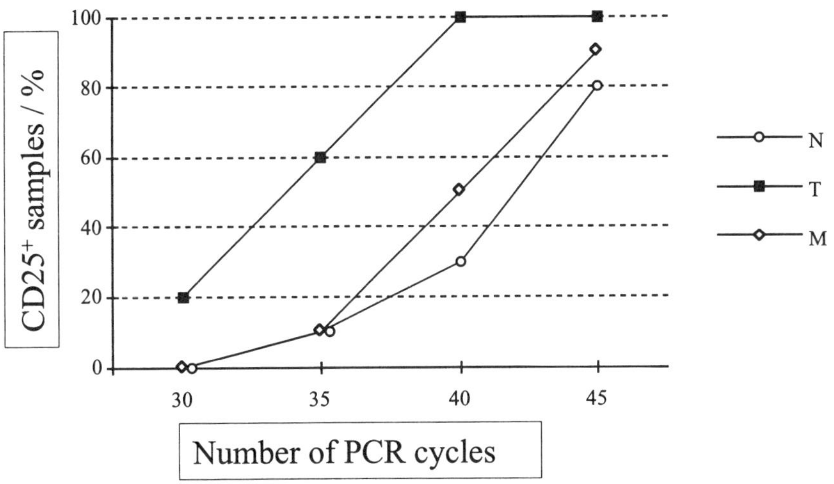

Fig. 2. Percentage of $CD25^+$ samples determined by RT-PCR after 30, 35, 40, and 45 cycles in normal colonic mucosa (N), primary carcinomas (T), and liver metastases (M) (for each, $n = 10$). Reprinted with permission from **ref.** *(7)*.

three tissue types showed no staining after 100 days of exposure. In the remaining 7 patients, the metastatic tissue stained in one case weaker, in one case equally strong, and in five cases stronger than the corresponding primary tumor. In both primary and metastatic carcinomatous tissues, FasL[1] tumor-infiltrating lymphocytes (TILs) were visible. However, they were sparse and weakly stained, and a difference in their number between primary carcinomas and liver metastases was not evaluated. *In situ* hybridization on paraffin-embedded sections may be influenced by RNase-mediated mRNA degradation, particularly if the mRNA to be detected is expressed at a relatively low level, as for FasL.

3.5.3. Western Blot

In order to quantify FasL-protein expression, the lysates from N, T, and M of six patients were analyzed by immunoblotting. Specific bands for FasL were detectable in two out of six lysates from normal colonic mucosa, in two out of six lysates from primary tumors, and in all lysates derived from metastatic tissue. In two patients, all three tissues expressed FasL-protein. In one instance, the intensity of the FasL band was stronger in the primary tumor, and in another instance weaker than in the metastatic tissue. In summary, in five out of six investigated patients, the metastatic tissue showed stronger FasL expression than the corresponding primary tumor.

3.5.4. Immunohistochemistry

Immunohistochemistry with anti-human FasL Mab was performed in sections of tumors and metastases of five patients. The staining of positive cells was diffuse and cytosolic, and was homogeneously distributed among most epithelial cells. The normal colonic mucosa stained weakly (4/5) or moderately (1/5). The primary tumors were stained moderately (3/5) or strongly (2/5), as were the metastases (2/5 and 3/5, respectively), with no discernible differences between the two groups. Immunoblotting clearly showed, that the FasL-protein expression in normal tissue is rare. Thus, the weak immunohistochemical staining of the normal mucosa was interpreted to be caused by unspecific binding of the antibody to the mucus.

3.5.5. Comparison of Results Obtained by All Methods

The results for the expression of FasL obtained by the four methods are summarized in **Table 1**. The detection of protein expression was more frequent with immunoblotting than with immunohistochemistry, and RT-PCR was more sensitive than *in situ* hybridization. The two more sensitive methods (immunoblotting and RT-PCR) yielded approx the same percentage of FasL-positivity in all types of specimens—namely, 30–50% in the normal mucosa, 30% in

Table 1
Frequency of FasL Detection in Normal Colonic Mucosa (N), Primary Tumors (T), and Liver Metastases (M) by Four Different Methods

Detection method	N	T	M
Immunohistochemistry	0/5	2/5	3/5
Immunoblotting	2/6	2/6	6/6
in situ hybridization	1/12	4/12	6/12
RT-PCR	5/10	3/10	9/10

primary tumors, and 90–100% in the metastases. Between the four methods used for analysis of FasL mRNA and protein, the percentage of FasL+ samples varied in primary tumors from 30–40%, and in metastases from 50–100%. However, all procedures yielded a higher frequency of FasL detection in the metastases than in the primary tumors. By contrast, the frequency of CD25 detection was similar in the three tissues, i.e., these cells contributed similarly in all three tissues to the overall expression of FasL.In the present work, no strict quantification of the amount of FasL protein or mRNA was carried out: the frequency of detection was used instead, in which expression was considered significant if it surpassed the detection level of each procedure. In summary, these results indicated that FasL is more frequently expressed at both the mRNA and protein levels in liver metastases than in their primary colorectal carcinomas.

3.6. Summary

Quantitative gene-expression analyses should be performed by Northern-blot analysis, when sufficient amounts of cells or tissue samples are available. Because of the scarcicity of tumor material, RT-PCR is frequently an alternative solution. However, there are limitations in RT-PCR analysis as a method for quantifying gene expression. The most promising procedure is real-time RT-PCR. It is, however, possible to obtain semi-quantitative information with conventional RT-PCR, without a requirement for expensive equipment, provided that certain considerations are taken into account (*see* **Notes 6** and **7**). Although standard RT-PCR has limitations as a method for quantification of gene expression, by combining data from additional techniques such as Western blot, *in situ* hybridization or immunohistochemistry, a reliable picture of the level of specific gene expression can be obtained (*see* **Note 8**).

4. Notes

1. RNAClean gives about fivefold higher RNA yields than the CsCl isolation method.

2. The amount of total RNA that can be isolated from the different tissues varies remarkably. From 200 mg of normal colonic mucosa, approx 100–250 μg of total RNA can be extracted with the RNAClean protocol. From neoplastic tissue (primary carcinoma and liver metastases), usually 4–5 times more total RNA can be isolated, i.e., 500–1.000 μg from 200 mg of tissue. For this reason, it is recommended to harvest sufficient amounts of normal colonic mucosa for the RNA isolation (0.5 to 1 g). Much smaller tissue specimens from the primary tumor and the metastases are enough to obtain sufficient amounts for the further analyses (100–200 mg resulting in 250–500 μg of total RNA, which is sufficient for cDNA synthesis, several RT-PCR analyses, and Northern-blot analysis).
3. When minute amounts of a specific mRNA need to be precisely quantitated, real-time RT-PCR is currently one of the methods of choice. However, the equiptment is currently expensive and not widely available in research laboratories. This method is discussed in detail elsewhere in the present volume.
4. Competitive PCR, discussed in detail elsewhere in the present volume, allows quantification of a specific cDNA. In this technique, an artifical template that has the same primer sites but yields a shorter PCR product than the analyzed transcript is added to the PCR. Dilutions of known concentration of this internal competitive DNA template are co-amplified together with a fixed aliquot of the total cDNA sample. The dilution of the competitive standard yielding a PCR product band with the same intensity as the band for the specific target cDNA indicates the amount of the investigated transcript. Neither this procedure, nor real-time PCR, however, circumvent the inherent inconsistency of the RT reaction.
5. To control for the variability of the reverse transcription reaction, competitive RT-PCR can be performed, where a competitive standard in RNA rather than DNA form is employed. The RNA standard is introduced prior to the RT step, and therefore competes with the target mRNA throughout the RT-PCR process. Various techniques can be employed to construct an RNA standard, some involving modification of the target cDNA after cloning into a plasmid containing the T7 RNA polymerase promoter. An RNA copy of the modified cDNA insert can then be synthesized by in vitro transcription from the plasmid using T7 RNA polymerase. Kits that permit the construction of competitor RNA are available (e.g., from Ambion) and the results can be evaluated with standard curve methodology ***(6)***.
6. When obtaining semi-quantitative information from RT-PCR, it is important to exclude genomic contamination, since genomic DNA will compete with the cDNA during PCR amplification. Also, equalize the amount of RNA template used for each RT-PCR, and determinate the overall amplification efficiency in each sample by performing RT-PCR for a housekeeping gene. For each PCR, determinate the range of cycle numbers in which the amount of PCR product increases linearly with increasing cycle number. This range will be influenced by the abundance of the analyzed mRNA transcript in the template: in our samples, for the housekeeping gene PDH, we found the linear range was between 25 and 35 cycles; for CD25, between 30 and 40 cycles, and for the weakly expressed transcript of FasL, it was between 35 and 45 cycles. The RT-PCR analyses for

each transcript must be performed in the middle of these determined ranges in order to obtain semi-quantitative information on gene expression (PDH: 30 cycles, CD25: 35 cycles, and FasL: 40 cycles).

7. Some investigators employ semi-quantitative evaluation of PCR band intensities in an effort to quantify gene expression. In view of the low quantitative precision of RT-PCR, the evaluation of signals via densitometry after normalization to a housekeeping gene is not recommended. This type of "quantification" may produce a false sense of accuracy, and does not allow for the nonlinear kinetics of PCR, the variation in amplification efficiency for different target templates, and the variation in expression of some housekeeping genes. In the present protocol, semi-quantitative assessment was used instead.
8. Because of the limitations of conventional RT-PCR as a quantitative technique, it is recommended that RT-PCR analysis is combined with other detection methods. The combination of integral analyses such as RT-PCR or Western blot with *in situ* detection methods such as *in situ* hybridization or immunohistochemistry is particularly helpful to obtain approximate information about the level of expression in vivo. *in situ* techniques also provide accurate localization of expression in specific cell types. These methods may show considerable differences in the percentage of positive samples or the expression level as a result of different sensitivities, accessibility of the detected antigen, and variations in stability between the specific protein and mRNA. However, if several methods demonstrate a more frequent detection of one gene or its product in one particular tissue type as in the present example, this collective result gives indirect but reliable information about the level of expression of this gene in the tissue.

References

1. O'Connell, J., O'Sullivan, G.C., Collins, J.K., and Shanahan, F. (1996) The Fas counterattack: Fas-mediated T-cell killing by colon cancer cells expressing Fas ligand. *J. Exp. Med.* **184,** 1075–1082.
2. Hanski, C., Klußmann, E., Wang, J., Böhm, C., Ogorek, D., Hanski, M.L., et al. (1996) Fucosyltransferase III and sialyl-Lex expression correlate in cultured colon carcinoma cells but not in colon carcinoma tissue. *Glycoconjugate J.* **13,** 727–733.
3. Sambrook, J., Fritsch, E.F., and Maniatis, T., eds. (1989) Molecular cloning, a laboratory manual. Laboratory Press, Cold Spring Harbor, NY
4. Rolfs, A., Schuller, I., Finckh, U., and Weber-Rolfs, I. (1992) *PCR: Clinical diagnostics and research.* Springer-Verlag, Berlin.
5. Foss, H.D., Demel, G., Anagnostopoulos, I., Aranjo, I., Hummel, M., and Stein, H. (1997) Uniform expression of cytotoxic molecules in anaplastic large cell lymphoma of Null/T-cell phenotype and in cell lines derived from anaplastic large cell lymphomas. *Pathobiology* **65,** 83–90.
6. Tsai, S.J. and Wiltbank M.C. (1996) Quantification of mRNA using competitive RT-PCR with standard-curve methodology. *BioTechniques* **21,** 862–866.
7. Mann, B., Gratchev, A., Bohm, G., et. al. (1999) FasL is more frequently expressed in liver metastases of colorectal cancer than in matched primary carcinomas. *Br. J. Cancer.* **79,** 1262–1269.

VII

Genetic Analysis

20

Detection of Single Nucleotide Polymorphisms Using a Non-Isotopic RNase Cleavage Assay

Frank Waldron-Lynch, Claire Adams, Michael G. Molloy, and Fergal O'Gara

1. Introduction

The common cardiovascular, respiratory, gastrointestinal tract, renal, and immune diseases share common etiologies, because they are multifactorial in origin, with both genetic and environmental components contributing to their clinical phenotype. Recently, the necessity for the identification and analysis of the inheritance of single-nucleotide polymorphisms (SNPs) in these complex diseases has become apparent. First, to identify genes contributing to susceptibility to these diseases, and second, to determine whether some regulatory SNPs may influence the severity or course of a disease.

Currently, two main experimental strategies exist to identify the genes involved in susceptibility to complex diseases. The existence of microsatellite marker maps, which span the human genome, and automated sequencing techniques, has made it possible to perform a total genome scan for susceptibility loci. Usually, families with one or more affected sibling are studied. Chromosomal regions potentially involved in susceptibility are identified by their preferential transmission to affected offspring. Although a total genome search identifies the chromosomal area potentially involved in disease susceptibility, the actual resolution of the mapping may be quite low (3–20 cM). Therefore, susceptibility genes are mapped to regions of between 5 and 1000 kb by linkage disequilibrium mapping, in which certain alleles, usually a microsatellite, are associated with the disease. When linkage disequilibrium has been ascertained for a marker locus, the area surrounding the marker must be fine-mapped to the

From: *Methods in Molecular Biology, vol. 193: RT-PCR Protocols*
Edited by: J. O'Connell © Humana Press Inc., Totowa, NJ

region in strongest disequilibrium with the disease. This is achieved by the characterization of SNPs, which occur approx every 1–2 kb in the genome *(1)*.

The second—and until recently—the most common technique for the genetic analysis of mutifactorial diseases is the candidate gene approach. The candidate gene technique involves the selection of genes for genetic analysis, whose products are believed to play a role in the pathogenesis of the disease. Essentially, the gene must be found to be in linkage disequilibrium with the disease by using SNPs or microsatellites. Importantly, some markers may be found in regulatory regions of a gene, thereby influencing gene transcription, and thus potentially influencing susceptibility to a disease.

The role of SNPs in complex disease has also been investigated by case-control disease-association studies. Interestingly, it has been found in rheumatoid arthritis (RA) that SNPs can correlate with clinical outcome, yet do not influence susceptibility to the disease. These findings will potentially have substantial implications for the management of the disease, allowing for the genetic definition and prediction of clinical phenotypes prior to their full expression *(2)*.

1.2. NIRCA™

An RNase cleavage assay can be performed as a simple and sensitive technique for the detection and analysis of SNPs. A non-isotopic RNase cleavage assay can be performed using a commercially available kit (NIRCA™; Ambion, Austin, TX). The method is robust and reliable, and requires a minimum of specialist equipment and training compared with other mutation detection methods. We have used the protocols in this chapter to investigate SNPs in the tumor necrosis factor-α (TNF-α) 3' untranslated region (TNF-α 3'UTR) in RA-affected individuals *(3)*. TNF-α is a pro-inflammatory cytokine implicated in the pathogenesis of RA and other chronic inflammatory conditions. The protocols described here are rapid, allowing the screening of a DNA target within two working days from collection of sample to detection of SNP.

Essentially, the NIRCA™ has seven steps: isolation of DNA, primary PCR, secondary PCR, RNA transcription, RNA hybridization, RNase cleavage, and agarose gel analysis (**Fig. 1**). A nested PCR strategy is used to amplify target DNA. The secondary PCR reaction uses primers that incorporate the T7 and SP6 RNA polymerase promoter sites. The PCR product from the second PCR is used as template for RNA polymerase to generate large quantities of sense and anti-sense single-stranded RNA. These individual aliquots of sense and anti-sense RNA are then hybridized to a complementary RNA probe that is known not to encode any SNPs. If any SNPs are present in the experimental sample, an RNA heteroduplex will form. The now double-stranded RNA is digested with RNase (specific for single-stranded RNA) leading to the cleav-

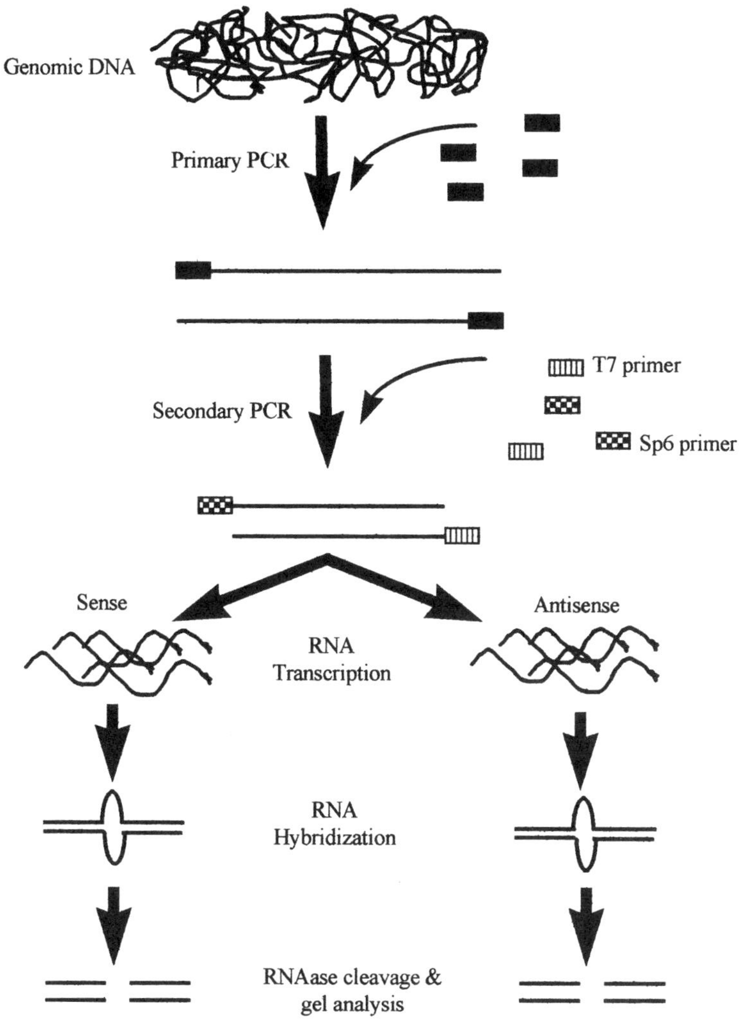

Fig. 1. Schematic diagram of NIRCA™.

age of any RNA hetroduplexes. The RNase cleavage products are then resolved by simple agarose gel electophoresis *(4)*.

Two methods are described here for the isolation of DNA. The buccal method is extremely useful, since it is rapid, noninvasive, uses no hazardous reagents, and does not require specialized medical training. However, far greater amounts of DNA are obtained from whole blood, and this method provides the option of establishing cell lines, thus negating the need for rebleeding.

The nested PCR protocol has two main advantages. Firstly, it increases the accuracy of the NIRCA™. Secondly, the primary PCR product provides a DNA sequencing template to confirm the identification of an SNP if one is detected. Several minor modifications to the manufacturer's protocol have been made—the most important to the format of the NIRCA™. The modified format of the NIRCA™ described in **Subheading 3.6.** was found to minimize the likelihood of identification errors occurring during the assay, thereby allowing the screening of up to twenty individual samples in an assay. In addition, the agarose gel electrophoresis is simplified to a standard protocol.

2. Materials

2.1. General

1. Sterile deionized water and mineral oil.
2. Sterile Eppendorf tubes; 0.5-mL PCR reaction tubes; 0.5-mL, 100-mL, and 500-mL pipets and tips; 10-mL pipets.
3. Microfuge and tabletop centrifuge.
4. Ultraviolet (UV) spectrophotometer and quartz curvets.
5. Perkin-Elmer Cetus 9200 termocycler (Perkin-Elmer, Foster City, CA.).

2.2. Blood Collection

1. 10-mL or 50-mL sterile syringes; butterfly needles; 10-mL or 5- mL sterile polypropylene tubes.
2. Preservative-free lithium heparin (Welcome, UK).

2.3. Buccal Cell Collection

1. Sterile 0.9% saline.
2. 50-mL polypropylene tubes.

2.4. DNA Extraction from Blood

1. Nucleon II DNA extraction kit (Scotlab, Scotland).
2. 50-mL and 5-mL sterile polypropylene tubes.
3. Two sterile 1l flasks and one sterile 0.1l flask.
4. Cell-preparation solution containing 10 m*M* Tris-HCl, pH 7.6, 320 m*M* Sucrose, 5 m*M* $MgCl_2$ and 1% Triton X-100.
5. Cell lysis solution containing 400 m*M* Tris-HCl, pH 7.6, 60 m*M* ethylenediaminetetraacetic acid (EDTA), 150 m*M* NaCl, and 1%sodium dodecyl sulfate (SDS) solution.
6. 2 *M* NaOH.
7. Chloroform at −20°C.
8. Pure ethanol at −20°C.
9. 70% ethanol at room temperature.
10. Shaking water bath.

2.5. DNA Extraction from Buccal Cells

1. 5% Chelax solution (Biorad).

2.6. Primary PCR Amplification

1. 100 ng/µl stock of purified genomic DNA (*see* **Note 1**).
2. The sequences of the primers used to amplify the TNF 3' untranslated region (UTR) are as follows: T11B: 5'-GCCGAGTCTGGGGAGGTCTAC-3' and T16: 5'-AGCCTGTGGAGACAGGACTTA-3'. The T11B primer is biotinylated. Dilute oligonucleotide primers (± biotinylation) in sterile deionized water to 10 m*M* (*see* **Note 2**).
3. 100-m*M* solutions of dATP, dTTP, dCTP, and dGTP (Promega, Madison, WI)
4. 400-µM working stock solutions of dATP, dTTP, DCTP, and dGTP, diluted in sterile deionized water and stored at −20°C.
5. Taq DNA polymerase 5 U/µl; 10X Taq DNA polymerase buffer containing 500 m*M* KCl, 100 m*M* Tris-HCl, pH 7.6, 1% Triton-X; 25 m*M* $MgCl_2$ solution (Promega).
6. 100 ng/µL positive control DNA.

2.7. PCR Purification

2.7.1. Using Dynabeads

1. 1 Sterile 1l flasks.
2. Streptavidin-coated Dynabeads (Dynal Ltd., Wirral).
3. Binding and washing buffer: 10 m*M* Tris-HCl, pH 7.5, 1 m*M* EDTA, 2 *M* NaCl.
4. Magnetic separator.

2.7.2. Using QIAquick Kit

1. QIAquick PCR purification kit (Qiagen Ltd., Dorking).
2. TE buffer: 10 m*M* Tris-HCl, pH 7.5, 1 m*M* EDTA.

2.8. Secondary PCR

1. 2 µL of purified primary PCR product.
2. The sequences of the nested primers used for the secondary PCR amplification of the TNF 3' UTR are: sense primer MD1(T7): 5'-GATCTAATCGACTCAT ATGGGGCCTACTGATTG-3'; anti-sense primer MD2 (SP6): 5'-GAT CATTTAGGTGACACTATAGTAGGATGTGGAGAGAGGACTTA-3'.
   The T7 and SP6 RNA polymerase promoter sequences are in bold, and additional consensus sequences for promoter function are underlined. The specific TNF 3'UTR sequences are in normal text. Dilute the primers in sterile deionized water to 5 m*M* (*see* **Note 2**).
3. Taq DNA polymerase, Taq DNA polymerase 10X buffer, MgCl2 stock solution, deoxynucleotide 5' triphosphate (dNTP) stock solutions, and positive control DNA are the same as for the primary PCR.

2.9. NIRCA™

1. MisMatch Detect™ NIRCA kit (>120 reactions) (Ambion, Austin, TX).
2. DNA from a normal individual (no SNP in the TNF 3'UTR) and an SNP-positive individual to generate positive and negative control RNA transcripts.

2.10. Agarose Gel Electrophoresis

1. Ultra-Pure agarose (Gibco-BRL).
2. 10X TBE: 114 g of Tris base, 55 g of Borate, and 10.3 g of EDTA per L.
3. Mismatch gel-loading solution (contains ethidium bromide).
4. Horizontal minigel running apparatus (Biorad).
5. Power pack.
6. UV transilluminator.

3. Methods

3.1. Specimen Collection

3.1.1. Blood

1. Wash syringes with preservative-free heparin (*see* **Note 3**).
2. Standard aseptic venesection technique is used for blood sample collection. Butterfly needles are recommended if 50-mL vol of blood or greater are required because they were found to cause the least discomfort.
3. Blood samples are transferred to 10-mL or 50-mL polypropylene tubes and stored at 4°C.

3.1.2. Buccal Cells

1. 13-mL of 0.9% sterile saline is aliquoted into 50-mL polypropylene tubes. 0.9% sterile saline may be conveniently obtained from intravenous (iv) bags found in all hospitals.
2. Pour saline into mouth and swish vigorously for 30 s. Spit back into the 50-mL polypropylene tube.
3. Transfer immediately to a 15-mL polypropylene tube and store at 4°C.

3.2. DNA Extraction

3.2.1. From Blood

1. Extract DNA from whole blood using the Nucleon II DNA extraction kit according to the manufacturer's protocol with the following modification: for the cell lysis buffer, make up 1% SDS and autoclave separately. Then add to an autoclaved solution of: 400 m*M* Tris-HCl, pH 7.6, 60 m*M* EDTA, and 150 m*M* NaCl to complete the lysis buffer.
2. Mix 10 mL of blood with 40 mL of cell preparation solution, and pellet the cells by centrifugation. Great care must be taken not to disturb the cell pellet when

removing the supernatant. Lyse the cells in lysis buffer.

3. Precipitate the DNA by addition of 2 vol of pure ethanol at −20°C. The genomic DNA precipitate may be removed conveniently by spooling onto a 1-mL plastic pipet tip. Wash the precipitate twice with 70% ethanol, and allow to dry by evaporation. Resuspend the DNA in 100–250 µL of sterile deionized water for storage at −20°C.
4. Measure the DNA concentration with a UV spectrophotometer at A_{260nm} using a quartz curvet.

3.2.2. From Buccal Cells

1. Spin the 15-mL polypropylene tube containing the specimen at 1000*g* for 15 min. Remove the supernatant, taking care not to dislodge the pellet, and place the tube on ice.
2. Add 50 µL of 5% Chelax solution (shake well before use). Resuspend the pellet by pipetting up and down with a 1-mL pipet tip.
3. Aliquot the resuspended solution into an Eppendorf tube and place in a boiling water bath for 10 min.
4. Place the tubes on ice for 2 min and then centrifuge at 13,000*g* for 30 s to pellet the Chelax. Transfer 400 µL of the DNA-containing supernatant to a new Eppendorf tube for storage at −20°C.
5. Measure the DNA concentration spectrophotometrically.

3.3. Primary PCR Amplification

1. Aliquot 500 ng of purified DNA into a PCR tube and place on ice.
2. Prior to making the PCR master mix, dilute an aliquot of the 25 m*M* $MgCl_2$ stock solution to 15 m*M* and the Taq DNA polymerase from 5 U/µL to 1 U/µL using sterile deionized water.
3. Prepare a PCR master mix for the requisite number of samples. The master mix per PCR reaction is as follows:

Stocks	Volume
Genomic DNA 100 ng/µL	5 µL
T11B primer 10 m*M*	1 µL
T16 primer 10 m*M*	1 µL
dATP 400 µ*M*	2.5 µL
dCTP 400 µ*M*	2.5 µL
dGTP 400 µ*M*	2.5 µL
dTTP 400 µ*M*	2.5 µL
$MgCl_2$ 15 µ*M*	5 µL
dH_2O	20.5 µL
10X Taq DNA polymerase buffer	5 µL
Taq DNA polymerase 1 U/µL	2.5 µL
Total Volume	**50 mL**

Sufficient master mix should be made to amplify the positive control DNA and the negative control (water instead of template). Hold all reagents on ice while making the master mix.

4. Cover the PCR reactions with 20 µL of mineral oil. Firmly cap all tubes and place them in the thermocycler. The following cycle is used for amplification of the TNF 3'UTR: an initial denaturation step of 94°C for 5 min, then 30 cycles of: 94°C for 1 min, 55°C for 1 min, and 72°C for 1 min; a final extension at 72°C for 5 min. Samples should be held at 4°C or stored at −20°C.

3.4. PCR Purification

1. Use both the Dynabead and QIAquick purification kits as per the manufacturers' protocol, with the following modifications:
2. Use 80 µL of Dynabeads per sample; process the sample through all the washing steps to remove the unused PCR reaction components; resuspened in 20 µL of sterile deionized water.
3. With the QIAquick method, the purified PCR products are resuspended in 30 µL of sterile deionized water (*see* **Note 5**).

3.5. Secondary PCR

1. Aliquot 2 µL of each primary PCR into a new PCR tube and place on ice.
2. Prepare the master mix for the secondary PCR; this is the same as for the primary PCR except that 23.5 µL of sterile deionized water is added per reaction to adjust for the different volume of template DNA (2 µL as opposed to 5 µL for the primary PCR). Also, the internal (nested) primers MD1(T7) and MD2 (SP6) will be used for the secondary PCR. Make up sufficient master mix to amplify the controls, and a DNA template that will produce an RNA complementary probe for the NIRCA™.
3. The PCR conditions are the same as for the primary PCR.
4. Completed PCRs may be stored at −20°C for several months and will withstand several freeze-thaw cycles.

3.6. NIRCA™

1. The following protocol is to screen a target region of DNA from one individual's sample (S1) to a complementary RNA probe from another individual known not to carry any SNPs in the region (N) and a mismatch control. This protocol may be scaled up if more samples are to be screened. It is important to keep careful note of all samples during the experiment to ensure that accurate results are obtained. A suggested format is provided below.
2. Aliquot DNA from the secondary PCR (taken from beneath the mineral oil) into four PCR tubes, and label as follows:

S1 (T7)	S1 (SP6)	N (T7)	N (SP6)
2 µL	2 µL	4 µL	4 µL

   Place the tubes on ice.

3. Prepare the master mixes to screen a single sample with a complementary RNA probe and mismatch control, as follows:

T7 Master Mix		**SP6 Master Mix**	
Stocks	Volume	Stocks	Volume
2.5 m*M* dNTP mix	6 µL	2.5 m*M* dNTP mix	3 µL
dH_2O	12 µL	dH_2O	12 µL
Transcription buffer	3 µL	Transcription buffer	3 µL
T7 RNA pol.	3 µL	SP6 RNA pol.	3 µL
Total volume	**24 mL**	**Total volume**	**24 mL**

The master mixes are most conveniently made up in sterile PCR tubes.

4. RNA transcription reaction: add master mix to each sample tube containing template DNA as follows:

T7 transcripts		**SP6 transcripts**	
Sample	T7 master mix	Sample	SP6 master mix
S1(T7)	8 µL	S1(Sp6)	8 µL
N(T7)	16 µL	N(SP6)	16 µL

Gently mix and spin. Incubate for 1 hour at 37°C in a thermocycler. Add equal volumes (10 µL or 20 µL) of mismatch hybridization solution to each tube, gently mix, and incubate at 95°C for 3 min in the thermocycler.

5. Set up the hybridization reactions to form RNA duplexes. Into the T7 (sense) and SP6 (anti-sense) single-stranded RNA transcripts generated for sample S1, add the complementary single-stranded RNA probe generated from the control sample N as follows:

T7 + **SP6** (sense + anti-sense)
10 µL of S1(T7) + 20 µL of N(SP6)

SP6 + **T7** (anti-sense + sense)
10 µL of S1(SP6) + 20 µL of N(T7)

Mismatch control
10 µL of N(T7) + 10 µL of N(SP6)

The final volume of the experimental sample is 40 µL. The mismatch control is constructed by hybridization of the sense and anti-sense RNA strands of the N sample, which is known not to have any SNPs in this region. Gently mix the samples and incubate at 95°C for 3 min in the thermocycler to denature the single-stranded RNA. Remove the tubes and allow to cool to room temperature, during which hybridization of the complementary strands takes place. The hybridized samples may be stored at −20°C for several months and can withstand several freeze-thaw cycles.

6. Subject the hybridizations to RNase digestion in sterile PCR tubes using the RNase solutions provided in the NIRCA™ kit. Incubate the RNase digests at 37°C for 45 min, and terminate the reactions by adding 4.5µL of the NIRCA™ mismatch-detect gel-loading buffer (containing ethidium bromide).

3.7. Agarose Gel Electrophoresis

1. Prepare a 2% Ultra Pure agarose gel in 100 mL of 1X Tris-boric acid-EDTA (TBE). Heat the mixture until the agarose dissolves and then cool and cast the

gel, ensuring that the wells will hold a 20-µL vol. The gel will set in 1–2 hours depending on room temperature.

2. Fill the gel electrophoresis rig with 1X TBE until the gel is fully immersed. Pipet 20 µL of each sample into a well, ensuring to change the pipet tip between each sample. Run a complementary mismatch control with each set of samples.
3. Run the gel on constant voltage at 70 V until the dye front has reached the end of the gel. Visualize and document the cleavage products using an UV transilluminator.

4. Notes

1. The following formulas are useful for the calculation of the amount of DNA in the initial samples and the primer solutions. Spectrophotometrically, an absorbance of 1.0 at 260 nm is approx equivalent to 50 µg/mL of DNA. Once the DNA concentration is known, use the following formula to determine how to dilute the stock to obtain the wanted DNA concentration: known stock DNA concentration X volume required (X µL) = wanted DNA concentration X final wanted volume (µL); solve for X (the volume of the stock DNA solution required to make the wanted DNA solution).
2. Primer design is an important factor in developing a reliable NIRCA™. The Oligo 5.0 primer design program (National Bioscience, Inc.) was used to design primers. The primer pair used for the primary PCR must be able to reliably and accurately amplify large quantities of the target sequence from complex mixtures. Ideally, the primer pair used for the secondary PCR should be completely nested within the primary-amplified DNA. However, this may not always be possible, depending on the sequence of the target DNA. Alternatively, a secondary PCR primer can be designed that is partially nested, and partly overlaps the corresponding primary PCR primer site (e.g., the MD2 primer used here).
3. The use of preservative-free heparin enables the isolated peripheral-blood lymphocytes (PBL) to remain viable for transformation in order to establish stable cell lines, if this is required.
4. All procedures in these protocols should be carried out using standard aspectic techniques to prevent DNase and RNase contamination. In particular, contamination with RNase will adversely affect the transcription reaction and therefore the likelihood of obtaining accurate results.
5. DNA sequencing is the gold standard for the identification of SNPs. After subsequent purification, the primary PCR provides DNA of suitable quality for sequencing. Biotinylated primers are required only if manual sequencing is being used. Otherwise, the primary PCR products may be sequenced directly—e.g., using an ABI Prism dye terminator sequencing kit and an ABI Prism 310 genetic analyzer (both Perkin-Elmer, Warrington).

Acknowledgments

The authors wish to thank P. Higgins for his expert technical assistance. The work was supported by a grant from the Health Research Board of Ireland (to Fergal O'Gara and Michael G. Molloy).

References

1. Vyse, T.J., and Todd, J.A. (1996) Genetic analysis of autoimmune disease. *Cell* **85,** 311–318.
2. Ruuls, S.R., and Sedgwick, J.D. (1999) Unlinking tumor necrosis factor biology from the major histocompatibility complex: lessons from human genetics and animal models. *Am. J. Hum. Genet.* **65,** 294–301.
3. Waldron-Lynch, F., Adams, C., Shanahan, F., Molloy, M.G., and O'Gara, F. (1999) Genetic analysis of the 3' untranslated region of the tumor necrosis factor shows a highly conserved region in rheumatoid arthritis affected and unaffected subjects. *J. Med. Genet.* **36,** 214–216.
4. Golderwick, M.M., Kimball, G.R., Lin, Q., Martin, S. S., and Tseng, J.Y. (1996) NIRCA: A rapid robust method for screening for unknown point mutations. *Biotechiques.* **21,** 106–112.

VIII

RT-PCR in Immunology

21

Detection of Clonally Expanded T-Cells by RT-PCR-SSCP and Nucleotide Sequencing of T-Cell Receptor β-CDR3 Regions

Manae Suzuki Kurokawa, Kusuki Nishioka, and Tomohiro Kato

1. Introduction

T-cell clonal expansion has been detected in the peripheral blood or the disease-affected sites in patients who suffer from various disease states such as infections, autoimmune diseases, malignancy, and post-transplantation complications *(1–7)*. Since T-cells begin to proliferate clonally when they recognize their specific antigens, the clonal T-cell expansion is considered to be a result of an antigen-specific immune response. For analysis of such antigen-specific T-cells, it is common to use their specific antigens if they are known. However, there are many diseases, such as rheumatoid arthritis, in which the pathogenic antigens are unknown. In these circumstances, the detection of clonally expanded T-cells is an effective method to evaluate whether antigen-specific immune responses are involved, since few clonally expanded T-cells are detected in healthy individuals *(4)*. In addition, the characterization of any clonally expanded T-cells that are detected would further promote the understanding of the disease mechanisms.

T-cells recognize their specific antigens through their T-cell receptors (TCRs). The TCR is a heterodimer which consists of either α and β chains (αβ TCR) or γ and δ chains (γδTCR). Approximately 90% of T-cells express αβ TCR in humans, which recognizes processed antigenic peptides within the major histocompatibility complex (MHC). The remaining 10% of T-cells express γδ TCR, some of which have been reported to recognize a whole antigenic molecule directly without the involvement of the MHC. The α and γ chains consist of variable (V), joining (J), and constant (C) regions, whereas the β and δ chains consist of V, diverse (D), J, and C regions. Multiple genes

From: *Methods in Molecular Biology, vol. 193: RT-PCR Protocols*
Edited by: J. O'Connell © Humana Press Inc., Totowa, NJ

encode the respective regions. For example, in the case of the human TCR β chain, approx 150 V genes, two D genes, 13 J genes, and two C genes exist on the chromosome (**Fig. 1A**). Mature TCR genes are constructed by selecting and joining one gene segment from each of the V, D, J, and C genes (rearrangement of TCR genes, **Fig. 1B**). Thus, the TCR repertoire is highly diverse because of this rearrangement. Further, some nucleotides are either added or deleted randomly at the V-D and D-J junctions, resulting in a further increase of the diversity of the TCR repertoire. Together, these rearrangements and alterations result in a TCR diversity that is estimated to be greater than 10^9. Furthermore, since the TCR β chains are expressed in an allelic exclusion manner, only one type of TCR β chain is expressed on a single T-cell. By this mechanism, the nucleotide sequences of the junctional regions of V, D, and J, the so called complementarity-determining region 3 (CDR3), differ from T-cell to T-cell. From these observations, the T-cells that carry the same TCR nucleotide sequences are believed to have been derived from a single T-cell. Thus, high expression of a particular TCR-mRNA among bulk T-cell-derived mRNAs may reflect the clonal expansion of a T-cell that carries the particular TCR. For detection of this highly expressed TCR-mRNA, several methods have been established. Representative methods include the combination of reverse-transcriptase-polymerase chain reaction (RT-PCR) and either nucleotide sequencing ***(8–12)***, separation by the single-strand conformation polymorphism (SSCP) ***(4, 5, 7, 13–18)***, or CDR3-length analysis ***(19, 20)***.

In this chapter, we introduce our protocol for the detection of clonally expanding T-cells by RT-PCR of TCR β genes, and subsequent SSCP analysis or nucleotide sequencing. This method specifically amplifies genes that encode TCR β CDR3 regions from T-cell-derived cDNAs. In the SSCP analysis, the TCR β genes amplified by familial PCR are denatured, and electrophoresed through a non-denaturing gel. The denatured (single-stranded) DNAs form unique single-strand conformations in non-denaturing conditions, depending on the sequences of the CDR3 regions. Electrophoresis results in single-stranded DNAs with the identical sequence migrating to the same position, whereas those with different CDR3 sequences migrate to different positions. Thus, highly expressed TCR β gene sequences produce sharp bands on the gel, whereas the remaining infrequent TCR β gene sequences produce a broad smear.

The SSCP analysis is a simple and effective method to detect clonally expanded T-cells. However, SSCP analysis is not quantitative. When nucleotide sequencing is used, the TCR β genes amplified by the familial PCR are randomly cloned into a plasmid vector. Afterwards, 30–40 of the cloned TCR genes are analyzed. The expanded T-cell clonotypes are detected as a multiple

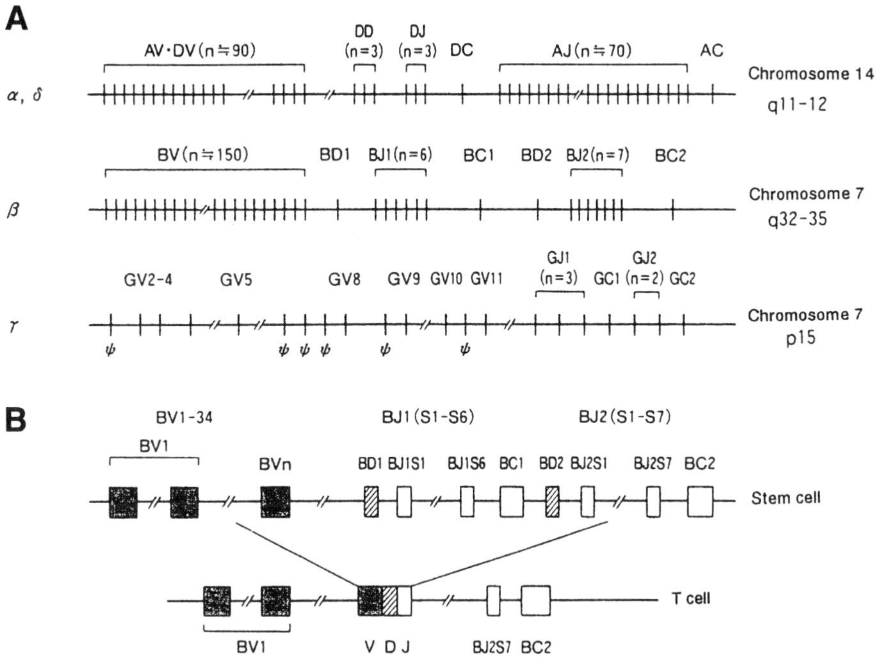

Fig. 1. Rearrangement of the TCR genes **(A)** An approximate location of TCR-encoding genes is shown. The δ gene-encoding region is included in the αgene-encoding region, so that the rearrangement of the α gene results in loss of the δ genes. The number of the genes is shown in parentheses, ψ represents pseudogenes. **(B)** The form of the rearrangement of a typical TCRβ gene is shown. Only one gene is selected from each of the V genes, from the two D genes, and from the 13 J genes, and connected together. Nucleotides are randomly added or deleted at this junctional region of V-D-J, so that the repertoire of TCR sequences becomes highly diverse (CDR3 region). Finally, one of two C genes is selected and connected to the end of the J gene to construct a mature TCR gene. Modified from **ref.** ***(26)***.

number of identical sequences. This procedure is laborious, but establishes the frequency of each clonally expanded T-cell.

The protocol is divided into four sections: total RNA extraction and cDNA synthesis, PCR-amplification of a particular TCR V β family, RT-PCR SSCP, and nucleotide sequencing. In addition, we introduce a method to amplify the TCRβCDR3-gene fragments from single-cell-derived cDNAs (single-cell PCR) ***(21)***, since this method can determine the frequency of each clonally expanded T-cell by cell count, and not by mRNA expression.

2. Materials

The materials described in this section are those that we use regularly. However, equivalent materials are available from various companies.

2.1. RNA Extraction and cDNA Synthesis

1. RNAzol™ B (Tel-Test, Inc., Friendswood, Texas), homogenizer, chloroform, isopropanol, a solution of 75% ethanol, diethyl pyrocarbonate (DEPC; Wako, Osaka, Japan).
2. Superscript II reverse transcriptase (RT) (Gibco-BRL, Gaithersburg, MD), supplied with 5X React Buffer and 0.1 *M* dithiothreitol (DTT); ribonuclease inhibitor (ToYoBo, Osaka, Japan); dinucleotide 5' triphosphate (dNTP), (dATP, dGTP, dCTP, dTTP; ToYoBo); and random hexanucleotide primers (Gibco-BRL).

2.2. PCR Amplification

1. Thermal Cycler (Takara, Otsu, Japan).
2. Taq DNA polymerase (Takara, supplied with 10x Taq buffer and dNTPs).
3. Primers for amplifying each TCR Vβ family *(22)*.

2.3. SSCP

1. Loading dye (10 mL of formamide, 10 mg of xylene cyanol, 10 mg of bromophenol blue, 200 μL of 0.5 *M* ethylenediaminetetraacetic acid [EDTA]).
2. Acrylamide; *bis*-acrylamide; TBE buffer (pH 8.3); glycerol; nitrocellulose membrane (Immobilon-S, Millipore Intertech, Bedford, MA); 20X SSPE; 50X Denhardt's solution; sodium dodecyl sulfate (SDS).
3. Biotinylated internal Cβ probe with the following sequence: 5'-AACAAGCGTG TTCCCACCCGAGGTCGCTGTGTT-3'.
4. Prepare a hybridizing solution: 2X SSPE, Denhardt's, 0.5% SDS.
5. Plex luminescence kit (Millipore Intertech).

2.4. Cloning andNucleotide Sequencing

1. PCR-Script™ Amp cloning kit (Stratagene, La, Jolla, CA).
2. Wizard plus SV Miniprep DNA purification system (Promega, Madison, WI).
3. Automated DNA sequencer: Applied Biosystems 377 (Perkin-Elmer/Applied Biosystems, Foster, CA); dye-primer DNA sequencing kit (Perkin-Elmer/Applied Biosystems).

2.5. Single-Cell PCR and Nucleotide Sequencing

1. Fluorescein-5-isothiocyanate (FITC)-conjugated antibody specific for each particular TCR Vβ family to be analyzed (Coulter, Hialeah, FL).
2. A cell sorter (ALTRA, Coulter).
3. Single-cell RNA extraction buffer: 40 m*M* Tris-HCl, pH 8.5, 60 m*M* KCl, 3 m*M*

$MgCl_2$, 10 m*M* DTT, 2 mg/mL Linear acrylamide, 0.5% Nonidet-P 40, and 0.1 U/mL ribonuclease inhibitor.
4. Single-cell RT buffer: 40 m*M* Tris-HCl, pH 8.5, 60 m*M* KCl, 3 m*M* $MgCl_2$, 10 m*M* DTT, 0.2 mg/mL bovine serum albumin (BSA), 7.5 m*M* dT15 primer, 0.3 m*M* dNTPs, and 4 U/mL reverse transcriptase.
5. Primers for (semi-)nested familial PCR: internal Cβ-antisense primers: Cβ1: 5'-GGGTGGGAACACCTTGTTCAGGT-3'; Cβ2: 5'-GGGTGGGAACACGTT TTTCAGGT-3'.

3. Methods

3.1. RNA Extraction and cDNA Synthesis

1. Extract total RNA from either cells or tissue samples. Lyse the tissue or cell samples in the RNAzol B lysis buffer, by homogenization if necessary, and purify the RNA according to the manufacturer's instructions.
2. To convert the extracted mRNA to cDNA, prepare the RT reaction as follows: 4 µL of 5X React Buffer, 1.5 µL of 100 pmol/µL random hexamer, 1 µL of 10 m*M* dNTPs (2.5 m*M* each), 2 µL of 0.1 *M* DTT, 1 µL of ribonuclease inhibitor, 1 µL of RT, and 9.5 µL of sample RNA solution (up to 20 µg). The total volume of the reaction is 20 µL. The use of ribonuclease inhibitor is essential to avoid degradation of the template RNAs.
3. Incubate the mixture at 42°C for 3 h.

3.2. PCR-Amplification of the Genes for the TCRβCDR3 Regions

Amplify the TCRβgenes of each Vβ family separately, using Vβ-specific primers and a common Cβ primer. The Vβ primer should amplify only one Vβ family, avoiding crossreaction with the other families. Efficient primers have been described by several groups ***(20,22,23)***. We regularly use the primers reported by Choi et al. ***(22)***, and perform the PCR-amplification as follows.

1. Adjust the concentration of the template cDNA to 0.05–0.1 µg/µL. Prepare the PCR reaction as follows: 2.5 µL of 10X Taq buffer, 2.0 µL of dNTPs (2.5 m*M* each), 1.0 µL of 50 pmol/µL Vβ-specific primer, 1.0 µL of 50 pmol/µL Cβ-specific primer, 0.2 µL of 5 U/µL Taq DNA polymerase, 1.0 µL of 0.05–0.1 µg/µL template cDNA, and 17.3 µL of sterile bi-distilled water. The final volume of the reaction is 25 µL. Overlay the reaction mix with a small amount of mineral oil to avoid evaporation.
2. Place the tube into a thermal cycler and incubate at 94°C for 1 min, followed by 35 cycles of: 94°C for 1 min, 58°C for 2 min, and 72°C for 3 min. A final incubation is performed at 72°C for 5 min.
3. Electrophorese approx 5 µL (1/5) of the PCR product through a 1.5% agarose gel to check the amplification. The TCR PCR product appears as a band of approx 200 bp (**Fig. 2**) (*see* **Note 1**).

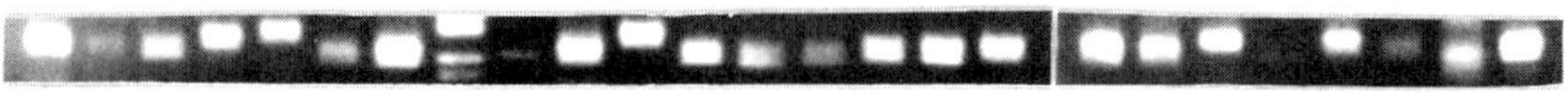

Fig. 2. PCR-amplification of TCRβCDR3 genes. A representative result of PCR-amplified TCRβgenes is shown. The CDR3 regions of TCRβgenes encoding each of Vβ 1–20 are amplified from peripheral blood by familial PCR. 22 Vβ family or sub-family genes are amplified in total (Vβ1, 2, 3, 4, 5.1, 5.2, 6, 7, 8, 9, 10, 11, 12, 13.1, 13.2, 14, 15, 16, 17, 18, 19, and 20). The size of the PCR products amplified by the chosen primers is approx 200 bp. M1, M2:DNA size markers. Size of the appearing band of M2 is 221 bp.

3.3. SSCP

The PCR products, obtained from the reactions described above, are next subjected to non-denaturing polyacrylamide gel electrophoresis (PAGE). Since the concentrations of the PCR products are different between the Vβ families, and between different experiments, the amount to be loaded onto the gel should be adjusted by diluting the PCR products in the SSCP loading dye.

1. Denature the diluted samples at 90°C for 2 min, and then maintain at room temperature.
2. Apply 3 µL of the denatured sample to a lane of a 4% polyacrylamide gel containing 10% glycerol.
3. Electrophorese the samples at 25°C at a voltage and for a time depending on the gel unit being used.
4. Blot the resolved PCR products onto a nitrocellulose membrane. Crosslink the DNA to the membrane by UV irradiation for 30 s.
5. Place the membrane in a plastic bag, and wash with 20 mL of hybridizing solution at 42°C for 2 h.
6. Incubate overnight at 42°C in 10 mL of hybridizing solution containing 1 pmol/µL biotinylated internal Cβ probe.
7. Wash with 200 mL of 0.2X SSPE/0.5% SDS solution at 55°C for 10 min.
8. Detect the bound Cβ probe using the Plex luminescence kit. Representative results of RT-PCR-SSCP are shown in **Fig. 3** (*see* **Note 2**).

3.4. Nucleotide Sequencing

The PCR product of one Vβ family contains various nucleotide sequences because of the diversity of the CDR3 region. This diversity makes direct sequencing extremely difficult. Thus, for the nucleotide sequencing, we choose to subclone the PCR product into a plasmid vector and prepare sequencing template corresponding to a single nucleotide sequence.

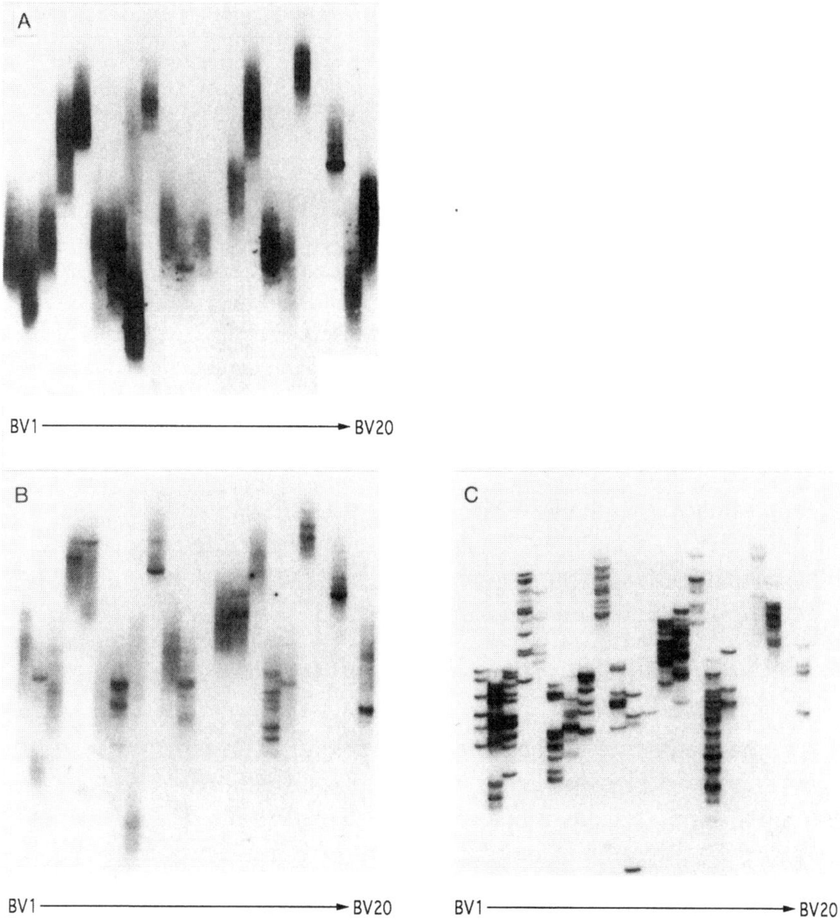

Fig. 3. T-cell clonality analysis by RT-PCR-SSCP. Representative SSCP results of T-cell clonality analysis are shown. TCRβCDR3 genes are amplified by familial PCR, and separated according to their SSCs. 22 Vβ family or subfamily genes are analyzed in total (Vβ1, 2, 3, 4, 5.1, 5.2, 6, 7, 8, 9, 10, 11, 12, 13.1, 13.2, 14, 15, 16, 17, 18, 19, and 20). (**A**) Peripheral blood obtained from a healthy subject produces a smear-like result, indicating an extremely heterogeneous T-cell clonality. (**B**) Peripheral blood of an RA patient shows several sharp bands, indicating the presence of distinct clonally expanded T-cells. (**C**) Synovial fluid of an RA patient shows a large number of sharp bands, indicating the oligoclonal T-cell expansion at the site of the disease.

1. Prior to the subcloning, purify the TCR-PCR product to remove excess primers and by-products. For this purpose, electrophorese the PCR product on a 1.5% agarose gel, and excise and purify the specific band.

2. Following purification, subclone the PCR product using a PCR-Script™ Amp cloning kit. Polish the insert gene and ligate to the PCR-Script plasmid vector.
3. Transform the plasmid carrying the PCR product insert into an *E. coli* host. The polishing, ligation, and transformation can all be performed using the kit.
4. After the transformation, select 30–40 colonies randomly, and culture separately overnight at 37°C in 1 mL of Luna Bertani (LB) medium containing ampicillin.
5. Purify the insert-containing plasmids using the Wizard-plus SV Miniprep DNA purification system.
6. Determine the nucloetide sequences of the cloned TCR fragments. Because the PCR-Script contains M13–20 primer sites, the M13 primers can serve as the primers for di-deoxy sequencing of the cloned insert. We use a 377 DNA sequencer and a sequencing kit that utilizes M13 primers. The nucleotide sequences of the prepared templates are analyzed according to the manufacturer's protocol for the sequencing system. The resultant sequences are read, and the deduced amino acid sequences are derived.

A representative result of T-cell clonality analysis using RT-PCR/nucleotide sequencing is shown (**Fig. 4**). Expanded clonotypes are detected as a multiple number of an identical sequence, whereas non-expanded clonotypes are identified as multiple unique sequences (*see* **Note 3**).

3.5. Method for Single Cell PCR and Subsequent Nucleotide Sequencing

Using the methods described above, the frequency of the T-cell clonotypes are estimated by expression levels of each TCR mRNA. No evidence has been presented to support the assumption that the T-cell clonotype frequency determined by the expression of individual TCR mRNA corresponds to the actual frequency of individual T-cell clonotypes. To address this problem, we established a method to determine the T-cell clonotype frequency by direct cell counts. The method utilizes single-cell PCR techniques. Briefly, T-cells with a particular Vβ family are sorted at 1 cell/well. Total RNA is extracted from each well and converted to cDNA. The TCRβCDR3 gene is amplified from each single cell-cDNA using either semi-nested or nested PCR, followed by nucleotide sequencing.

1. Isolate mononuclear cells from source samples and stain with the FITC-conjugated antibody specific for the particular TCR Vβ family to be analyzed. Using a cell sorter, the positive cells are sorted at a frequency of 1 cell/well onto a microtiter plate that contains 20 µL of single-cell RNA extraction buffer in each well.
2. Incubate the plates for 10 min at 25°C
3. Add 20 µL of single-cell RT buffer to each well and incubate for 1 h at 37°C for cDNA synthesis.
4. PCR-amplify TCR β gene transcripts using either nested- or semi-nested PCR. Perform the primary PCR using 1.0 µL of 10X Taq buffer, 0.5 µL of dNTPs (2.5

*F (%)	BV3	NDN	BJ	
8 (24.2)	CASSL	TGR	TGELF	J2S2
2 (6.0)	CAS	TTLPGLVW	F	J2S1
2 (6.0)	CASS	PMKGG	TDTQY	J2S3
2 (6.0)	CASSL	GG	AKNIQY	J2S4
1 (3.0)	CASS	LGLMD	TEAF	J1S1
1 (3.0)	CASS	SPRQGD	TEAF	J1S1
1 (3.0)	CAS	RPGAAD	TEAF	J1S1
1 (3.0)	CASS	LAPNL	YGYT	J1S2
1 (3.0)	CASS	AGTGTL	YGYT	J1S2
1 (3.0)	CASSL	GGAG	NQPQH	J1S5
1 (3.0)	CAS	RTRGS	SYNSPLH	J1S6
1 (3.0)	CAS	TSD	SYNEQF	J2S1
1 (3.0)	CASS	SRLAT	SYNEQF	J2S2
1 (3.0)	CAS	SRQP	TGELF	J2S2
1 (3.0)	CASS	QAGGP	DTQY	J2S3
1 (3.0)	CASSL	GSA	DTQY	J2S3
1 (3.0)	CASS	QSGYIS	DTQY	J2S3
1 (3.0)	CASS	NRGY	TDTQY	J2S3
1 (3.0)	CASSL	FGV	YEQY	J2S7
1 (3.0)	CAS	RPSTSGT	YEQY	J2S7
1 (3.0)	CASS	PDYL	YEQY	J2S7
1 (3.0)	CAS	WQGT	YEQY	J2S7
1 (3.0)	CASS	PPDGS	SYEQY	J2S7

Fig. 4. T-cell clonality analysis by nucleotide sequencing. A representative result of T-cell clonality analysis by nucleotide sequencing. TCRβCDR3 genes are amplified by familial PCR, and subcloned in preparation for nucleotide sequencing. Sequences of deduced amino acids are shown. The specific sequences that are detected as multiple copy numbers are that of clonally expanded T-cells. A result derived from an RA patient is shown. *F = Frequency of each clonotype. A total of 33 subcloned TCRβgene transcripts were analyzed. Reprinted with permission *(12)* from BMJ publishing group.

m*M* each), 1.0 μL of 5 pmol/μL Vβ-specific primer, 1.0 μL of 5 pmol/μL common Cβ primer, 0.2 μL of 5 U/μL Taq DNA polymerase, 1.0 μL of template cDNA, and 5.3 μL of sterile bi-distilled water. The total reaction volume is 10 μL. The primers for the primary PCR are the same as those used previously in **Subheading 3.2.** *(22)*. Overlay an appropriate amount of mineral oil.

5. Perform thermal cycling at 94°C for 30 s, followed by 35 cycles of: 94°C for 30 s, 58°C for 30 s, and 72°C for 30 s. The reaction is maintained at 72°C for 30 s.
6. The secondary PCR is performed using the two internal Cβ-anti-sense primers (Cβ1 and Cβ2) and either the same Vβ-specific primer as used in the primary PCR above (for semi-nested PCR), or an internal Vβ-sense primer (for fully nested PCR). Prepare a pre-mix containing per reaction: 9.0 μL of 10X Taq buffer, 3.5 μL of dNTPs (2.5 m*M* each), 1.0 μL of 50 pmol/ μL (internal) primer, 0.5 μL of 50 pmol/μL internal Cβ1-primer, 0.5 μL of 50 pmol/μL internal Cβ2-

A

BV	NDN	BJ		(total 24)	
CASS	FPGGG	YNEQF	J2S1	4	(16.7%)
CA	IRKDRAT	NTGELF	J2S2	3	(12.5%)
CAS	TTTGFW	ETQY	J2S5	3	(12.5%)
CASS	PTGA	TDTQY	J2S3	3	(12.5%)
CASSL	EGGGSD	EQY	J2S7	2	(8.3%)
CASSL	SSGGSA	EQF	J2S1	1	(4.2%)
CASS	WRGN	EQY	J2S7	1	(4.2%)
CAS	GTSQG	TQY	J2S5	1	(4.2%)
CASS	FSGAV	YNEQF	J2S1	1	(4.2%)
CASS	IDRDP	NQPQH	J1S5	1	(4.2%)
CAS	RRLAGN	NEQF	J2S1	1	(4.2%)
CASS	WGGT	NTGELF	J2S2	1	(4.2%)
CASS	SGHQG	EQY	J2S7	1	(4.2%)
CASS	FELAS	NEQF	J2S1	1	(4.2%)

B

BV	NDN	BJ		(total 24)	
CASS	FPGGG	YNEQF	J2S1	2	(8.3%)
CA	IRKDRAT	NTGELF	J2S2	2	(8.3%)
CAS	TTTGFW	ETQY	J2S5	2	(8.3%)
CASSL	EGGGSD	EQY	J2S7	1	(4.2%)
CASSL	GTVR	DTQY	J2S3	1	(4.2%)
CASS	FSTCSA	NYGYT	J1S1	1	(4.2%)
CASS	FRE	AKNIQY	J2S4	1	(4.2%)
CASS	FGQGEG	PQH	J1S5	1	(4.2%)
CAS	GLPGQSILGS	FGSG	J1S2	1	(4.2%)
CASS	KTDGP	GELF	J2S2	1	(4.2%)
CASS	VWTGT	NSPLH	J1S6	1	(4.2%)
CASS	FWSA	GYT	J1S2	1	(4.2%)
CASS	RSLSLRDPHDK	AF	J1S1	1	(4.2%)
CASSL	ASGG	GELF	J2S2	1	(4.2%)
CA	TPRWDS	NTGELF	J2S2	1	(4.2%)
CASS	PGQG	NQPQH	JIS5	1	(4.2%)
CASS	FGPN	YEQY	J2S7	1	(4.2%)
CASSL	TGG	TGELF	J2S2	1	(4.2%)
CASS	PRERGSW	QPQH	J1S5	1	(4.2%)
CAS	RFGIRGD	QPQH	JIS5	1	(4.2%)
CASSL	TAGQ	NEKLF	J1S4	1	(4.2%)

Fig. 5. Comparison of the frequency of clonally expanded T-cells in a rheumatoid joint evaluated by TCR frequency and by cell frequency. Synovial fluid mononuclear cells (SFMCs) from an RA patient were divided into two aliquots, one of which was used for TCR frequency analysis and the other for cell frequency analysis (single-cell PCR). Results by these methods were compared. (**A**) To examine TCR frequency, Vβ8+ TCRβ gene transcripts were amplified by PCR from SFMC-derived bulk cDNAs. The PCR products were subcloned into a plasmid vector and 24 subcloned

primer, 0.5 µL of 5 U/µL of Taq DNA polymerase, and 75.5 µL of sterile bi-distilled water.

7. Add 90 µL of the pre-mix to the corresponding primary PCR reaction, mix well, and subject to thermal cycling as for the primary PCR above, except that the annealing temperature is changed from 580C to 640C.
8. Subclone the PCR products from each of the secondary PCRs separately into the plasmid vector for nucleotide sequencing as described in **Subheading 3.4.**
9. Determine T-cell clonality by analyzing the sequences of the TCR CD3 region derived from approx 20–30 single T-cells.

A representative result is shown in **Fig. 5**. Since this method is not affected by bias in RT-PCR amplification of the different TCR CD3 mRNA variants, we can determine the frequency of expanded T-cell clonotypes more accurately by directly counting the frequency of specific sequences derived from single T-cells (**Fig. 5 A,B**).

3.6. Conclusion

The protocols for the analysis of T-cell clonality using RT-PCR and either subsequent SSCP or nucleotide sequencing have been described. These methods can be used to detect an expanded T-cell clonotype. Single cell PCR and subsequent nucleotide sequencing is also useful for the detection and the determination of the precise frequency by cell count. Furthermore, these methods can determine, by nucleotide sequencing, the profile of the amino acid sequences of TCRβCDR3, which is known to interact primarily with the antigenic peptide presented by an MHC molecule *(24,25)*. Our results have demonstrated that these methods generate much information on T-cell responses, which provide a more detailed analysis of disease mechanisms.

4. Notes

1. If no PCR product is obtained, there are a number of possible reasons. Optimum Taq DNA polymerase activity and appropriate Cβ primer sequences are critical

---

(continued) genes were randomly selected for nucleotide sequencing. Deduced amino-acid sequences of the junctional regions are shown. (**B**) To examine cell frequency, Vβ8+ T-cells derived from the SFMCs were sorted at a frequency of 1 cell/well and TCRβ genes were amplified by RT-PCR (single-cell PCR). The PCR products were subcloned into plasmid vectors separately. Deduced amino-acid sequences of their junctional regions are shown. Both approaches exhibit good agreement: both methods identified the same 3 clonally most expanded T-cell variants (the top 3 in A and B). However, 2 variants found to be clonally expanded by bulk RT-PCR analysis (variants 4 and 5 in A) were not found to be clonally expanded by the single-cell RT-PCR method in this study. These differences may be due to sampling variations. Reprinted with permission from **ref.** *(21)*.

for amplification, so check these. Positive control cDNA can be prepared from human peripheral-blood lymphocytes and should be amplified concurrently in every experiment. If the positive control yields a PCR product, the problem is likely to be caused by insufficient template in the test sample. If the amount of the original sample is very small, amplification by semi-nested PCR using equal amounts of the internal Cβ1 and Cβ2 primers is helpful. All of the reagents used for PCR amplification should be checked and changed if found to be contaminated or inactive.

2. If no band, or a smear-like gel pattern is detected by SSCP, there may be a problem with the probe. Check and change the biotinylated internal Cβ probe. Degradation or sequence differences will adversely affect detection. Also, excessive crosslinking by ultraviolet (UV) irradiation may damage the target DNA, thus preventing hybridization of the probe.
3. For identifying the expansion of T-cell clonotypes in each sample, it is vital to prevent contamination by previously amplified sequences. The genes of dominant clonotypes in previous experiments may persist as contaminants in the current experiment, both during RT-PCR and subcloning. If contamination is suspected, change all the reagents for RT-PCR and subcloning and repeat the experiment. To reduce the risk of contamination, it is recommended to aliquot all the reagents into amounts necessary for a single experiment and to discard the aliquot after each experiment.

Acknowledgment

The authors thank Ms. Sayaka Yukii and Ms. Yumiko Ajiri for their assistance in preparing the manuscript.

References

1. Murali-Krishna, K., Altman, J. D., Suresh, M., Sourdive, D. J., Zajac, A. J., Miller, J. D.et al. (1998) Counting antigen-specific CD8 T-cells: a reevaluation of bystander activation during viral infection. *Immunity* **8,** 177–187.
2. Novak E.J., Liu A.W., Nepom G.T., and Kwok W.W. (1999) MHC class II tetramers identify peptide-specific human CD4(+) T-cells proliferating in response to influenza A antigen. *J. Clin. Investig.* **104,** 63–67.
3. Shanmugam, A., Copie–Bergman, C., Caillat, S., Bach, J. F., and Tournier-Lasserve, E. (1995) In vivo clonal expression of T lymphocytes specific for an immunodominant N-terminal myelin basic protein epitope in healthy individuals. *J. Neuroimmunol.* **59,** 165–172.
4. Yamamoto, K., Sakoda, H., Nakajima, T., Kato, T., Okubo, M., Doi, M., et al. (1992) Accumulation of multiple T-cell clonotypes in the synovial lesions of patients with rheumatoid arthritis revealed by a novel clonality analysis. *Int. Immunol.* **4,** 1219–23.
5. Mato, T., Masuko, K., Misaki, Y., Hirose, N., Ito, K., Takemoto, Y., et al. (1997) Correlation of clonal T-cell expansion with disease activity in systemic lupus erythematosus. *Int. Immunol.* **9,** 547–554.

6. Zarour, H. M., Kirkwood, J. M., Kierstead, L. S., Herr, W., Brusic, V., Slingluff, C. L., Jr, et al. (2000) Melan-A/MART-1(51–73) represents an immunogenic HLA-DR4-restricted epitope recognized by melanoma-reactive CD4(+) T-cells. *Proc. Natl. Acad. Sci. USA* **97,** 400–405.
7. Masuko, K., Kato, S., Hagihara, M., Tsuchida, F., Takemoto, Y., Izawa, K., et al. (1996) Stable clonal expansion of T-cells induced by bone marrow transplantation. *Blood* **87,** 788–799.
8. Kato, T., Kurokawa, M., Masuko-Hongo, K., Sasakawa, H., Sekine, T., Ueda, S., et al. (1997) T-cell clonality in synovial fluid of a patient with rheumatoid arthritis. Pursistent but fluctuant oligoclonal T-cell expansions. *J. Immunol.* **159,** 5143–5149.
9. Suzuki, S., Masuko-Hongo, K., Hoger, T. A., Nguyen, M.H., Sasakawa, H., Kurokawa, M., et al. (1998) Contribution of the T-cell receptor BJ gene to recognition of the P91A tumor antigen in DBA/2 mice. *Cancer Immunol. Immunother.* **46,** 93–103.
10. Murata, H., Kita, Y., Sakamoto, A., Matsumoto, I., Matsumura, R., Sugiyama, T., et al. (1995) Limited TCR repertoire of infiltrating T-cells in the kidneys of Sjogren's syndrome patients with interstitial nephritis. *J. Immunol.* **155,** 4084–4089.
11. Sekine, T., Kato, T., Masuko-Hongo, K., Nakamura, H., Yoshino, S., Nishioka, K., et al. (1999) Type II collagen is a target antigen of clonally expanded T-cells in the synovium of patients with rheumatoid arthritis. *Ann. Rheum. Dis.* **58,** 446–450.
12. Kurokawa, M., Kato, T., Masuko-Hongo, K., Ueda, S., Kobata, T., Okubo, M., et al. (1999) Characterization of T-cell clonotypes that accumulated in multiple joints of patients with rheumatoid arthritis. *Ann. Rheum. Dis.* **58,** 546–553.
13. Okubo, M., Kurokawa, M., Ohto, H., Nishimaki, T., Nishioka, K., Kasukawa, R., et al. (1994) Clonotype analysis of peripheral blood T-cells and autoantigen-reactive T-cells from patients with mixed connective tissue disease. *J. Immunol.* **153,** 3784–3790.
14. Yamamoto, K., Masuko, K., Takahashi,S., Ikeda, Y., Kato, T., Mizushima, Y., et al. (1995) Accumulation of distinct T-cell clonotypes in human solid tumors. *J. Immunol.* **154,** 1804–1809.
15. Tanaka, A., Iwabuchi, S., Takatori, M., Ohno, A., Yamada, H., Hashimoto, N., et al. (1997) Clonotypic analysis of T-cells in patients with autoimmune and viral hepatitis. *Hepatol.* **25,** 1070–1076.
16. Hoger, T. A., Jacobson, S., Kawanishi, T., Kato, T., Nishioka, K., and Yamamoto, K. (1997) Accumulation of human T lymphotropic virus (HTLV)-I-specific T-cell clones in HTLV-I-associated myelopathy/tropical spastic paraparesis patients. *J. Immunol.* **159,** 2042–2048.
17. Matsumoto, I., Tsubota, K., Satake, Y., Kita, Y., Matsumura, R., Murata, H., et al. (1996) Common T-cell receptor clonotype in lacrimal glands and labial salivary glands from patients with Sjogren's syndrome. *J. Clin. Investig.* **97,** 1969–1977.
18. Sumida, T., Matsumoto, I., Murata, H., Namekawa, T., Matsumura, R., Tomioka, H., et al. (1997) TCR in Fas-sensitive T-cells from labial salivary glands of patients with Sjogren's syndrome. *J. Immunol.* **158,** 1020–1025.
19. Goronzy, J. J., Bartz-Bazzanella, P., Hu, W., Jendro, M. C., Walser-Kuntz, D. R.,

and Wayand, C. M. (1994) Dominant clonotypes in the repertoire of peripheral CD4+ T-cells in rheumatoid arthritis. *J. Clin. Investig.* **94,** 2068–2076.
20. Eiraku, N., Hingorani, R., Ijichi, S., Machigashira, K., Gregersen, P. K., Monteiro, J., et al. (1998) Clonal expansion within CD4+ and CD8+ T-cells subsets in human T lymphotropic virus type 1-infected individuals. *J. Immunol.* **161,** 6674–6680.
21. Kurokawa, M., Furukawa, H., Yabe, T., Matsui, T., Toda, M., Hamada, C., et al. (1999) Frequency of clonally expanded T-cells evaluated by PCR from a single cell. *J. Immunol. Meths.* **224,** 203–208.
22. Choi, Y., Kotzin, B., Herron, L., Callahan, J., Marrack, P., and Kappler, J. (1989) Interaction of Staphylococcus aureus toxin "superantigens" with human T-cells. *Proc. Natl. Acad. Sci. USA* **86,** 8941–8945.
23. Sumida, T., Yonaha, F., Maeda, T., Tanabe, E., Koike, T., and Yoshida, S. (1992) T-cell receptor reperoire of infiltrating T-cells in lips of Sjogren's syndrome patients. *J. Clin. Investig.* **89,** 681–685.
24. Garboczi, D.N., Ghosh, P., Utz, U., Fan, Q.R., Biddison, W.E., and Wiley, D.C. (1996) Structure of the complex between human T-cell receptor, viral peptide and HLA-A2. *Nature* **384,** 134–141.
25. Garcia, K. C., Degano, M., Stanfield, R. L., Brunmark, A., Jackson, M. R., Peterson, P. A., et al. (1996) An alphabeta T-cell receptor structure at 2.5 A and its orientation in the TCR-MHC complex. *Science* **274,** 209–219.
26. Kazùhiko, Y. (1994) *Autoimmune Diseases.* Yodosha Co., Ltd., Tokyo, Japan, pp. 36–47, pp. 70–87.

22

Generation of scFv from a Phage Display Mini-Library Derived from Tumor-Infiltrating B-Cells

Nadège Gruel, Beatrix Kotlan, Marie Beuzard, and Jean-Luc Teillaud

1. Introduction

1.1. Anti-Tumor Antibodies and Tumor-Infiltrating B-Cells in Cancer Patients

B lymphocytes that infiltrate solid tumors, known as tumor-infiltrating lymphocyte B-cells (TIL-B-cells) , are present in human tumors from different histological types *(1,2)* and tumor-specific antibodies can be detected in many cancer patients. For instance, anti-idiotypic antibodies have been derived from patients with B lymphoma *(3)* and anti-neuroblastoma antibodies have been detected in some neuroblastoma patients *(4)*. In most cases, these antibodies are characterized as auto-antibodies binding to self intracellular proteins. Some of these antibodies, such as anti-*p53* antibodies, could be used as markers of relapsing tumors, of tumors not yet detected, or even as effector molecules in adjuvant immunotherapy *(5,6)*. However, the analysis of the repertoire and of the specificity of antibodies present in cancer patients has been limited by considerable technical difficulties. Traditional methods such as hybridoma technology or Epstein Barr Virus (EBV) transformation of B-cells introduce an important selection bias due to their low efficacy. They allow B-cell lines to be generated which are highly unstable and poor immunoglobulin (Ig) producers. However, molecular engineering now makes it possible to study the antibody repertoire in great detail and to derive human recombinant antibody fragments. The random combination of cloned heavy- and light-chain variable regions (VH and VL) and their expression as single-chain Fv (scFv) or Fab fragments on the surface of bacteriophages *(7–9)* is a powerful method (termed "phage display") to study and select antibodies in patients with different diseases.

From: *Methods in Molecular Biology, vol. 193: RT-PCR Protocols*
Edited by: J. O'Connell © Humana Press Inc., Totowa, NJ

Several anti-tumor human scFv fragments have been isolated using phage display *(10–12)*. Lymphocytes isolated from tumor-draining lymph nodes or from tumors represent a valuable source of antibodies that could help define new molecules and/or new therapeutic targets. Recently, anti-*p53* scFvs have been derived from patients with colorectal cancer using phage display *(12)*. These patients exhibited anti-*p53* antibodies in their serum. Some of the antibody fragments showed medium-high affinity to wild-type *p53* and recognized the N-terminus of human *p53*.

1.2. Building up Phage Display Mini-Libraries from Overexpressed Variable Regions

Among solid tumors, high-grade breast medullary carcinoma (MBC) is heavily infiltrated by B lymphocytes and plasmocytes. Despite its high histological grade, it is associated with a relatively good prognosis *(13–15)*. It is therefore a tumor of choice for testing the utility of the phage-display strategy to study the Ig repertoire expressed by TIL-B-cells *(15)* and to screen for tumor-specific scFv fragments. In addition , most of the MBC patients exhibit a high frequency of *p53* mutations and have anti-*p53* antibodies in their serum *(16)*.

This chapter describes a method that can be used to study the antibody repertoire of TIL-B-cells isolated from MBC and to generate a phage-display scFv library to isolate antibody fragments of interest. This method can be used for building up a phage-display library from any solid human tumor. The amplification and the cloning of VH, Vκ, and Vλ derived from B-cells infiltrating an MBC using specific degenerate oligonucleotide primers *(9, 17)*, as well as the sequencing of the cloned regions, are examined. A strategy to overcome the need for building up a very large phage-display library is then presented. Based on the sequencing data, only the most represented VH and VL regions are cloned, randomly assembled, and expressed on the phage surface as scFvs, representing a "mini-library." It is postulated that a mini-library contains a higher number of homologous VH-VL association than large combinatorial libraries, thus representing more accurately the pairing of VH-VL that takes place in TIL-B-cells in vivo. In addition, a mini-library is easy to generate and manipulate. It can be rapidly screened to select antibody fragments using appropriate antigens such as purified proteins, and cells.

2. Materials

2.1. Tumor Sample Preparation

1. Tumor specimens are obtained aseptically during surgery; a tumor biopsy sample can also be used. Comply with all local regulatory procedures for using human-derived tissues.

2. RPMI-1640 with 2.0 g/L $NaHCO_3$ (Seromed, Berlin, Germany).
3. 200 m*M* L-glutamine (Seromed).
4. 10,000 IU/mL penicillin/10 mg/mL streptomycin (Seromed).
5. Petri dishes (Nunc, Rockslide, Denmark).
6. Scalpel.
7. Deoxyribonuclease (Sigma, Saint Louis, MO).
8. Collagenase (Sigma).
9. Hyaluronidase (Sigma).
10. Sterile gauze.
11. Fetal calf serum (FCS) (Hyclone, Logan, UT).

2.2. VH and VL Repertoire Study

2.2.1. VH and VL Cloning

1. Phosphate-buffered saline (PBS): 5.84 g NaCl, 4.72 g Na_2HPO_4, 2.64 g NaH_2PO_4, $2H_2O$. Adjust pH to 7.2. Complete to 1 L with sterile, double-distilled water and autoclave.
2. Total RNA preparation kit (Trizol reagent, Gibco-BRL, Paisley, Scotland).
3. Chloroform.
4. Isopropanol.
5. 75% (v/v) ethanol.
6. Ultrapure agarose (Gibco-BRL).
7. 10X Tris-boric acid-EDTA (TBE) (Quantum, Illkirch, France).
8. 10 mg/mL ethidium bromide (Quantum).
9. 6X loading buffer: 4 g sucrose, 25 mg bromophenol blue. Complete to 10 mL with sterile, double-distilled water.
10. OneStep RT-PCR Kit (Qiagen, Hilden, Germany).
11. PCR primers for cloning the human rearranged immunoglobulin variable-gene regions (VH, Vκ, Vλ) (**Table 1**) ***(9, 16)***.
12. Ultrapure low-melting point agarose (Gibco-BRL).
13. 50× Tris-acetate-EDTA (TAE) buffer (Quantum).
14. Gel purification extraction kit (Qiagen).
15. Klenow DNA polymerase with 10X buffer (New England Biolabs, Beverly, MA).
16. Ultrapure deoxynucleotide 5' triphosphate (dNTP) set (Amersham Pharmacia Biotech, Buckinghamshire, England).
17. T4 polynucleotide kinase with 10X buffer and 10 m*M* dATP (Amersham Pharmacia Biotech).
18. QIAquick PCR product purification kit (Qiagen).
19. T4 DNA ligase with 10X buffer (New England Biolabs).
20. pUC18 (SmaI/BAP) plasmid (Amersham Pharmacia Biotech).
21. *E. coli* TG1 strain (F'traD36 $lacI^q$ D(lacZ)M15 $proA^+B^+$/supE D(hsdM-mcrB)5 ($r\kappa^- m\kappa^- McrB^-$) thi D(lac-proAB).
22. 2X TY medium (Bio 101, Carlsbad, CA).
23. 20% (w/v) glucose solution.
24. 100 mg/mL stock solution of ampicillin (Boehringer Mannheim, Mannheim, Germany).

Table 1
PCR Primers Used for Cloning the Human Rearranged Immunoglobulin Variable Gene Regions (VH, Vκ, and Vλ)*

JHFor primer :	
HuJH1-2	5'-TGA GGA GAC GGT GAC CAG GGT GCC-3'
HuJH3	5'-TGA AGA GAC GGT GAC CAT TGT CCC-3'
HuJH4-5	5'-TGA GGA GAC GGT GAC CAG GGT TCC-3'
HuJH6	5'-TGA GGA GAC GGT GAC CGT GGT CCC-3'
VHBack primers :	
HuVH1	5'-CAG GTG CAG CTG GTG CAG TCT GG-3'
HuVH2	5'-CAG ATC ACC TTG AAG GAG TCT GG-3'
HuVH3-5-7	5'-CAG GTG CAG CTG GTG GAG TCT GG-3'
HuVH4	5'-CAG GTG CAG CTG CAG GAG TCG GG-3'
HuVH6	5'-CAG GTA CAG CTG CAG CAG TCA GG-3'
HuJκFor primers :	
HuJκ1-4	5'-ACG TTT GAT TTC CAC CTT GGT CCC-3'
HuJκ2	5'-ACG TTT GAT CTC CAG CTT GGT CCC-3'
HuJκ3	5'-ACG TTT GAT ATC CAC TTT GGT CCC-3'
HuJκ5	5'-ACG TTT AAT CTC CAG TCG TGT CCC-3'
HuVκBack primers :	
HuVκ1-4	5'-GAC ATC CAG ATG ACC CAG TCT CC-3'
HuVκ2a	5' GAT GTT GTG ATG ACT CAG TCT CC-3'
HuVκ2b	5'-GAT ATT GTG ATG ACC CAG ACT CC-3'
HuVκ3	5'-GAA ATT GTG TTG ACG CAG TCT CC-3'
HuVκ5	5'-GAA ACG ACA CTC ACG CAG TCT CC-3'
HuVκ6a	5'-GAA ATT GTG CTG ACT CAG TCT CC-3'
HuVκ6b	5'-GAT GTT GTG ATG ACA CAG TCT CC-3'
HuJλFor primers :	
HuJλ1	5'-ACC TAG GAC GGT GAC CTT GGT CCC-3'
HuJλ2–3	5'-ACC TAG GAC GGT CAG CTT GGT CCC-3'
HuJλ7	5'-ACC GAG GAC GGT CAG CTG GGT GCC-3'
HuVλBack primers :	
HuVλ1a-5–9	5'-CAG TCT GTG CTG ACT CAG CC-3'
HuVλ1b	5'-CAG TCT GTG TCG ACG CAG CCG CC-3'
HuVλ2	5'-CAG TCT GCC CTG ACT CAG CCT GC-3'
HuVλ3a	5'-TCC TAT GTG CTG ACT CAG CCA CC-3'
HuVλ3b	5'-TCT TCT GAG CTG ACT CAG GAC CC-3'
HuVλ3c	5'-TCT TAT GAG CTG ACA CAG CTA CC-3'
HuVλ4a	5'-CAG CCT GTG CTG ACT CAA TCA TC-3'
HuVλ4b	5'-CAG CTT GTG CTG ACT CAA TCG CC-3'
HuVλ4c	5'-CTG CCT GTG CTG ACT CAG CCC CC-3'
HuVλ6	5'-AAT TTT ATG CTG ACT CAG CCC CA-3'
HuVλ7–8	5'-CAG ACT GTG GTG ACT CAG GAG CC-3'
HuVλ10	5'-CAG GCA GGG CTG ACT CAG CCA CC-3'

*Primers are designed according to Marks et al. *(9)* and to the V-base database (http://www.mrc-cpe.cam.ac.uk/imt-doc/restricted/ALIGNMENTS.html).

25. AmpliTaq DNA polymerase with 10X buffer and 25 m*M* $MgCl_2$ (PE Applied Biosystems, Foster City, CA).
26. Primers :
    LMB2 Forward primer, 17 mer M13 sequencing primer
    5'-GTAAAACGACGGCCAGT-3'
    LMB3 Backward primer, 17 mer M13 reverse primer
    5'-CAGGAAACAGCTATGAC-3'
27. Sterile toothpicks.
28. Glycerol (autoclaved) (Sigma).

2.2.2. Sequencing

1. QIAprep Spin Miniprep kit (Qiagen).
2. Dye Terminator Cycle Sequencing Ready Reaction Kit (PE Applied Biosystems).
3. Primers LMB2 and LMB3 (*see* **Subheading 2.2.1.**).
4. DyeEx Spin kit (Qiagen).
5. Deionized formamide (Sigma).
6. 50 m*M* ethylenediaminetetraacetic acid (EDTA) at pH 8.0.
7. Urea (Sigma).
8. 19:1 acrylamide/*bis*-acrylamide (Biorad, Hercules, CA).
9. Amberlite (Sigma).
10. 10X TBE solution.
11. 10% (w/v) solution of ammonium persulfate (Quantum).
12. TEMED (Quantum).

2.3. scFv Generation

1. PCR primers corresponding to the 5' and 3' VH, Vκ, Vλ sequences over-represented in the variable regions cloned into pUC18 (*see* **Table 1**).
2. 5' and 3' primers with adequate restriction sites (5' : Sfi I/Nco I ; 3' : Not I) for PCR II and PCR III, hybridizing to the 5' and 3' VH, Vκ, Vλ sequences most represented among the variable regions cloned into pUC18.
3. Software packages for sequence analysis and comparison are commercially availablc. Databases for human antibody variable regions allow further comparisons, for example: http://imgt.cnusc.fr:8104/; http://immuno.bme.nwu.edu/; http://www.mrc-cpe.cam.ac.uk/imt-doc/public/INTRO.html.

2.4. Phage Display and Isolation of Specific scFv

2.4.1. Cloning into pHEN I Phagemid

1. Restriction enzymes with 10X buffers and 100X acetylated bovine serum albumin (BSA) (*Not* I,*Nco* I) (New England Biolabs).
2. Calf Intestinal Phosphatase (CIP) (New England Biolabs).
3. Hybaid recovery purification kit (Hybaid, Ashford, England).
4. pHEN I phagemid (**Fig. 1**).
5. 245 × 245 × 20 mm bacterial plates (Nunc).
6. Glycerol (autoclaved) (Sigma).

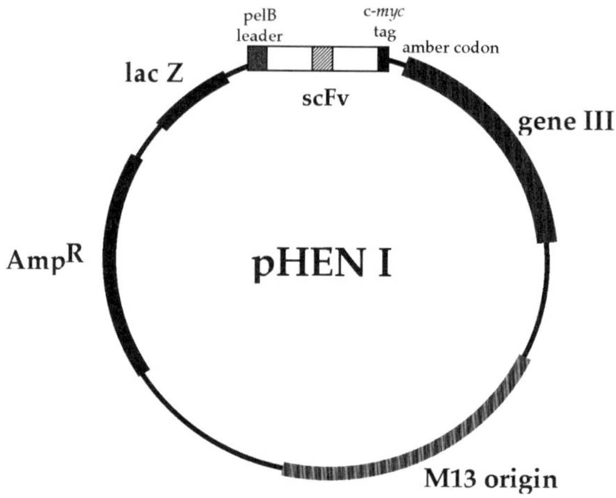

Fig. 1. PHEN1 map.

2.4.2. Phage Production

1. VCSM13 helper phage (Stratagene, La Jolla, CA).
2. 100 mg/mL stock solution kanamycin (Boehringer Mannheim).
3. 20% PEG/2.5 M NaCl: 200*g* of polyethylene glycol (PEG 8000), 146.1 g NaCl. Complete to 1 L with sterile, double-distilled water and autoclave.

2.4.3. Selection of Phage scFvs on Immobilized Antigen

1. 75 × 12 mm immunotubes (Nunc).
2. Skimmed milk powder.
3. PBS: 0.1% (v/v) Tween 20.
4. 100 m*M* triethylamine (autoclaved) (Sigma).
5. 1 *M* Tris-HCl at pH 7.4, (autoclaved).

2.4.4. Phage ELISA

1. Sterile toothpicks.
2. 96-microwell V-bottom plates (Nunc).
3. Gas permeable sealing membranes for microtiter plates "Breathe-easy" (Diversified Biotech, Boston, MA).
4. VCSM13 helper phage (Stratagene).
5. ELISA maxisorp immunoplate (Nunc).
6. Mouse anti-M13 antibody (Amersham Pharmacia Biotech).
7. Goat-anti-mouse IgG (H+L)-AP antibodies (Southern Biotechnology, Birmingham, AL).
8. PNPP (p-nitrophenyl phosphate) substrate tablets (Sigma) (product:yellow, read at 405 nm).

9. Substrate buffer:48.5 mL diethanolamine, 350 mL distilled water. Agitate, and add 62.5 µL 4 *M* $MgCl_2$. Adjust pH to 9.8 with 32% HCl and complete to 500 mL with distilled water. Store at 4°C, in the dark.
10. DNA sequencing primers:
    pHEN I For primer, 20 mer
    5'-GTCTATGCGGCCCCATTCAG-3'
    Linker For primer, 23 mer
    5'-CGATCCGCCACCGCCAGAGCCAC-3'
    LMB3 Back primer (*see* **Subheading 2.2.2.**).

3. Methods

The overall strategy is presented in **Fig. 2**.

3.1. Tumor Sample Preparation

1. A small tumor specimen (1.5 cm in diameter) is obtained aseptically during surgery and transferred to the laboratory within a few hours in cold RPMI-1640 medium containing 2 m*M* L-glutamine, 100 IU/mL penicillin and 100 µg/mL streptomycin (RPMI-1640-PS).
2. Put the tumor piece into a sterile petri dish containing RPMI-1640-PS, and cut the tumor in small pieces (1 mm in diameter) (*see* **Note 1**).
3. Transfer the tumor suspension to a sterile 50-mL conical tube.
4. Complete the volume to 50 mL with RPMI-1640-PS and centrifuge (10 min, 180*g*, room temperature).
5. Discard the supernatant carefully after centrifugation. Resuspend the tumor cell pellet and complete to 50-mL with RPMI-1640-PS.
6. Add 3860 IU of deoxyribonuclease, 35,000 IU of collagenase, and 125 IU of hyaluronidase.
7. Incubate the mixture for 1–1.5 h at 37°C. The suspension should be turbid at the end of incubation.
8. Filter the tumor suspension through a sterile gauze and collect the cells into a 50 mL sterile conical tube. Complete to 50-mL with RPMI-1640-PS.
9. Centrifuge (10 min, 180*g*, room temperature).
10. Discard the supernatant carefully and complete to 50-mL with RPMI 1640-PS.
11. Centrifuge (10 min, 180*g*, room temperature).
12. Discard the supernatant and resuspend the tumor cell pellet in 2–3 mL RPMI-1640-PS containing 10% heat-inactivated FCS.
13. Evaluate the cell suspension viability using a trypan blue dye exclusion assay.
14. Analyze the single-cell suspension by making a cytocentrifuge smear using 500,000–700,000 cells/sample; perform a May-Grunwald Giemsa staining.
15. Cells can be directly used for RNA preparation or frozen in liquid nitrogen until further use. They can be also examined for the presence of specific markers such as CD19, CD20, or sIg, etc.) using flow cytometry.

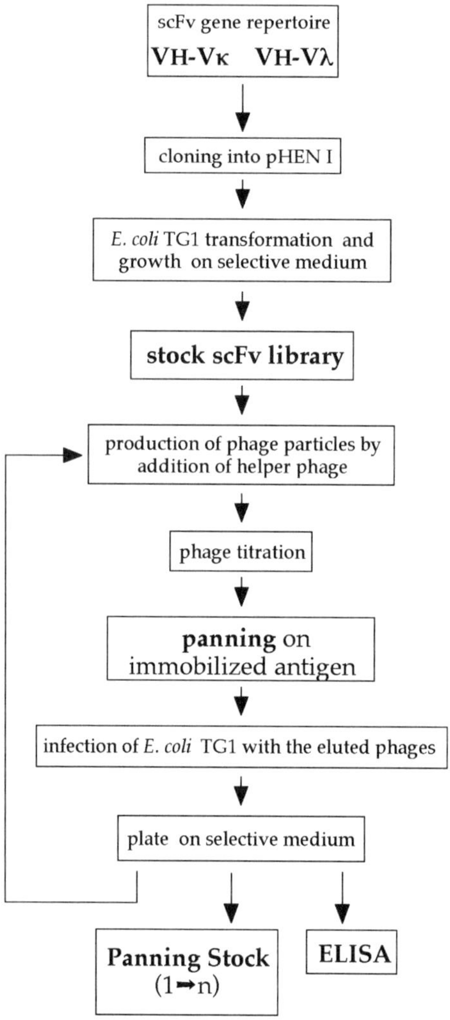

Fig. 2. Strategy for the selection of scFvs derived from tumor-infiltrating B-cells using phage display.

3.2. VH and VL Repertoire Study

3.2.1. Total RNA Preparation (see **Note 2)**

1. Take 5×10^6–1×10^7 cells isolated from the tumor piece and wash twice with PBS (10 min, 180*g*, 4°C). Discard the supernatant.
2. Lyse cells by adding 0.2 mL/10^6 cells of Trizol and resuspending the cell pellet. Vortex.
3. Add 0.1 V chloroform ; vortex several times and incubate for 5 min on ice.

4. Centrifuge to separate total RNA from DNA and proteins (15 min, 13,000*g*, 4°C).
5. Save the upper phase (aqueous phase containing RNA).
6. Precipitate RNA from the aqueous phase by adding 1 V isopropanol; incubate for 15 min on ice.
7. Centrifuge (15 min, 13,000*g*, 4°C) and carefully discard the supernatant.
8. Wash the RNA pellet with 1 mL of 75% ethanol (10 min, 13,000*g*, 4°C). Discard the supernatant.
9. After a brief spin, discard the remaining liquid, and evaporate for 10 min at RT.
10. Dissolve the RNA in 50 µL of RNase-free double-distilled water.
11. Determine the concentration and quality of the RNA preparation by optical density (OD) measurements at wavelengths λ = 260 nm and λ = 280 nm. Optimal 260:280 OD ratio should be between 1.7 and 1.9.
12. Check the RNA by 1% agarose gel electrophoresis: denaturate the RNA at 65°C for 5 min (10 µL of RNA [1 µg] + 1 µL of 1% SDS + 9 µL of distilled water); put on ice before running with 5 µL of loading buffer.

3.2.2. RT-PCR (see ***Note 3)***

RT-PCR can be performed using the OneStep RT-PCR kit from Qiagen. This kit allows for both reverse transcription and PCR amplification in a "one-step" reaction. To amplify the human immunoglobulin VH-JH, Vκ-Jκ and Vλ-Jλ encoding cDNAs, specific primers are designed according to Marks et al., 1991 ***(9)*** and to the V-base database (http://www.mrc-cpe.cam.ac.uk/imt-doc/public/INTRO.html). 20-, 23- or 24-bp oligo-nucleotide primers are synthesized and HPLC-purified (**Table 1**).

Before setting up the PCRs, prepare equimolar mixtures (final concentration: 10 µ*M*) of the 4 JHFor primers (HuJH1–2, HuJH3, HuJH4-5, HuJH6); of the 5 VHBack primers (HuVH1, HuVH2, HuVH3-5-7, HuVH4, HuVH6); of the 4 HuJκFor primers (HuJκ1–4, HuJκ2, HuJκ3, HuJκ5); of the 7 HuVκBack primers (HuVκ1–4, HuVκ2a, HuVκ2b, HuVκ3, HuVκ5, HuVκ6a, HuVκ6b); of the 3 HuJλFor primers (HuJλ1, HuJλ2-3, HuJλ7); of the 12 HuVλBack primers (HuVλ1a-5–9, HuVλ1b, HuVλ2, HuVλ3a, HuVλ3b, HuVλ3c, HuVλ4a, HuVλ4b, HuVλ4c, HuVλ6, HuVλ7–8, HuVλ10).

1. Prepare the following RT-PCR mix on ice for each primer pair :

	VH or Vκ or Vλ
5*X* OneStep RT-PCR buffer with 12.5 m*M* $MgCl_2$:	10 µL
10 m*M* dNTP:	2 µL
10 µ*M* forward primer:	5 µL
10 µ*M* backward primer:	5 µL
Q solution:	10 µL
Template RNA:	10 ng
OneStep RT-PCR enzyme mix:	2 µL
RNase-free double distilled water to:	50 µL

2. Place the PCR tubes in the thermal cycler only after it has reached 50°C.
3. RT-PCR conditions: reverse transcription: 50°C for 30 min PCR: 95°C for 5 min; this heating step inactivates the RT, denaturates the cDNA and activates the Taq DNA polymerase. 94°C for 1 min, 60°C for 1 min, 72°C for 1 min Repeat for 30 cycles. Incubate at 72°C for 5 min at the end.
4. Analyze the PCR products by 1% agarose gel electrophoresis in TBE (8 μL of PCR product + 2 μL of loading buffer). Expected sizes are around 370 bp for VH, 340 bp for Vκ and 335 bp for Vλ **(Fig. 3)**.

3.2.3. Purification of PCR Products

1. Perform 1.5 % low melting agarose gel electrophoresis in TAE using all the PCR products obtained.
2. Excise the bands of the correct size, and purify with the gel extraction purification kit (Qiagen).
3. Gel-quantify the amount of purified PCR products using quantitated DNA markers.

3.2.4. Blunt-End cloning of VH, Vλ and Vκ into E. coli TG1

3.2.4.1. Klenow Treatment and Phosphorylation of Purified PCR Products

1. Prepare the Klenow reaction mix on ice:

	VH or Vκ or Vλ
10X Klenow buffer:	2 μL
Klenow DNA polymerase:	3.25 IU
Sterile, double-distilled water to:	10 μL

2. Add 10 μL of purified PCR product to the relevant tube and incubate for 15 min at room temperature.
3. Add 10 μL of 1.25 m*M* dNTP solution to each tube and incubate for another 15 min at room temperature.
4. Prepare the phosphorylation reaction mix on ice:

	VH or Vκ or Vλ
10X kinase buffer:	5 μL
10 m*M* dATP:	5 μL
Polynucleotide kinase:	10 IU
Sterile, double-distilled water to:	30 μL

5. Add 21 μL of the Klenow reaction to the phosphorylation mix and incubate for 1 h at 37°C.
6. Purify DNA products with QIAquick PCR product purification kit (Qiagen).

3.2.4.2. Blunt-End Ligation into pUC18 Plasmid

1. Prepare the following ligation mix on ice:

	VH or Vκ or Vλ
10X T4 DNA ligase buffer:	2 μL
pUC18 plasmid (× mol):	250 ng
DNA insert:	3× mol

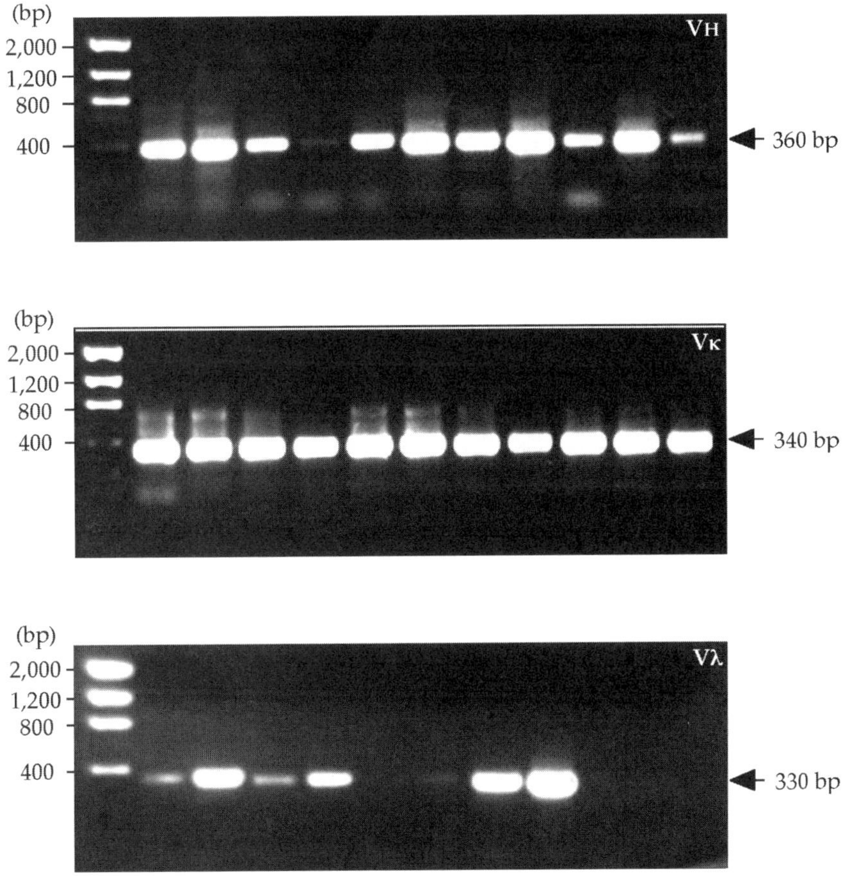

Fig. 3. RT-PCR amplification of VH, Vκ, and Vλ regions expressed by cells present in MBC. PCR products were run on 1% agarose gels. Mol-wt markers (in bp) are indicated on the left, and the length of the VH, Vκ, and Vλ is indicated on the right.

T4 DNA ligase:	2.5 IU
Sterile, double-distilled water to:	20 µL

2. Incubate overnight at 16°C.

3.2.5. Electroporation and PCR Screening

1. Use 2 µL of ligation product to transform 40 µL of electrocompetent *E. coli* TG1 (200 ohm, 25 µF, 2.5 kV).
2. Resuspend the transformed bacteria immediately in 1 mL of 2X TY-1% glucose (2X TY-Glu), and grow for 1 h at 37°C with shaking (250 rpm).
3. Plate on 2X TY-100 µg/mL ampicillin-1% glucose (2X TY-Amp-Glu) plates and grow overnight at 37°C.

4. Screen colonies by PCR.
5. Prepare the following PCR mix :

	VH or Vκ or Vλ
10X Taq DNA polymerase buffer:	2 μL
25 m*M* $MgCl_2$:	1.3 μL
5 m*M* dNTP:	1 μL
10 pmol/μL forward LMB2 primer:	1 μL
10 pmol/μL backward LMB3 primer:	1 μL
Taq DNA polymerase:	1 IU
Sterile, double-distilled water to:	20 μL

6. Aliquot 1 mL 2X TY-Amp-Glu culture medium into 5-mL sterile tubes.
7. Pick up individual bacterial colonies from the overnight culture with sterile toothpicks and rinse them into the PCR tubes containing the PCR mix. Then put toothpicks into the 5 mL tubes containing culture medium and let the bacteria grow overnight at 37°C with shaking (250 rpm).
8. Start the PCR: 94°C for 5 min (bacterial lysis and denaturation); perform 30 cycles of: 94°C for 1 min, 50°C for 1 min, 72°C for 1 min; keep at 72°C for 5 min at the end.
9. Analyze PCR products on 1% agarose gel in TBE (expected sizes are about 400 bp for VH, 370 bp for Vλ, and 380 bp for Vκ).
10. Based on the PCR results, incubate the cultures containing insert-positive clones (inoculated in **step 7**) overnight: freeze and store the positive clones after adding 150 μL of glycerol to 850 μL of the bacterial suspension.

3.2.6. Double-Stranded DNA Sequencing of VH, Vκ, and Vλ

3.2.6.1. DNA Preparation and Sequencing Reaction

1. Inoculate 5 mL 2X TY-Amp-Glu with positive bacterial clones and grow overnight at 37°C with shaking (250 rpm).
2. Purify pUC18 plamid DNA using QIAprep Spin Miniprep Kit (Qiagen).
3. Analyze plasmid DNA minipreps by spectrophotometry (OD at λ= 260 nm) and 1% agarose gel electophoresis.
4. Prepare the following PCR-sequencing mix:

Terminator-ready reaction mix:	8 μL
Plasmid DNA:	2 μg
Primer (LMB2 or LMB3):	3.2 pmol
Sterile, double-distilled water to:	20 μL

5. PCR : 96°C for 10 s, 50°C for 5 s, 60°C for 4 min. Perform 25 cycles.
6. Purify the sequencing product using DyeEx Spin Kit (Qiagen).
7. Dry the pellet.
8. Prepare the following mix :
   5 V deionized formamide.
   1 V 50 m*M* EDTA at pH 8.0.

9. Add 4 µL to each tube and agitate vigorously to dissolve the pellet. Centrifuge (short spin).
10. Denature the samples at 90° for 2 min, then transfer immediately onto ice.

3.2.6.2. Sequencing Gel Electrophoresis

1. Dissolve 40 *g* of urea with 24 mL of distilled water and 12 mL of 40% stock acrylamide/*bis*-acrylamide solution.
2. Deionize the mix with 1 g of amberlite. Mix for 10 min at room temperature.
3. Filter the mix, add 8 mL of 10× TBE and complete to 80 mL with sterile, double-distilled water.
4. Just before pouring the gel, add 400 µL of freshly prepared 10% ammonium persulfate solution and 45 µL of TEMED.
5. When the gel has set, load the samples and perform electrophoresis at a suitable voltage and time, depending on the gel unit being used.
6. After electrophoresis, analyze the data with Editor program®.

3.3. Generation of scFv

3.3.1. Amplification of Selected VH, Vκ, and Vλ with Specific Primers (Based upon DNA Sequence Analysis)

1. Prepare a PCR mix (PCR I) for each of the VH, Vκ and Vλ pUC18 plasmid clones. Based on sequencing results, use primers that will hybridize to the most highly represented variable regions.

	VH or Vκ or Vλ
10X Taq DNA polymerase buffer:	5 µL
25 m*M* $MgCl_2$:	3.25 µL
5 m*M* dNTP:	2.5 µL
10 pmol/µL forward primer:	2.5 µL
10 pmol/µL backward primer:	2.5 µL
Sterile, double-distilled water to:	50 µL

2. Put the tubes under UV light (λ= 254 nm) for 10 min to destroy any contaminant DNA.
3. Add 2.5 µL of the plasmid DNA (5–50 ng) and 5 IU of Taq DNA polymerase.
4. PCR conditions: 94°C for 2 min. Perform 30 cycles of: 94°C for 1 min, 60°C for 1 min, and 72°C for 1 min Incubate at 72°C for 5 min at the end.
5. Analyze the PCR products on 1% agarose gel in TBE (expected sizes are approx 360 bp for VH, 340 bp for Vκ, and 330 bp for Vλ).
6. Purify the PCR products with QIAquick PCR product purification kit (Qiagen).

3.3.2. PCR II

The secondary PCR (PCR II) is performed with primers adding *Sfi* I/*Nco* I restriction sites at the 5' VH and *Not* I at the 3' Vκ or Vλ ends. Primers coding for the linker (GGGGS)3 are used at the 3' VH (ScVHFor) and 5' Vκ (ScVκBack) or Vλ ends. These primers are complementary to each other over 15 bp.

1. Prepare the PCR II mix for each selected VH, Vκ, and Vλ clones:

	VH or Vκ or Vλ
10X Taq DNA polymerase buffer:	5 μL
25 m*M* MgCl2:	3.25 μL
5 m*M* dNTP:	2.5 μL
10 pmol/μL forward primer:	2.5 μL
10 pmol/μL backward primer:	2.5 μL
PCR I product:	15 ng
Taq DNA polymerase:	5 IU
Sterile, double-distilled water to:	50 μL

2. PCR conditions: 94°C for 1 min, followed by 10 cycles of: 94°C for 1 min, 50°C for 1 min, and 72°C for 1 min; incubate at 72 °C for 5 min at the end.
3. Analyze each PCR II product by 1% agarose gel electrophoresis in TBE. Expected sizes are 400 bp for VH, 380 bp for Vκ, and 370 bp for Vλ.
4. Treat the PCR II products with Klenow DNA polymerase:

PCR II product:	42 μL
Klenow DNA polymerase:	5 IU
2.5 m*M* dNTP:	5 μL
Sterile, double-distilled water to:	50 μL

5. Incubate for 15 min at room temperature.
6. Purify the Klenow-treated PCR II products using low-melting agarose gel in TAE, and gel-extraction purification kit (Qiagen).
7. Quantitate the DNA by 1% agarose gel electrophoresis with quantitated DNA markers.

3.3.3. Assembly PCR (PCR III)

1. Prepare the following (PCR III) mix:

10X Taq DNA polymerase buffer:	10 μL
25 m*M* $MgCl_2$:	6.5 μL
5 m*M* dNTP:	5 μL
VH product:	10 ng
Vκ or Vλ product:	10 ng
Taq DNA polymerase:	5 IU
Sterile, double-distilled water to:	90 μL

2. Perform 10 cycles of: 94°C for 1 min, 50°C for 1 min, 72°C for 1 min.
3. Add the SfiVHBack and NotVκFor primers (5 μL); perform 15 cycles of: 94°C for 1 min, 72°C for 2 min.
4. Analyze the PCR III products on a 1% agarose gel in TBE (expected size : 850 bp).
5. Run PCR products on a 1% low melting agarose gel in TAE and purify using gel-extraction purification kit (Qiagen).

3.4. Phage Display and Isolation of Specific scFv

3.4.1. scFv Cloning into pHEN I Phagemid

1. ScFv cDNA and pHEN I phagemid are digested with *Not* I and *Nco* I.

	ScFv or pHEN I
DNA:	1 µg
10X Buffer:	5 µL
100X BSA:	0.5 µL
Not I:	10 IU
Sterile, double-distilled water to:	50 µL

   Incubate for 2 h at 37°C.

2. Add 10 IU of *Nco* I, and incubate for an additional 2 h at 37°C.
3. Dephosphorylate pHEN I with 10 IU of CIP at 37°C for 30 min.
4. Digested samples are purified on a 1% low-melting agarose gel in TAE, using the Hybaid purification kit and quantified using DNA markers. ScFv and pHEN I DNA are then ligated. The ligation product is electroporated into *E. coli* TG1.
5. Prepare the ligation reaction as follows:

pHEN I (× mol):	100 ng
Insert DNA:	3× mol
10X DNA ligase buffer:	2 µL
T4 DNA ligase:	200 IU
Sterile, double-distilled water to:	20 µL

   Incubate for 12 h at 16°C.

6. Use 2 µL of ligation product to transform 40 µL of electro-competent *E. coli* TG1 (200 ohms, 25 µF, 2.5 kV).
7. Resuspend the transformed bacteria immediately in 1 mL of 2X TY-Glu, dilute in 10 mL of 2X TY-Glu and incubate at 37°C for 1 h with shaking (250 rpm).
8. Plate on a 245 × 245 × 20 mm 2X TY-Amp-Glu plate and grow overnight at 30°C.
9. After overnight culture, scrape the colonies off the plate into 2 mL of 2X TY-Amp-Glu and add 10% glycerol before freezing at –80°C. These bacteria represent the stock library.

3.4.2. Phage Production (see **Note 4**)

1. Inoculate 100 mL of culture medium (2X TY-Amp-Glu) with 100 µL of bacteria from the stock library. Grow overnight at 37°C with shaking (250 rpm).
2. Dilute a 500 µL aliquot of the overnight culture in 50 mL fresh culture medium (2X TY-Amp-Glu) and grow at 37°C until OD_{600nm} reaches 0.5.
3. Add 2×10^{10} pfu of VCSM13 helper phage (Stratagene) (phage/bacteria ratio is about 20/1), and incubate at 37°C for 30 min, without shaking.
4. Centrifuge (20 min, 3500*g*, 4°C) (to remove the glucose-containing medium) and resuspend the bacteria pellet in 100 mL of 2X TY-Amp-25 µg/mL kanamycin (2X TY-Amp-Kan).

5. Grow overnight at 30°C with shaking (250 rpm).
6. Pellet the bacteria by centrifugation (20 min, 3500g, 4°C). Collect the phage-containing supernatant.
7. Precipitate phage particles by adding 0.2 V (20 mL) of 20% PEG 8000-2.5 *M* NaCl and incubating the mix for 1 h on ice, with gentle stirring.
8. Centrifuge (30 min, 3500*g*, 4°C), discard the supernatant, and resuspend the phage pellet in 10 mL of PBS.
9. Centrifuge (10 min, 3500*g*, 4°C) to remove any residual bacterial debris or aggregated phage and transfer the supernatant to a 15-mL conical tube. Precipitate phage again with 2 mL of 20% PEG 8000-2.5 *M* NaCl. Mix gently and incubate on ice for 30 min.
10. Centrifuge (30 min, 3500*g*, 4°C).
11. Discard the supernatant and resuspend the phage in 2 mL of PBS.
12. Centrifuge (2 min, 13,000*g*, 4°C) to remove any residual bacterial debris or aggregated phage.
13. Store the phage-containing supernatant at 4°C.

3.4.3. Phage Titration

1. Inoculate 5 mL of 2X TY-Glu with a single TG1 colony (from plate) and grow overnight at 37°C with shaking (250 rpm).
2. Dilute 50 µL of the bacteria overnight culture in 5 mL 2X TY-Glu and incubate at 37°C until $OD_{600\ nm}$ is about 0.5.
3. In the meantime, dilute the phage supernatant (ten-fold dilutions from 10^{-3} to 10^{-11}) in 2X TY.
4. Inoculate 100 µL of the TG1 cells (from **step 2**) with 1 µL of phage from each dilution (**step 3**) and incubate at 37°C for 30 min with gentle shaking (130 rpm). Include a negative control with 1 µL of 2X TY.
5. Plate on 2X TY-Amp-Glu plates and grow overnight at 37°C.
6. Phage titration is assessed by using the following formula: titer (pfu/mL) = number of colonies/plate × 10^3 × dilution factor
   Only plates with a number of colonies that can be easily counted are considered. The titer of the phage preparation is the mean value of the titer determined for each of these plates.

3.4.4. Selection of Phage scFvs on Immobilized Antigen

1. Coat a 75 × 12 mm Nunc Immunotube with 4 mL of antigen (1–1000 µg/mL) in PBS overnight at room temperature, with rotating.
2. Wash 3× with PBS and saturate with 2% skimmed milk in PBS at 37°C for 2 h.
3. Wash 3× with PBS.
4. Add 10^{13} phage (in a volume as small as possible) to 4 mL of 2% skimmed milk-PBS, place in the antigen-coated tube, and incubate for 2 h at room temperature.
5. Remove non specifically bound phages by extensive washing with PBS-0.1% Tween 20 (10×) and then PBS (10×).

6. Elute the bound phage by adding 1 mL of 100 m*M* triethylamine. Incubate for 10 min at room temperature with rotating and neutralize immediately with 0.5 mL of 1 *M* Tris-HCl, pH 7.4 (*see* **Note 5**).
7. Save 0.75 mL of the eluate, and store the other 0.75 mL at –80°C in 15% glycerol (final).
8. For the next round of selection, incubate the eluate (0.75 mL) with 10 mL of exponentially growing (OD_{600nm} = 0.5) *E. coli* TG1 (30 min, 37°C, without shaking).
9. Centrifuge the infected TG1 (15 min, 3500g, 4°C), resuspend the pellet with 1.5 mL 2X TY and plate on a 245 × 245 × 20 mm 2X TY-Amp-Glu plate. Grow overnight at 37°C.
10. Scrape the bacterial colonies into 2 mL of 2X TY-Amp-Glu-15% glycerol.
11. Freeze 1.9 mL at –80°C and use 0.1 mL for production of phages (see Subheading 3.3.2.). After precipitation by PEG-NaCl and titration, the phages are used for the next round of panning. The panning procedure is conducted at least 4×.

3.4.5. Phage ELISA

1. Inoculate 50 mL of culture medium (2X TY-Amp-Glu) with 50 µL of bacteria from each bacterial frozen stock and grow overnight at 37°C with shaking (250 rpm).
2. Dilute a 500 µL aliquot of the overnight culture in 50 mL fresh culture medium (2X TY-Amp-Glu) and grow at 37°C until OD_{600nm} is approx 0.5.
3. Dilute the bacteria in 2X TY-Amp-Glu medium (from 10^{-6} to 10^{-14}) and plate 100 µL of each dilution onto 2X TY-Amp-Glu plates. Grow overnight at 37°C.
4. Toothpick individual colonies and incubate each colony into 100 µL/well of 2X TY-Amp-Glu in a 96-well V-bottomed microplate. Be careful to avoid cross-contamination. Include control wells without bacteria. Seal with Breathe-easy sterile films (*see* **Note 6**).
5. Incubate overnight at 37°C with shaking (250 rpm). If the control wells contain bacteria, discard the plate and repeat the experiment. Carefully label the plate (= plate A).
6. Transfer 2 µL of each well into a new 96-well V-bottomed microplate containing 175 µL/well of 2X TY-Amp-Glu and grow for an additional 2 h at 37°C, with shaking (250 rpm).
7. Prepare a mix of 50% 2X TY-Amp-Glu/glycerol (v/v) in a sterile container. Add 25 µL to each well of microplate A, and store at –80°C. The microplate A can be stored for several mo.
8. Add 25 µL/well of 2X TY-Amp-Glu containing 10^9 pfu of helper phage. Incubate for 30 min at 37°C, without shaking.
9. Centrifuge the plate (10 min, 180*g*, room temperature).
10. Carefully discard the supernatants by pipeting using aerosol resistant tips and resuspend the pellets with 200 µL/well of 2X TY-Amp-Kan. Grow overnight at 30 °C.

11. Centrifuge the plate (20 min, 180*g*, 4°C). In a new 96-well V-bottomed microplate, mix 100 µL of the phage-containing supernatants with 100 µL of 4% skimmed milk-PBS.
12. Incubate for 15–30 min at room temperature.
13. Coat a microtitration ELISA plate with 50 µL/well of antigen (10 µg/mL in PBS).
14. Incubate for 2 h at RT or overnight at 4°C.
15. Saturate the plate with 200 µL/well of 2% skimmed milk-PBS for 2 h at room temperature or overnight at 4°C.
16. Transfer 100 µL of the mix (*see* **step 9**) into the microtitration enzyme-linked immunosorbent assay (ELISA) plate and incubate for 2 h at room temperature.
17. Carefully discard the supernatant (to avoid phage dissemination) and wash 3× with PBS-0.1% Tween 20.
18. Add 100 µL/well of mouse anti-M13 antibody (Pharmacia) (5 µg/mL in 2% skimmed milk-PBS). Incubate for 1 h at room temperature.
19. Wash 3× with PBS-0.1% Tween 20.
20. Add 100 µL/well of goat anti-mouse IgG (H+L)-AP (Southern Biotechnology) (1/1000 in 2% skimmed milk-PBS) and incubate for 1 h at room temperature.
21. Wash 3× with PBS-0.1% Tween 20 before adding 100 µL/well of AP-substrate.
22. Read the plate at OD 405 nm.
23. Grow overnight cultures of the bacterial clones of microplate (A) (*see* **step 4**) corresponding to the positive wells (*see* **Subheading 3.2.6.1.**). Add 15% glycerol, and freeze and store at –80°C.
24. Phagemid pHEN1 of each positive clone is then sequenced, using pHEN I For, Linker For and LMB3 primers (*see* **Subheading 2.4.5.**).

4. Notes

1. Tumor sample preparation: fat and necrotic tissues should be removed from the tumor piece before cutting it in small pieces. Be careful to avoid any bacterial contamination.
2. Total RNA extraction: work on ice with gloves and with RNase-free solutions to avoid any RNA degradation.
3. cDNA synthesis, PCR assay and cloning: all reagents should be handled on ice. Restriction enzymes and T4 DNA ligase should be kept in the –20°C freezer. Sterile, aerosol resistant tips should be used to avoid DNA cross-contamination.
4. Handle phage preparation very carefully. Phage contamination of a laboratory is very difficult to eradicate. NEVER flick phage-containing microtiter plates or NEVER discard phage-containing supernatant in the laboratory sink. Each solution containing phage should be discarded in capped tubes.
5. Selection of phage scFvs using immobilized antigen: do not incubate phage with triethylamine for more than 10 min as phage are sensitive to alkaline solutions.
6. ELISA: to avoid any cross-contamination during the bacterial culture using 96-well V-bottomed microplates, make sure that the gas-permeable sealing membrane is perfectly adhered to the microtiter plate.

Acknowledgments

This work was supported by grants from the Association pour la Recherche sur le Cancer (ARC), the Institut Curie, the INSERM. B. Kotlan was supported by fellowships from the Association pour la Recherche sur le Cancer (ARC), from the French Embassy in Budapest (Hungary) and from the Société Française du Cancer (SFC). The authors wish to thank Dr. B. Zafrani (Département d'Anatomo-Pathologie, Institut Curie) for providing tumor samples and the scientific advice of Pr. W. H. Fridman.

References

1. Barbuto, J.A., Verastegui, E.L., and Hersh, E.M. (1993) The use of transformed T-cell lines for clonal expansion of human B-cells from peripheral blood, spleen, and tumor-infiltrating lymphocytes. *Hybridoma* **12,** 115–125.
2. Punt, C.J., Barbuto, J.A., Zhang, H., Grimes, W.J., Hatch, K.D. and Hersh, E.M. (1994) Anti-tumor antibody produced by human tumor-infiltrating and peripheral blood B lymphocytes. *Cancer Immunol. Immunother.* **38,** 225–232.
3. Hsu, F.J., Caspar, C.B., Czerwinski, D., Kwak, L.W., Liles, T.M., Syrengelas, A., et al. (1997) Tumor-specific idiotype vaccines in the treatment of patients with B-cell lymphoma : long-term results of a clinical trial. *Blood* **89,** 3129–3135.
4. Dalmau, J., Graus, F., Cheung, N.K., Rosenblum, M.K., Ho, A., Canete, A., et al. (1995) Major histocompatibility proteins, anti-Hu antibodies, and paraneoplastic encephalomyelitis in neuroblastoma and small cell lung cancer. *Cancer* **75,** 99–109.
5. Crawford, L.V., Pim, D.C., and Bulbrook, R.D. (1982) Detection of antibodies against the cellular protein *p53* in sera from patients with breast cancer. *Int. J. Cancer* **30,** 403–408.
6. Lubin, R., Schlichtholz, B., Teillaud, J.-L., Garay, E., Bussel, A., Wild, C.P. et al. (1995) *p53* antibodies in patients with various types of cancer: assay, identification, and characterization. *Clin. Cancer Res.* **1,** 1463–1469.
7. Mc Cafferty, J., Griffiths, A.D., Winter, G., and Chiswell, D.J. (1990) Phage antibodies: filamentous phage displaying antibody variable domains. *Nature* **348,** 552–554.
8. Clakson, T., Hoogenboom, H.R., Griffiths, A.D., and Winter, G. (1991) Making antibody fragments using phage display libraries. *Nature* **352,** 624–628.
9. Marks, J.D., Hoogenboom, T.P., Bonnert, J., McCafferty, J., Griffiths, A.D., and Winter, G. (1991) By-passing immunization. Human antibodies from V-gene libraries displayed on phage. *J. Mol. Biol.* **222,** 581–597.
10. Zhang, H., Lake, D.F., Barbuto, J.A., Bernstein, R.M., Grimes, W.J., and Hersh, E.M. (1995) A human monoclonal anti-melanoma single-chain Fv antibody derived from tumor-infiltrating lymphocytes. *Cancer Res.* **55,** 3584–3591.
11. Clark, M.A., Hawkins, N.J., Papaioannou, A., Fiddes, R.J., and Ward, R.L. (1997) Isolation of human anti-c-erbB-2 Fabs from a lymph node-derived phage display library. *Clin. Exp. Immunol.* **109,** 166–174.
12. Coomber, D.W., Hawkins, N.J., Clark, M.A., and Ward, R.L. (1999) Generation

of anti-*p53* Fab fragments from individuals with colorectal cancer using phage display. *J. Immunol.* **163,** 2276–2283.
13. Moore, O.S., and Foote, F.W. (1949) The relatively favourable prognosis of medullary carcinoma of the breast. *Cancer* **2,** 635–640.
14. Gaffey, M.J, Frierson, H.F., Mills, S.E, et al. (1990) Medullary carcinoma of the breast: Identification of lymphocyte subpopulations and their significance. *Mod. Pathol.* **6,** 721–728.
15. Kotlan, B., Gruel, N., Zafrani, B., Furedi, G., Foldi, J., Petranyi, G.G., et al. (1999) Immunoglobulin variable region usage by B lymphocytes infiltrating a human breast medullary carcinoma. *Immunol. Lett.* **65,** 143–151.
16. de Cremoux, P., Salomon, A.V., Liva, S., Dendale, R., Bouchind'homme, B., Martin, E., et al. (1999) *p53* mutation as a genetic trait of typical medullary breast carcinoma. *J. Nat'l Cancer Inst.* **91,** 641–643.
17. Tomlinson, I.M., Walter, G., Marks, J.D., Llewelyn, M.B., and Winter, G. (1992) The repertoire of human germline VH sequences reveals about fifty groups of VH segments with different hypervariable loops. *J. Mol. Biol.* **227,** 776–798.

23

Generation of Murine scFv Intrabodies from B-Cell Hybridomas

Chang Hoon Nam, Sandrine Moutel, and Jean-Luc Teillaud

1. Introduction

1.1. Intracellular Use of Antibodies

The intracellular expression of recombinant single-chain Fv (scFv) molecules—termed intrabodies—has potential advantages for therapeutic use *(1,2)*. These single polypeptide chains are made of one heavy-chain variable region (VH) linked through a flexible spacer, usually a repeated motif of 3× GGGGS (although some others have also been described) *(3)*, to one light chain variable region (VL). The VH-linker-VL sequence can be inverted with no loss of the binding. Shortened linkers have been used to produce scFvs of divalent or bispecific forms *(3,4)*.

scFvs can be directed to various subcellular compartments to target oncogenes, tumor-suppressor gene products, cell-surface receptors, and viral proteins *(1)*. They have been mostly derived from neutralizing monoclonal antibodies (MAbs), and have provided the basis for phenotypic knockouts ("Antibody-Mediated Knockout", AMKO) via the binding and sometimes the relocalization of the targeted antigen *(5,6)*. Restoration of function, e.g. transcriptional activity, may also be achieved *(7)*.

Two sources of recombinant scFvs are available. The screening of phage libraries permits the isolation of scFvs directed against a variety of antigens such as tumor-associated antigens, viral proteins, membrane receptors, or cytokines *(8–10)*. One limitation of this approach is the frequent selection of antibody fragments with low or intermediate affinities. Also, one must be able to select a scFv with the desired functional property, which may prove to be difficult. This approach requires very large libraries and extensive screening

From: *Methods in Molecular Biology, vol. 193: RT-PCR Protocols*
Edited by: J. O'Connell © Humana Press Inc., Totowa, NJ

work. An alternative is to derive scFvs from well-characterized mouse or rat hybridomas. In most cases, the scFv will exhibit properties similar to the parental antibody, with the same or a decreased affinity for the epitope. Once cloned, the N-terminus or the C-terminus part of scFvs can be fused to sequences that allow to target various cell compartments. For instance, SKDEL sequence allows the retention of recombinant antibodies within the endoplasmic reticulum (ER) ***(11)***. Similarly, the Nuclear Localization Signal (NLS) sequence derived from the SV40 T antigen is used to target the nucleus ***(7)*** and the CAAX box motif ***(12)*** derived from the C-terminus part of p21ras is used to target the inner face of the plasma membrane ***(13)***. Many sequences have been extensively characterized over the last decade, which makes it possible to target recombinant molecules to almost any cell compartment. As a consequence, the ability of scFvs to be expressed in precise cell locations triggered intensive efforts to develop intracellular immunization strategies against a variety of molecules.

1.2. Generation of scFv from Hybridomas

The generation of scFvs from hybridomas can be readily performed as described in this chapter. Using cDNA obtained by reverse transcription of hybridoma total RNA, a three-step PCR is performed with appropriate VH and VL primers. It permits the amplification and assembly of VH and VL regions. Once assembled, the scFv-encoding cDNA is cloned into an inducible prokaryotic expression vector that contains a leader sequence (such as pelB) ***(14)***, allowing the scFv translocation into the periplasm. A *c-myc* tag made of 11 amino-acids derived from the *c-myc* sequence is fused to the C-terminus part of the scFv. It allows the detection of the recombinant protein in various assays using a MAb, 9E10 ***(15)***. The recombinant protein can be then purified from bacterial-culture supernatant or from periplasm. It is advisable not to use the 9E10 MAb-based affinity purification procedure, as it will lead to a low yield of scFv. A poly-histidine tag (6 × His), located at the C-terminus part of the scFv allows an efficient purification by Immobilized Metal-Affinity Chromatography (IMAC).

However, VH or VL cloning or cloning of the correct variable region (i.e., responsible for antigen specificity) may be difficult. First, some hybridomas produce antibodies with variable regions that belong to the miscellaneous groups of variable regions as defined by Kabat et al. ***(16)***, or that contain point mutations at their 5' and/or 3' ends. Thus, primers used for PCR, although optimized to avoid any bias ***(17,18)***, can exhibit mismatches that prevent them from hybridizing these particular variable regions. A number of strategies have been developed to solve this problem. N-terminal sequencing of the MAb makes it possible to design new sets of primers that perfectly match the 5' ends of VH or VL ***(19)***. Alternatively, "anchored" PCR permits determination of the 5' and 3' VH or VL sequences ***(20)***. However, these two approaches are time-consuming and expensive. By contrast, the use of pairs of primers that specifically hybridize to VH or VL leader sequences *(21)* and to CH1 or Cκ (or Cλ) constant domains, respectively, permits the rapid determination of sequences from the 5' and 3' ends of the variable regions that could not be otherwise amplified. A specific set of primers that perfectly match the VH or the VL regions are then designed and used for further amplification and assembly. Second, many hybridomas generated by fusing B cells with P3-X63-Ag8.653 or Sp2/O myeloma/hybridoma cells express mRNA encoding a rearranged abortive variable region. This aberrant Vκ transcript (MOPC abVκ) originates from the parental mouse myeloma MOPC-21, from which these fusion partners have been derived. Sets of oligonucleotides designed to amplify mouse Vκ variable regions cannot discriminate between the functional gene derived from the fused B cell and the aberrant transcript. A simple method can be used to specifically inhibit the amplification of this aberrant gene with a peptide nucleic acid (PNA) matching its CDR3 *(22)*. PNAs are analogs of oligodeoxynucleotides *(23)*, which can form highly stable complexes with cDNA, but cannot function as primers for DNA polymerase *(24)*.

After scFv cloning into a prokaryotic expression vector, the addition of sequences that encode localization peptides is performed by molecular engineering using adequate restriction sites. Some examples are provided here. Many other possibilities exist to target various molecules depending on their location *(25,26)*. A leader sequence should be inserted into the 3' part of the scFv construct if the targeted molecule is localized in the ER, the Golgi or the cell membrane. The leader sequence may be the leader sequence of the cloned VH, although some others have been successfully used. In addition, the insertion of a localization sequence can be performed only after the cloning of the scFv construct into a eukaryotic expression vector. The choice of the promoter strength that will control scFv expression in mammalian cells, is important. Some scFvs aggregate and eventually precipitate in the reducing environment of the cytosol. The amount of scFv produced may be critical to avoid this major drawback of intrabodies. Another approach to improve scFv stability is to add a Cκ sequence to the C-terminus part of the recombinant molecule *(22)*. The Cκ domain will tend to homodimerize, and to generate divalent molecules with better avidity for the target antigen. However, some scFv will aggregate and precipitate once it is expressed in the cytosol. It is then advisable to test several scFvs, if available. If not, the selection of other scFvs or of mutated scFvs, which exhibit a good stability in the reducing environment of cytosol, can be assayed. These approaches are currently being explored by different groups, either by a so-called "rational" design of VH and VL sequences *(27)* or by an in vivo selection using the double-hybrid system in yeasts and mammalian cells *(28)*.

2. Materials

2.1. Hybridoma Cell Culture

1. Culture medium: Dulbecco's modified Eagle's medium (DMEM) + 10% fetal calf serum (FCS) + 1 m*M* sodium pyruvate + 2 m*M* L-glutamine + 100 IU/mL penicillin, 100 µg/mL streptomycin.
2. 96-well tissue-culture flasks.
3. IsoStrip (Boehringer Mannheim, Indianapolis, IN).

2.2. VH and Vκ Cloning and scFv Generation from B-Cell Hybridomas

2.2.1. RNA Extraction

1. QIAshredder homogenizer kit (Qiagen, Hilden, Germany).
2. RNeasy Midi RNA extraction kit (Qiagen).
3. 14.5 *M* β-mercaptoethanol.
4. Sterile, RNase-free pipet tips.
5. Ethanol (70% in RNase-free water).

2.2.2. cDNA Preparation

1. Ready-To-Go You-Prime First-Strand cDNA Synthesis Beads (Amersham Pharmacia Biotech, Piscataway, NJ).
2. pd(N)6 Random hexamer primers (Amersham Pharmacia Biotech).

2.2.3. PCR and scFv Assembly

1. AmpliTaq DNA polymerase with 10X buffer and 25 m*M* $MgCl_2$ (PE Applied Biosystems, Foster City, CA).
2. HotStarTaq DNA polymerase (Qiagen).
3. Pfu DNA polymerase (Stratagene, La Jolla, CA).
4. Ultrapure dNTP set (Amersham Pharmacia Biotech).
5. DNA polymerase I Klenow fragment (New England Biolabs, Beverly, MA).
6. Low/High-gelling temperature agarose (electrophoresis grade).
7. Tris-acetate- EDTA (TAE): 40 m*M* Tris-HCl, pH 8.3, 20 m*M* acetic acid, 1 m*M* EDTA.
8. Ethidium bromide (10 mg/mL).
9. Primers (*see* **Tables 1–5**) ***(17,18)***.
10. Qiaquick PCR purification kit (Qiagen).
11. Qiaquick gel extraction kit (Qiagen).
12. Geneclean II kit (BIO 101, La Jolla, CA).
13. Sterile, double-distilled water.

2.3. Sequence Addition for Intracellular Expression and Targeting

1. Ultrapure agarose (Gibco-BRL).
2. Tris-boric acid-EDTA (TBE) (Quantum, France).

3. 6X loading buffer:15% glycerol, 25 mg bromophenol blue, adjust the vol to 10 mL with distilled sterile water.
4. Wizard PCR Prep kit (Promega, Madison, WI).
5. Restriction endonucleases (New England Biolabs).
6. T4 DNA ligase (New England Biolabs).

Table 1
Oligonucleotides used for VH PCR I Amplification

VH1Back : 5'-AGGTSMARCTGCAGSAGTCWGG- 3'
VH1For : 5' -TGAGGAGACGGTGACCGTGGTCCCTTGGCCCC- 3'

* R=A or G, Y=C or T, K=G or T, M=A or C, S=G or C, W=A or T.

Table 2
Oligonucleotides used for Mouse Vκ PCR I Amplification

Vκ MB-I :	5'-GAC ATT GTG ATG W*CA CAG TCT CCR T-3'
Vκ MB-II:	5'-GAT RTT GTG RTG ACY CAR ACT CCA-3'
Vκ MB-III:	5'-RAC ATT GTG CTG ACM CAR TCT CCW G-3'
Vκ MB-IV:	5'-SAA AWT GTK CTS ACC CAG TCT CCA-3'
Vκ MB-V:	5'-GAY ATY MAG ATG ACM CAG TCT MCA-3'
Vκ MB-VI:	5'-CAA ATT GTT CTC WCC CAG TCT CCA-3'
or alternatively, Vκ 2Back :	
	5'-GACATTGAGCTCACCCAGTCTCCA-3'
Vκ 2 For (mix with equimolar quantity):	
5'-CCG TTT GAT TTC CAG CTT GGT GCC- 3'	
5'-CCG TTT TAT TTC CAG CTT GGT CCC-3'	
5' -CCG TTT TAT TTC CAA CTT TGT CCC-3'	
5'-CCG TTT CAG CTC CAG CTT GGT CCC-3'	

* R=A or G, Y=C or T, K=G or T, M=A or C, S=G or C, W=A or T.

Table 3
Oligonucleotides used for Mouse V PCR II Amplification

SfiVH back:	5'-TAC TCG CGG CCC AAC CGG CCA TGG CCC AGG T*SM ARC TGC AGS AGT C-3'
ScVH For:	5'-AGA GCC ACC TCG GCC TGA ACC GCC TCC ACC TGA GGA GAC GGT GAC CG-3'

* R=A or G, Y=C or T, K=G or T, M=A or C, S=G or C, W=A or T.

Table 4
Oligonucleotides used for Mouse Vκ PCR II Amplification

ScVK MB-I/III:	5'-GGC GGA GGT GGC TCT GGC GGT GGC GGA TCG R*AC ATT GTG MTG WCA CAR T-3'
ScVK MB-II:	5'-GGC GGA GGT GGC TCT GGC GGT GGC GGA TCG GAT RTT GTG RTG ACY CAR ACT CCA-3'
ScVK MB-IV:	5'-GGC GGA GGT GGC TCT GGC GGT GGC GGA TCG SAA AWT GTK CTS ACC CAG TCT CCA-3'
ScVK MB-V:	5'-GGC GGA GGT GGC TCT GGC GGT GGC GGA TCG GAY ATY MAG ATG ACM CAG TCT MCA-3'
ScVK MB-VI:	5'- GGC GGA GGT GGC TCT GGC GGT GGC GGA TCG CAA ATT GTT CTC WCC CAG TCT CCA-3'
NotVK For (mix with equimolar quantity):	
	5'-GAT ATG AGA TAC TGC GGC CGC CCG TTT GAT TTC CAG CTT GGT GCC-3'
	5'-GAT ATG AGA TAC TGC GGC CGC CCG TTT TAT TTC CAG CTT GGT CCC-3'
	5'-GAT ATG AGA TAC TGC GGC CGC CCG TTT TAT TTC CAA CTT TGT CCC-3'
	5'-GAT ATG AGA TAC TGC GGC CGC CCG TTT CAG CTC CAG CTT GGT CCC-3'

* R=A or G, Y=C or T, K=G or T, M=A or C, S=G or C, W=A or T.

Table 5
Oligonucleotides used for PCR III Amplification

SfiVH back :
5'-TAC TCG CGG CCC AAC CGG CCA TGG CCC AGG T*SM ARC TGC AGS AGT C-3'
Not VK for (mix with equimolar quantity) :
5'-GAT ATG AGA TAC TGC GGC CGC CCG TTT GAT TTC CAG CTT GGT GCC-3'
5'-GAT ATG AGA TAC TGC GGC CGC CCG TTT TAT TTC CAG CTT GGT CCC-3'
5'-GAT ATG AGA TAC TGC GGC CGC CCG TTT TAT TTC CAA CTT TGT CCC-3'
5'-GAT ATG AGA TAC TGC GGC CGC CCG TTT CAG CTC CAG CTT GGT CCC-3'

* R=A or G, Y=C or T, K=G or T, M=A or C, S=G or C, W=A or T.

*2.3.1. Addition of a Cκ Dimerization Sequence (see **Fig. 1**)*

1. The human plasma cell leukemia cell line ARH77 (American Type Culture Collection, ATCC, Rockville, MD).
2. PCR primers for cloning the human Cκ region:

VκF-Cκ2 primer, 36 mer:
5'-AGATGGTGCAGCCACAGTCCGTTTGATTTCCAGCTT-3'
hCκ Back, 18 mer :
5'-ACTGTGGCTGCACCATCT-3'
CκFNot, 39 mer :
5'- GATATGAGATACTGCGGCCGCCTCTCCCCTGTTGAAGCT-3'

2.3.2. Addition of Targeting Sequences

2.3.2.1. CAAX Box for scFv Targeting to the Inner Surface of the Plasma Membrane

The sequence corresponding to the CAAX box (KMSKDGKK KKKKSKTKCVIM) can be derived from the p147 eukaryotic expression CAAX box containing vector ***(12)***.

2.3.2.2. NLS (Nuclear Localization Signal) Targeting Sequence

The addition of the NLS sequence (PKKKRKV) is performed by a two-step PCR method. Sequence encoding the peptide MGWSCPKKKRKVGGG TATGVHSQ is added to the 5' end (or alternatively to the 3' end) of the scFv by using two partially overlapping complementary oligonucleotide primers.

1. PCR primers:
   * 5'NLS-1, 58 mer:
     *Bam*HI P K K K R K V
     5'- A CTA *GGA TCC* ATG GGA TGG AGC TGT **CCA AAA AAG AAG AGA AAG GTA** GGT GGA GGA ACA-3'
   * 5'NLS-2, 53 mer:
     5' VH
     5'- AAG GTA GGT GGA ACA GCT ACA GTT CAT TCG CAG **GTS MAR CTG CAG SAG TCW GG**-3'
2. pcDNA3 eukaryotic expression vector (Invitrogen, Carlsbad, CA).

2.3.2.3. Endoplasmic Reticulum and Leader Targeting Sequences

The leader sequence allows the nascent polypeptide chain to be translocated into the ER lumen. The ER SEKDEL sequence allows the retention of the protein within the ER. Any mouse VH leader sequence can be used. The addition of the leader sequence termed MHVP4 ***(16)*** is presented here:

The leader sequence is added to the 5' end of the scFv sequence using two oligonucleotides (B-MHVP4-1 and MHVP4-2). A first PCR is performed that permits the addition of the SEKDEL sequence to the 3' end of the scFv and the 3' region of the leader sequence (MHVP4-2). A second PCR is then performed to add the 5' region of the leader sequence (B-MHVP4-1).

1. PCR primers:

   Vκ-ER-For, 57 mer:
   *Eco*RI Stop SEKDEL *Not*I 3'Vκ
   5'-CG *GAA TTC* TTA **GAG CTC ATC GTT CTC GGA** *TGC GGC CGC* CCG TT TAT TTC CAG CTT G-3'
   B-MHVP4-1, 57 mer :
   *Bam*HI Leader sequence
   5'- A CTA *GGA TCC* ATG AAC TTT GGG CTC AGC TTG ATT TTC CTT GTC CTA ATT TTA AAA GG-3'
   MHVP4-2, 53 mer:
   Leader sequence 5'VH
   5'- CTT GTC CTA ATT TTA AAA GGT GTC CAG TGT CAG **GTS MAR CTG GAG SAG TCW GG**-3'

2. pcDNA3 eukaryotic expression vector (Invitrogen).

2.3.2.4. Mitochondria Targeting Sequence

The sequence corresponding to the MSVLTPLLLRGLTGSARR LPVPRAK IHSLPPEGK LQ mitochondria targeting peptide can be derived from the pCOX8.21 BMP vector ***(29)***.

2.3.2.5. Spacer for Fusion Molecules (Mouse IgG3 Hinge Region)

The hinge region of mouse IgG3 can be used as a spacer for fusion molecules (scFv and cytokine, scFv and enzyme). Two partially complementary primers which encode the Hinge region are used.

1. PCR primers:
   * Hinge *Not*I Back, 32 mer:
   5'- *G GCC GCA* CCC AAG CCC AGT ACC CCC CCA GGT G-3'
   * Hinge *Bam*HI For, 32 mer:
   5'- *GA TCC* ACC TGG GGG GGT ACT GGG CTT GGG TGC-3'
2. T4 polynucleotide kinase (New England Biolabs).
3. d(ATP) 100 m*M* (Amersham Pharmacia Biotech).

2.4. ScFv Expression and Purification

2.4.1. Electro-transformation of E. coli

1. *Not*I and *Nco*I restriction enzymes (New England Biolabs).
2. pSW1 prokaryotic expression vector ***(18)***.
3. T4 DNA ligase (New England Biolabs).
4. SOB media: 20 g bacto-tryptone, 5 g bacto-yeast extract, 0.5 g NaCl, 5 g $MgSO_4 \cdot 7H_2O$.
5. SOC media: Add 20 mL of sterile-filtered 1 *M* glucose solution/L of SOB media.
6. Electrocompetent bacteria (TG1) (*see* **Subheading 3.4.2.1.**).

7. 2X TY media: 16 g bacto-tryptone, 10 g bacto-yeast extract, 5 g NaCl.
8. Ampicillin (100 µg/mL) glucose (1% w/v) containing 2X TY agar plates.
9. Qiaprep spin miniprep kit (Qiagen).

2.4.2. PCR Screening

1. AmpliTaq DNA polymerase (PE Applied Biosystems).
2. Primers:
   LMB2 : 5'-GTA AAA CGA CGG CCA GT-3'
   LMB3 : 5'-CAG GAA ACA GCT ATG AC-3'

2.4.3. ScFv Cloning, Expression and Purification

1. Isopropyl-B-D-thiogalactopyramoside (IPTG) (stock solution: 1 *M*).
2. Sucrose (Sigma, Saint Louis, MO).
3. 9E10 anti-*c-myc* monoclonal antibody (MAb) (Clontech, Palo Alto, CA).
4. Goat anti-mouse IgG (H+L)-AP (Southern Biotechnology Associates,Inc., Birmingham, AL).
5. Talon Metal Affinity Resin (Clontech).
6. Nitrocellulose membrane.
7. Resuspension buffer I (for periplasmic fraction): 30 m*M* Tris-HCl, pH 7.0, 20% (w/v) sucrose, 1 m*M* EDTA. Resuspension buffer II (for osmotic shock fraction): 5 m*M* $MgSO_4$ ***(30)***.
8. Loading buffer (20 m*M* Tris-HCl, pH 8.0, 500 m*M* NaCl).
9. Washing buffer (20 m*M* Tris-HCl, pH 8.0, 500 m*M* NaCl, 20 m*M* imidazole).
10. Elution buffer (20 m*M* Tris-HCl, pH 8.0, 500 m*M* NaCl, 150 m*M* imidazole).

3. Methods

3.1. Subcloning of B-Cell Hybridoma and MAb Evaluation

Subcloning the hybridoma by limiting dilution is necessary before generating the scFv to avoid the risk of working with variant cells such as nonproducing cells or cells producing a mutated MAb that exhibit characteristics different from those of the parental MAb. Following cloning, test the hybridoma supernatant for antigen specificity and isotype (heavy and light chains) by enzyme-linked immunosorbent assay (ELISA) and IsoStrip assays.

1. Prepare 10 mL of cells at 600 cells/mL in DMEM culture medium.
2. Prepare three 96-well tissue-culture flasks with 150 µL of medium in each well except for the first columns.
3. Put 300 µL of cell preparation in the first column and perform successive two-fold dilutions.
4. Incubate for 7–12 d in a humidified 37°C, 8% CO_2 incubator.
5. Evaluate clones for antibody production with an effective technique such as ([ELISA], RIA, Western-Blot, dot-blot, or immunofluorescence). Expand one of the positive clones for VH and VL cloning.

3.2. VH and Vκ Cloning and scFv Generation from B-Cell Hybridomas

3.2.1. RNA Extraction from B-Cell Hybridomas

Follow the manufacturer's recommendations when using the RNeasy mini RNA isolation kit (Qiagen) (*see* **Notes 1, 2**).

3.2.2. cDNA Preparation

1. For first-strand cDNA synthesis, bring 2 μg of total RNA to a final vol of 30 μL by dilution with RNase-free water in a sterile 1.5-mL microcentrifuge tube.
2. Heat at 65°C for 10 min, then chill on ice for 2 min.
3. Transfer the RNA solution to a tube containing First-Strand Reaction Mix Beads.
4. Add 0.2 μg of $pd(N)_6$ primers and sterile, double-distilled water to a final volume of 33 μL.
5. Incubate at room temperature for 1 min.
6. Mix the contents of the tube by repeatedly pipetting the mixture up and down.
7. Centrifuge briefly with minifuge to spin down liquid.
8. Incubate the tube at 37°C for 60 min.
9. Store at –20°C until further use.

3.2.3. PCR Assembly

3.2.3.1. PCR I

3.2.3.1.1. Screen for Optimal Primers to be Used in PCR I

A preliminary PCR allows the search for optimal backward (or reverse) primers to be used for the PCR I amplification of the VH and Vκ region of the MAb to be cloned. The primers to be tested in this way are listed in **Tables 1** and **2**. Make up a PCR mix containing per reaction:

10X PCR buffer	2 μL
dNTP (5 m*M*)	1 μL
$MgCl_2$ (25 m*M*)	2 μL
Forward primer (10 μ*M*)	1 μL (*see* **Tables 1**, **2**)
Backward primer (10 μ*M*)	1 μL (*see* **Tables 1**, **2**)
Sterile double distilled water to	20 μL

1. Irradiate the mix on a short-wave UV (254 nm) transilluminator for 10 min.
2. Add 1 μL of the cDNA to the mix.
3. Heat at 94°C for 5 min in a PCR thermal cycler.
4. Add 0.2 μL of Taq DNA polymerase (5 U/μL).
5. Perform 30 PCR cycles as follows:
   94°C for 1 min, 60°C for 1 min, 72°C for 1 min 30 s.
6. Heat at 72°C for 5 min in the PCR thermal cycler.
7. Analyze 8 μL of PCR product on a 1% agarose gel. Expected sizes are about 350 bp (VH) and 320 bp (Vκ) (*see* **Notes 3** and **4**).

3.2.3.1.2. PCR I.

1. Make up 50 μL PCR mix in 0.5-mL microcentrifuge tubes containing:

10X PCR buffer	5 μL
dNTP (5 m*M*)	2.5 μL
$MgCl_2$ (25 m*M*)	3 μL (*see* **Note 5**)
Forward primer (10 μ*M*)	2.5 μL (*see* **Tables 1, 2**).
Selected backward primer (10 μ*M*)	2.5 μL (*see* **Subheading 3.2.3.1.1.**)
Sterile, double-distilled water to	50 μL

2. Irradiate the mix on a short-wave UV (254 nm) transilluminator for 10 min.
3. Add 5 μL of the cDNA to the mix (*see* **Note 6**).
4. Heat at 94°C for 5 min in the PCR thermal cycler.
5. Add 0.5 μL of Taq DNA polymerase (5 U/μL) (*see* **Note 7**).
6. Perform 30 PCR cycles as follows: 94°C for 1 min, 60°C for 1 min, 72°C for 1 min 30 s. Heat at 72°C for 5 min for a final elongation.
7. Analyze 8 μL of PCR product on a 1% agarose gel. Expected sizes are approx 350 bp (VH) and 320 bp (Vκ).
8. Use Qiaquick PCR purification kit for the purification of PCRI products. Store at –20°C until further use.

3.2.3.2. PCR II

1. Make up 100 μL PCR mix in 0.5 mL microcentrifuge tubes containing:

	VH	Vκ
10X PCR buffer	10 μL	10 μL
dNTP (5 m*M*)	5 μL	5 μL
$MgCl_2$ (25 m*M*)	6 μL	6 μL
ScVH for (VH) (10 μ*M*)	5 μL	–
Not Vκ for (Vκ) (10 μ*M*) (*see* **Subheading 2.2.3.**)	–	5 μL
Sfi VH back (VH) (10 μ*M*)	5 μL	–
ScVK MX* (Vκ) (10 μ*M*) (*see* **Subheading 2.2.3.**).	–	5 μL
Sterile, double-distilled water to	100 μL	100 μL

*This primer is selected from those listed in **Table 4**; the selected primer (I–VI) will correspond to the Vκ MB sequence (I–VI) that gave the optimal amplification in PCR I (*see* **Subheading 3.2.3.1.1.**)

2. Add 3–7 μL of PCR I product to the mix.
3. Heat at 94°C for 5 min in PCR thermal cycler.
4. Add 1 μL of Taq DNA polymerase (5 U/μL).
5. Perform 10 PCR cycles as follows:
6. 94°C for 1 min, 60°C for 1 min, 72°C for 2 min.
7. Heat at 72°C for 5 min in PCR thermal cycler.

8. Analyze 8 µL of PCR product on a 1% agarose gel.
9. Use Qiaquick PCR purification kit for purifying PCR II products (VH and Vκ).
10. Make up 50 µL Klenow reaction mix in a 1.5-mL microcentrifuge tube containing:

Klenow fragment	1 µL
10X *Eco* Pol buffer	5 µL
dNTP (5 m*M*)	2.5 µL
Purified PCRII product	42.5 µL

Incubate at room temperature for 15 min.

11. Purify PCR fragments by electrophoresis on a 1% low-gelling temperature agarose gel: excise the bands corresponding to the PCR products (expected sizes are about 410 bp [VH] and 370 bp [VK]) with a sterile razor blade and extract from the gel using the Qiaquick gel extraction kit or the Geneclean II kit (*see* **Note 8**).
12. Store at –20°C until further use.

3.2.3.3. PCR III (Assembly PCR)

1. Make up 100 µL PCR reaction mix in a 0.5-mL microcentrifuge tube containing:

10X PCR buffer	10 µL
dNTP (5 m*M*)	5 µL
$MgCl_2$(25 m*M*)	6 µL
Purified VH fragments	100 ng*
Purified Vκ fragments	100 ng*
Sterile, double-distilled water to	90 µL

* This quantity can be approximated by analysis on an 1% agarose gel by comparison to markers of known sizes and amounts.

2. Heat at 94°C for 5 min in PCR thermal cycler.
3. Add 1 µL of Taq DNA polymerase (5 U/µL).
4. Perform 10 PCR cycles as follows: 94°C for 1 min, 50°C for 1 min, 72°C for 1 min.
5. Then add: 5 µL of Sfi VH Back (10 *M*) and 5 µL of *Not* Vκ For (10 µ*M*).
6. Perform 15 PCR cycles as follows: 94°C for 1 min 30 s, 60°C for 1 min, 72°C for 2 min. Heat at 72°C for 5 min for a final elongation.
7. Analyze the PCR assembly product on a 1% agarose gel.
8. Purify the assembled fragment by electrophoresis on a 1% low-gelling-temperature agarose gel ; excise the band corresponding to the assembled product (approx 770 bp) with a sterile razor blade and extract from the gel using the Qiaquick gel extraction kit or the Geneclean II kit.

3.3. Sequence Addition for Intracellular Expression and Targeting

3.3.1. Cκ Sequence (Fig. 1)

1. Amplification of the scFv and of the Cκ. Prepare two PCR mixes as follows :

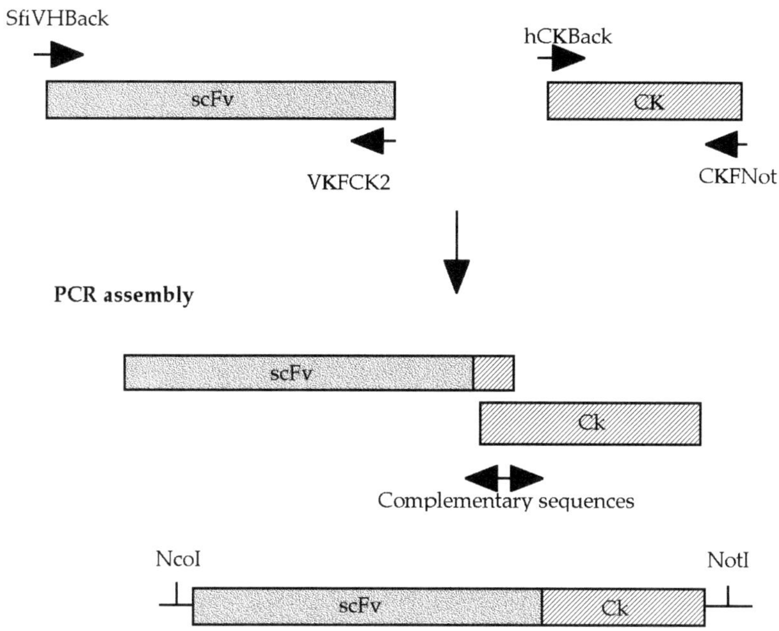

Fig. 1. Addition of a human Cκ domain to the C-terminus part of scFv.

	scFv	Cκ
10X PCR buffer	10 µL	10 µL
$MgCl_2$ (25 m*M*)	6.5 µL	6.5 µL
dNTP (5 m*M*)	5 µL	5 µL
SfiVH Back (10 µ*M*)	5 µL	–
VκF-Cκ2 (10 µ*M*)	5µL	–
scFv (miniprep)	50 ng	–
or		
HCκ Back (10 µ*M*)	–	5 µL
Cκ FNot (10 µ*M*)	–	5 µL
ARH77 cDNA	–	1 µL
AmpliTaq DNA polymerase	5 U	5 U
Sterile, double-distilled water to	100 µL	100 µL

2. Perform 30 PCR cycles as follows: 94°C for 2 min (before starting cycles) scFv: 94°C for 1 min, 60°C for 1 min, 72°C for 1 min.
   C : 94°C for 1 min, 53°C for 1 min, 72°C for 1 min.
3. Analyze the PCR products using a 1% agarose gel (8 µL PCR product + 2 µL loading buffer). Expected sizes are about 770 bp for the scFv and 320 bp for the C region.

4. Purify the PCR products with the Wizard PCR prep kit.
5. PCR assembly: prepare a PCR mix as follows:

10X PCR buffer	10 µL
$MgCl_2$ (25 m*M*)	6.5 µL
dNTP (5 m*M*)	5 µL
scFv purified PCR product	10 µL
Cκ purified PCR product	10 µL
AmpliTaq DNA polymerase	5 U
Sterile, double-distilled water to	100 µL

6. Perform 10 PCR cycles as follows: 94°C for 2 min (before starting cycles); 94°C for 1 min, 50°C for 1 min, 72°C for 1 min 30 s.
7. Add 5 µL of Sfi VH Back (10 µ*M*) and 5 µL of Cκ For Not (10 µ*M*).
8. Perform 15 PCR cycles as follows: 94°C for 1 min, 50°C for 1 min, 72°C for 1 min 45 s.
9. Analyze the PCR product on a 1% agarose gel (8 µL PCR product + 2 µL loading buffer). The expected size is 1080 bp.
10. Purify the PCR product with the Wizard PCR prep kit.
11. Digest the PCR products with *Nco* I and *Not* I for subsequent ligation into pSW1 (*see* **Subheading 3.4.1.**).

3.3.2. Addition of Targeting Sequences

All of these additional sequences are most conveniently added to the scFv sequence after the scFv has first been cloned into a plasmid vector (*see* **Subheading 3.4.1.**).

3.3.2.1. CAAX Box for scFv Targeting to the Inner Surface of Plasma Membrane

The CAAX box motif KMSKDGKKKKKKSKTKCVIM allows scFv targeting to the inner face of the plasma membrane. The CAAX box fragment is ligated at the 3' end of the scFv sequence as follows:

1. Isolate the CAXX box sequence by digesting the p147 eukaryotic expression vector *(12)* with *Eco*RI:

   *Eco*RI   CAAX box
   5'-G AAT TC A AAG ATG AGC AAA GAT GGT AAA AAG AAG AAA AAG
                                                  stop
   AAG TCA AAG ACA TGT GTA ATT ATG **TAA** ATA CAA TTT GTA CTT
                                                                        *Xba*I   *Sal*I
   TTT TCT TAA GGC ATA CTA GTA C*TC TAG A GT CGA C*CT GCA GGC
                                  *Eco*RI
   ATG CAA GCT G*GG AAT T*C A-3'

2. Ligate the CAAX box fragment into the pSW1 prokaryotic-expression vector

containing the scFv insert; the vector is digested with *Eco*RI and dephosphorylated (*see* **Subheading 3.4.1.**). The entire sequence can be cloned into a eukaryotic expression vector using adequate restriction sites.

3.3.2.2. NLS (Nuclear Localization Signal) Targeting Sequence

The addition of the NLS motif (PKKKRKV) to the C-terminus of the scFv sequence is obtained by performing two consecutive PCRs:

1. Prepare a first PCR mix as follows:

10X PCR buffer	10 µL
$MgCl_2$ (25 m*M*)	6.5 µL
dNTP (5 m*M*)	5 µL
5' NLS-2 (10 µ*M*)	5 µL
Vκ 2-For (10 µ*M*)	5 µL
scFv (miniprep)	50 ng
AmpliTaq DNA polymerase	5 U
Sterile, double-distilled water to	100 µL

2. Perform 30 PCR cycles as follows: 94°C for 2 min (before starting cycles); 94°C for 1 min, 60°C for 1 min, 72°C for 1 min.
3. Analyze the PCR product on a 1% agarose gel (8 µL PCR product + 2 µL loading buffer). The expected size is approx 760 bp.
4. Purify the PCR product with the Wizard PCR prep kit.
5. Prepare a second PCR mix as follows:

10X PCR buffer	10 µL
$MgCl_2$ (25 m*M*)	6.5 µL
dNTP (5 m*M*)	5 µL
5' NLS-1 (10 µ*M*)	5 µL
Not-Vκ -For (10 µ*M*)	5 µL
PCR I product	50 ng
AmpliTaq DNA polymerase	5 U
Sterile, double-distilled water to	100 µL

6. Perform 30 PCR cycles as follows: 94°C for 2 min (before starting cycles); 94°C for 1 min, 55°C for 1 min, 72°C for 1 min.
7. Analyze the PCR product by 1% agarose gel electrophoresis (8 µL of PCR product + 2 µL loading buffer). The expected size is 790 bp.
8. Purify the PCR product with the Wizard PCR prep kit.
9. Digest the PCR product with *Bam*H I and *Not* I and ligate into the pcDNA3 eukaryotic expression vector.

3.3.2.3. Endoplasmic Reticulum and Leader Sequences

The expression and retention of the scFv in the ER requires the addition of a leader and of the SEKDEL sequences. The leader sequence allows the nascent polypeptide chain to be translocated into the ER lumen. The ER SEKDEL sequence allows the retention of the protein within the ER. Two PCRs are performed:

1. Prepare the first PCR mix as follows:

10X PCR buffer	10 µL
$MgCl_2$ (25 m*M*)	6.5 µL
dNTP (5 m*M*)	5 µL
MHVP4-2 (10 µ*M*)	5 µL
Vκ 2-For (10 µ*M*)	5 µL
scFv (plasmid miniprep)	50 ng
AmpliTaq DNA polymerase	5 U
Sterile, double-distilled water to	100 µL

2. Perform 30 PCR cycles as follows: 94°C for 2 min (before starting cycles); 94°C for 1 min, 60°C for 1 min, 72°C for 1 min.
3. Analyze the PCR product on a 1% agarose gel (8 µL PCR product + 2 µL loading buffer). The expected size is approx 750 bp.
4. Purify the PCR product with the Wizard PCR prep kit.
5. Prepare the second PCR mix as follows:

10X PCR buffer	10 µL
$MgCl_2$ (25 m*M*)	6.5 µL
dNTP (5 m*M*)	5 µL
B-MHVP-1 (10 µ*M*)	5 µL
Vκ -ER-For (10 µ*M*)	5 µL
PCR I product	1 µL
AmpliTaq DNA polymerase	5 U
Sterile, double-distilled water to	100 µL

6. Perform 30 PCR cycles as follows: 94°C for 2 min (before starting cycles); 94°C for 1 min, 60°C for 1 min, 72°C for 1 min.
7. Analyze the PCR product on a 1% agarose gel (8 µL PCR product + 2 µL loading buffer). The expected size is 820 bp.
8. Purify the PCR product with the Wizard PCR prep kit.
9. Digest the PCR product with *Bam*H I and *Not* I and ligate into the pcDNA3 eukaryotic expression vector.

3.3.2.4. Mitochondria Targeting Sequence

To target mitochondria with scFv, the sequence encoding the subunit VIII of the human c-oxidase cytochrome is fused to the 5' end of the scFv insert. This motif contains a presequence eliminated during the translocation through the mitochondrial membrane.

Presequence N-Ter Cyt. VH
MSVLTPLLLRGLTGSARRLPVPRAK**IHSLPPEGKL** QVQLQ

The fragment encoding this peptide can be obtained by digesting the pCOX8.21 BMP eukaryotic expression vector with *Hind*III ***(29)***.

1. Add an *Nco* I restriction site to the 5' and 3' ends of the fragment by PCR.
2. Digest the fragment and the pSW1 plasmid containing the scFv insert with *Nco* I and dephosphorylate the digested pSW1 plasmid; ligate the fragment at the 5' end of the scFv sequence. Transform *E. coli* TG1 (*see* **Subheadings 3.4.1.** and **3.4.2.**).
3. Isolate TG1 clones and PCR-screen for inserts with the expected size (*see* **Subheading 3.4.3.1.**).
4. Prepare plasmid minipreps and check for the correct orientation of the mitochondria targeting sequence by double digestion with adequate restriction enzymes.
5. Transfer the entire insert (targeting sequence + scFv) into a eukaryotic expression vector.

3.3.2.5. Spacer for Fusion Molecules (Mouse IgG3 Hinge Region)

The hinge region of the mouse IgG3 can be used as a spacer between the scFv and another molecule (e.g. cytokine or enzyme) to form a fusion protein. The annealing of the partially complementary Hinge-*Not* I-Back and Hinge-*Bam*H I-For primers (*see* **Subheading 2.3.2.5.**) generates a hinge-linker fragment containing a *Not* I site at the 5' end and a *Bam*H I site at the 3' end. Since this linker will be subsequently ligated in between the two moieties of the fusion construct, both oligonucleotide strands of the linker must first be phosphorylated.

1. Prepare the following phosphorylation mix for each primer :

primer (10 pM)	50 µL (Hinge-*Not*I-Back or Hinge-*Bam*HI-For)
10X kinase buffer	10 µL
dATP (100 m*M*)	1 µL
T4 polynucleotide kinase	5 µL
Sterile, double-distilled water	34 µL

2. Incubate for 1 h at 37°C.
3. Mix the tubes containing the two phosphorylated primers.
4. Incubate for 10 min at 94°C; then slowly cool the tube to room temperature to allow annealing. Store at –20°C. Use 1 µL of the linker preparation for 1 µg of plasmid to be ligated.
5. Digest the pSW1 plasmid containing the scFv with *Not* I and *Eco*R I (*see* **Subheading 3.4.1.**). Digest the hinge product with *Not* I and *Bam*H I; digest the cDNA encoding the second moiety of the fusion protein with *Bam*H I and *Eco*R I; this cDNA will have been generated with a *Bam*H I site at the 5' end and an

*Eco*R I site at the 3' end. Ligate pSW1 with the two fragments to generate a plasmid-scFv-hinge-fusion cDNA hybrid.

3.4. ScFv Cloning, Expression, and Purification

3.4.1. scFv cDNA Cloning

For optimal digestion, the digestion of the PCR assembly product is performed using the buffer supplied by the manufacturer. A high concentration of restriction enzymes (*Not* I: 25-fold, *Nco* I: fivefold over the amount recommended, *see* **Subheading 4.1.3.3.**) is used for an efficient cutting (*see* **Note 9**).

1. Prepare the following mix :

Purified PCR product	50 µL
10X NEB buffer	10 µL
Nco I (10 U/µL)	5 µL
Sterile, double-distilled water	35 µL

2. Incubate for 4 h or overnight at 37°C.
3. Purify the digested product using Qiaquick PCR purification kit (final vol: 50 µL).
4. Prepare the following mix :

Purified *Nco* I digested PCR product	50 µL
10X NEB buffer 4	10 µL
100X BSA	1 µL
Not I (10 U/µL)	5 µL
Sterile, double-distilled water	35 µL

5. Incubate for 4 h or overnight at 37°C.
6. In parallel, digest the purified pSW1 plasmid with *Not* I/*Nco* I and remove 5' phosphate groups of the termini with calf intestinal phosphatase (CIP) to avoid self-ligation.
7. After digestion of pSW1 with *Nco* I/*Not* I, add 2 µL (20 U) of CIP directly to the tube and incubate for 1 h at 37°C.
8. Purify the digested products (pSW1 and PCR assembly products) by electrophoresis on a 1% low-gelling temperature agarose gel; extract from the gel using the Qiaquick gel-extraction kit or the Geneclean II kit.
9. Prepare the following ligation reactions:

	1	**2**	**3**
Digested pSW1 (100 ng)	100 ng	100 ng	100 ng
Digested PCR band (80–120 ng*)	–	80 ng	120 ng
Sterile, double-distilled water to	18 µL	18 µL	18 µL

* Molar ratio: plasmid/insert = 1/3 to 1/4.5

10. Incubate for 5 min at 50°C and leave for 1 min on ice.
11. Add 2 µL NEB T4 ligase buffer and 1 µL T4 DNA ligase (400 units/µL).
12. Leave overnight at 16°C; store at –20°C until further use.
13. Purify the ligated product using the Geneclean II kit.

3.4.2. E. Coli *Electrotransformation* **(31)**

3.4.2.1. Preparation of *E. coli* TG1 Electrocompetent Cells

1. Inoculate freshly grown *E. coli* single colony on agar plate to 5 mL SOB.
2. Grow overnight at 37°C with shaking at 250 rpm.
3. Add 5 mL of the *E. coli* TG1 overnight culture to 500 mL of SOB.
4. Grow at 37°C with shaking at 250 rpm until $OD_{600} = 0.5$ ($\pm$0.03).
5. Incubate on ice for 30 min.
6. Centrifuge at 3000*g* for 15 min at 4°C.
7. Resuspend the pellets in 500 mL of ice-cold sterile, double-distilled H_2O.
8. Centrifuge at 3000*g* for 15 min at 4°C.
9. Resuspend the pellets in 250 mL of ice-cold sterile, double-distilled H_2O.
10. Centrifuge at 3000*g* for 15 min at 4°C.
11. Resuspend the pellets in 20 mL of ice-cold 10% glycerol-sterile, double-distilled H_2O.
12. Centrifuge at 3000*g* for 15 min at 4°C.
13. Carefully remove the supernatant.
14. Resuspend the pellets in 200 µL of ice-cold 10% glycerol-sterile, double-distilled H_2O.
15. Use directly for electroporation or store 50 µL aliquots at –70°C.

3.4.2.2. Electrotransformation

1. Purify the ligated product (*see* **Subheading 3.4.1., step 13**) using the Geneclean II kit.
2. Thaw *E. coli* TG1 electrocompetent cells (*see* **Subheading 3.4.2.1.**) at room temperature and keep on ice.
3. Add 2 µL of the purified ligated product to 50 µL of *E. coli* TG1 electrocompetent cells (*see* **Note 10**).
4. Mix carefully by pipeting up and down. Do not create any bubbles.
5. Incubate for 1 min on ice.
6. Set the pulse generator of the electroporation unit to the 25 µF capacitor, 2.5 kV, and 200 Ω in parallel with the sample chamber.
7. Transfer the mixture of cells and ligated product to a cold 0.2-cm electroporation cuvet, and ensure that it fits to the bottom of the cuvet (shake if necessary) (*see* **Note 11**).
8. Apply a single pulse using the settings described in **step 5**. This should result in a pulse of 12.5 kV/cm with a time constant of 4.5–5 msec.
9. Immediately add 1 mL of prewarmed SOC medium (at 37°C) into the cuvet and gently but quickly resuspend the cells with a sterile Pasteur pipet.
10. Transfer the cell suspension to a 17 x 100 mm polypropylene tube and incubate at 37°C for 1 h with shaking at 225 rpm.
11. Plate appropriately diluted aliquots (10-fold, 10^3-fold, 10^5-fold, 10^7-fold) on TYE plates containing 100 µg/mL ampicillin and 1% glucose (TYE-AMP-GLU) to obtain single colonies.

3.4.2.3. PCR Screening

1. Single colonies of *E. coli* TG1 electroporated with pSW1 containing the scFv insert are obtained as described in **Subheading 3.4.2.2, step 11**.
2. Make up 20 µL PCR reaction mixes in 0.5-mL microcentrifuge tubes containing:

10X PCR buffer	2 µL
dNTP (5 m*M*)	1 µL
$MgCl_2$ (25 m*M*)	1 µL
LMB2 (10 µ*M*)	1 µL
LMB3 (10 µ*M*)	1 µL
Taq DNA polymerase	1 U
Sterile, double-distilled water to:	20 µL

3. Transfer single colonies into PCR reaction mixes with sterile toothpicks. In parallel, plate on TYE-AMP-GLU.
4. Heat at 94°C for 2 min in the PCR thermal cycler.
5. Perform 30 cycles as follows: 94°C for 1 min, 50°C for 1 min, 72°C for 1 min.
6. Heat at 72°C for 5 min for a final elongation.
7. Analyze 8 µL of PCR reaction on a 1% agarose gel. The expected size is approx 870 bp.

3.4.3. ScFv Expression and Purification

3.4.3.1. Expression of Soluble scFv in *E. coli*

1. Inoculate a positive colony (i.e., with an insert of approx 870 bp) from the TYE-AMP-GLU plate (*see* **Subheading 3.4.2.3. step 3**) into 6 mL of 2X TY medium containing 100 µg/mL ampicillin and 1% glucose (2X TY-AMP-GLU). Incubate at 37°C overnight.
2. Inoculate 300 µL of bacteria into 30 mL 2X TY-AMP-GLU(0.1%) (to give a 1/100 dilution).
3. Grow at 37°C with shaking at 250 rpm to OD_{600} = 0.9 (about 3 h 30 min).
4. Pellet bacteria at 3000*g* for 10 min at room temperature.
5. Resuspend the pellet in 30 mL of induction medium (2X TY medium containing 0.4 *M* sucrose, 100 µg/mL ampicillin and 1 m*M* IPTG) ***(32)***.
6. For harvesting scFv from bacterial culture supernatant:
   a. Grow bacteria at 29°C with shaking at 275 rpm overnight (16–20 h).
   b. Centrifuge the culture (10,800*g*, 15 min, at 4°C) and collect the supernatant.
   c. Filter the supernatant through a Sterivex-HV 0.45-µm filter unit (Millipore, Bedford, MA). Filtered supernatant can be used directly in various assays (immunofluorescence, Western-blot, dot-blot assays) or the scFv can be further purified if needed (*see* **Subheading 3.4.3.3.**).
7. For harvesting scFv from bacterial periplasm:
   a. Pellet bacteria at 3000*g* for 30 min at 4°C.

b. Resuspend in one-twentieth of the original volume with 30 m*M* Tris-HCl, pH 7.0, 20% (w/v) sucrose, 1 m*M* EDTA.
c. Leave on ice for 30 min while stirring.
d. Spin at 10,800*g* for 15 min at 4°C and collect the supernatant (periplasmic fraction) into a new tube.
e. Resuspend the pellet in one-twentieth of the original vol of 5 m*M* $MgSO_4$ (osmotic shock) and incubate for 30 min on ice with stirring.
f. Spin at 10,800*g* for 15 min at 4°C and collect the supernatant fraction; add to the periplasmic fraction. The scFv can be further purified from this preparation if needed (*see* **Subheading 3.4.3.3.**).

3.4.3.2. Dot-Blotting

1. Pre-wet a nitrocellulose filter for 10 min at room temperature in PBS-0.5% Tween 20.
2. Set the filter on the Dot-Microfiltration apparatus with 3 presaturated filter papers.
3. Load the scFv-containing preparation (periplasmic fraction or culture supernatant) into the Dot-Microfiltration apparatus, and vacuum until the fragments pass completely through the filter.
4. Saturate the filter for 30 min at room temperature in PBS-3% BSA-0.5% Tween 20.
5. Incubate the filter for 45 min at room temperature with anti-*c-myc* antibody (9E10) (4 μg/mL in PBS-3% BSA-0.5% Tween 20).
6. Wash the filter with PBS for 5 min (3×).
7. Incubate the filter for 1 h at room temperature with alkaline phosphatase-coupled goat anti-mouse IgG (H+L) (1/1000 dilution in PBS-3% BSA-0.5% Tween 20).
8. Wash the filter with PBS for 5 min (3×).
9. Add BCIP-NBT(5-bromo-4-chloro-3-indolyl phosphate/nitro blue tetrazolium) substrate. If the scFv expression is successful, an insoluble violet color should appear on the filter after a few minutes of incubation at room temperature.
10. Stop the reaction by rinsing the filter with sterile distilled water.

3.4.3.3. Purification of Soluble scFv

1. Dialyze the scFv fragments preparation (10-kDa cut-off) against loading buffer (20 m*M* Tris-HCl, pH 8.0, 500 m*M* NaCl) overnight at 4°C.
2. Prepare a column with 1-mL Talon™ resin (Clontech).
3. Equilibrate the column with 50 mL of loading buffer.
4. Load the scFv preparation onto the column.
5. Wash the column with 50 mL of washing buffer (20 m*M* Tris-HCl, pH 8.0, 500 m*M* NaCl , 20 m*M* imidazole).
6. Elute scFv with 20 mL of elution buffer (20 m*M* Tris-HCl, pH 8.0, 500 m*M* NaCl, 150 m*M* imidazole).
7. Collect 1-mL fractions.
8. Dialyze the scFv-containing fractions (as detected by ELISA or dot-blot) overnight at 4°C against PBS (10-kDa cut-off) to remove imidazole and NaCl.

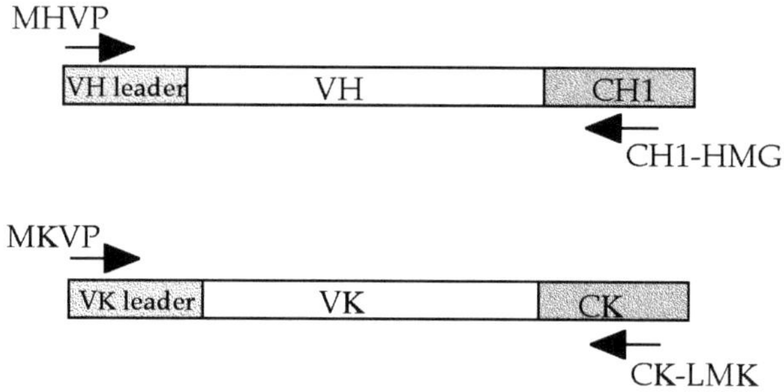

Fig. 2. Cloning of VH and VL regions with oligonucleotides hybridizing with the leader sequences and Constant CH1 and Cκ domains.

4. Notes

1. It is important to determine the adequate number of hybridoma cells to ensure a good RNA recovery yield. This depends on the H and L chain mRNA content of the hybridoma, which varies greatly between cell lines. 5×10^6 cells are generally processed. If RNA yield is not satisfactory, then prepare RNA from 2×10^7 cells per one RNeasy midi spin column.
2. Total RNA concentration is determined by measuring the absorbance at 260 nm (A_{260}=1.0 is equivalent to approx 40 µg/mL). This relation is valid only for measurements performed in water. Therefore, always perform several dilutions (from 1/10 to 1/50) of the RNA sample in RNase-free double-distilled water. The ratio between the OD readings at 260 nm and 280 nm (A_{260}/A_{280}) provides an estimate of the purity of RNA. Optimal ratios range from 1.5 to 1.9. Always check the RNA by running a 1% agarose gel before proceeding to the cDNA synthesis step. Purified RNA should be stored at –70°C in RNase-free double-distilled water to avoid degradation that may occur when stored at –20°C.
3. Amplification failure. The amplification of the VH or of the Vκ regions of some MAbs can be unsuccessful. These variable regions often belong to the "miscellaneous" VH and Vκ groups of the Kabat database ***(14)*** and PCR failure is likely caused by the presence of mismatches between the oligonucleotides used and the 5' or the 3' ends of the MAb VH or Vκ. An alternative strategy is to use primers that hybridize with VH or Vκ leader sequences (for the 5' end) ***(19)*** and with sequences located within the CH1 or C domains (for the 3' end), since the diversity of mouse VH and Vκ leader sequences is low (**Fig. 2**). Twelve pairs of MHVP leader/CH1 primers and eleven pairs of MκVP leader/Cκ primers can be tested (*see* **Tables 6–8**). This usually enables amplification of the MAb VH and Vκ regions. The corresponding PCR products are then sequenced, making it possible to design specific 5' and 3' oligonucleotide primers.
4. Aberrant transcript amplification: use of PNAs. Most of the myeloma fusion partners used to generate hybridomas are derived from the MOPC-21 cell line that

Table 6
Oligonucleotides Specific for Mouse VH Leader Sequences

MHVP1	5'-ACTAGTCGACATGAAATGCAGCTGGGTCATS*TTCTTC-3'
MHVP2	5'-ACTAGTCGACATGGGATGGAGCTRTATCATSYTCTT-3'
MHVP3	5'-ACTAGTCGACATGAAGWTGTGGTTAAACTGGGTTTTT-3'
MHVP4	5'-ACTAGTCGACATGRACTTTGGGYTCAGCTTGRTTT-3'
MHVP5	5'-ACTAGTCGACATGGACTCCAGGCTCAATTTAGTTTTCCTT-3'
MHVP6	5'-ACTAGTCGACATGGCTGTCYTRGSGCTRCTCTTCTGC-3'
MHVP7	5'-ACTAGTCGACATGGRATGGAGCKGGRTCTTTMTCTT-3'
MHVP8	5'-ACTAGTCGACATGAGAGTGCTGATTCTTTTGTG-3'
MHVP9	5'-ACTAGTCGACATGGMTTGGGTGGGAMCTTGCTATTCCTG-3'
MHVP10	5'-ACTAGTCGACATGGGCAGACTTACATTCTCATTCCTG-3'
MHVP11	5'-ACTAGTCGACATGGATTTTGGGCTGATTTTTTTTATTG-3'
MHVP12	5'-ACTAGTCGACATGATGGTGTTAAGTCTTCTGTACCTG-3'

* R=A or G, Y=C or T, K=G or T, M=A or C, S=G or C, W=A or T.

Table 7
Oligonucleotides Specific for Mouse Vκ Leader Sequences

Mκ VP1 :	5'-ACTAGTCGACATGAAGTTGCCTGTTAGGCTGTTGGTGCTG-3'
Mκ VP2 :	5'-ACTAGTCGACATGGAGW*CAGACACACTCCTGYTATGGGT-3'
Mκ VP3 :	5'-ACTAGTCGACATGAGTGTGCTCACTCAGGTCCTGGSGTTG-3'
Mκ VP4:	5'ACTAGTCGACATGAGGRCCCCTGCTCAGWTTYTTGGMWTCTTG-3'
Mκ VP5 :	5'-ACTAGTCGACATGGATTTWCAGGTGCAGATTWTCAGCTTC-3'
Mκ VP6 :	5'-ACTAGTCGACATGAGGTKCYYTGYTSAGYTYCTGRGG-3'
Mκ VP7 :	5'-ACTAGTCGACATGGGCWTCAAGATGGAGTCACAKWYYCWGG-3'
Mκ VP8 :	5'-ACTAGTCGACATGTGGGGAYCTKTTTYCMMTTTTTCAATTG-3'
Mκ VP9 :	5'-ACTAGTCGACATGGTRTCCWCASCTCAGTTCCTTG-3'
Mκ VP10:	5'-ACTAGTCGACATGTATATATGTTTGTTGTCTATTTCT-3'
Mκ VP11:	5'-ACTAGTCGACATGGAAGCCCCAGCTCAGCTTCTCTTCC-3'

* R=A or G, Y=C or T, K=G or T, M=A or C, S=G or C, W=A or T.

Table 8
Oligonucleotides Specific for Mouse CH1 and Cκ Constant Domains

CH1-HMG :	5'-GCCTCGAGTGGACAGGGATCCAGAGTTCCA-3'
Cκ -LMκ :	5'-GCCTCGAGTCGACACGACTGAGGCACCTCCAG-3'

produces an aberrant κ light-chain transcript, with a nonfunctional VJ recombination. The preferential amplification by RT-PCR of the aberrant transcript has been widely reported when trying to generate scFv from hybridomas. A method to specifically inhibit this amplification consists of hybridizing its CDR3 with a specific peptide nucleic acid (PNA) that forms highly stable complexes with cDNA, but cannot function as a primer for DNA polymerases. Thus, a PNA was designed to specifically block the PCR elongation of MOPC21 aberrant Vκ in conditions in which the functional Vκ from the hybridoma would be amplified from cDNA *(22)*. The PNA sequence is as follows: H_2N- CGTGTAAGC TCCCTA-H. A search made using the Kabat database showed that the amplification of only a few Vκ regions would be blocked using this PNA *(22)*. Prepare a 50-µL PCR mix containing Stoffel buffer (Perkin-Elmer), 3 m*M* $MgCl_2$, 250 µ*M* dNTPs, 0.25 µ*M* of backward and forward mouse Vκ PCR I primers, 10 µ*M* of PNA, 2 U AmpliTaq DNA polymerase (Stoffel fragment). Perform 30 PCR cycles as follows: 94°C for 2 min (before starting cycles); 94°C for 30 s, 65°C for 30 s, 50°C for 30 s, 60°C for 30 s. After completing the cycles, incubate at 60°C for 2 min.

5. The standard final concentration of $MgCl_2$ is 1.5 m*M*. However, if little or no PCR product is obtained, perform PCR with different final concentrations of $MgCl_2$ ranging from 1.5–5.0 m*M* (0.5-m*M* steps).
6. The amount of first-strand cDNA to be used in PCR I will depend on the relative abundance of the cDNA of interest in the final cDNA reaction mixture. Thus, test 1 µL, 5 µL, or 10 µL of cDNA in PCR I.
7. Point mutations that occur during PCR can be a major problem when generating scFv from a given hybridoma. Therefore, the use of a high-fidelity Taq DNA polymerase is recommended for PCR I, such as Pfu DNA polymerase (Stratagene).
8. Contamination may occur during purification of PCR II products from low-gelling-temperature agarose gels. First, to avoid contamination by DNases leading to PCR II product degradation, prepare all buffers (loading buffer, running buffer) using DNase-free double-distilled water; also wash the electrophoresis apparatus and combs thoroughly with DNase-free double-distilled water. Second, DNA contamination (plasmid, PCR product or phage) may occur. In that case, depurinate the electrophoresis apparatus and combs with 0.25 *M* HCl for at least 30 min before use and wash briefly with DNase-free double-distilled water *(33)*.
9. A fivefold overdigestion (as compared with calculated units needed for digesting a given amount of DNA at a given temperature by the manufacturer) with *Nco* I and a 25-fold overdigestion with *Not* I usually allows efficient digestion (more than 95%) of the PCR fragment. Supercoiled plasmids may require up to fivefold more *Not,* I enzyme for a complete digestion than linear DNA. Thus, digest the plasmid first with *Nco* I and then with *Not* I. Avoid *Nco* I digestion for longer than 4 h, since a nonspecific "star" activity can be observed using this enzyme.
10. Ligation efficiency is lower than 20%. Therefore, concentrate the ligated product by pooling several ligation reaction tubes (about five tubes) into one tube to obtain higher electroporation efficiency (especially for phage display); purify the pooled ligated product using the Geneclean II kit before electroporation.
11. Use a 0.2-cm electrode gap electroporation cuvet. To obtain higher eletroporation efficiency (for phage display), use a 0.1-cm electrode-gap electroporation cuvet. In this case, decrease the pulse-generator voltage down from 2.5 kV to about 2.1 kV.

Acknowledgments

This work was supported by grants from the BioAvenir program, the Association pour la Recherche sur le Cancer (ARC), the Ligue Nationale contre le Cancer (LNCC), the Institut Curie, the INSERM, the French MENRT (CRTG program) and from Aventis. Chang Hoon Nam is a PhD student supported by a fellowship from the Republic of Korea government (Korean Government Overseas Scholarship Program). Part of the work was performed in collaboration with Drs L. Bracco, C. Caron de Fromentel, I. Delumeau, F. Schweighoffer, and B. Tocqué (Aventis, Vitry sur Seine, France). The authors wish to thank Dr. O. Cochet and acknowledge the scientific advice of Pr. W. H. Fridman and Pr. D. Louvard.

References

1. Marasco, W.A. (1997) Intrabodies: turning the humoral immune system outside in for intracellular immunization. *Gene Ther.* **4,** 11–15.
2. Deshane, J., Grim, J., Loechel, S., Siegal, G.P., Alvarez, R.D. and Curiel, D.T. (1996) Intracellular antibody against erbB-2 mediates targeted tumor cell eradication by apoptosis. *Cancer Gene Ther.* **3,** 89–98.
3. Huston, J.S., Mudgett-Hunter, M., Tai, M.S., McCartney, J., Warren, F., Haber, E., et al. (1991) Protein engineering of single-chain Fv analogs and fusion proteins. *Methods Enzymol.* **203,** 46–88.
4. Kontermann, R.E., and Muller, R. (1999) Intracellular and cell surface displayed single-chain diabodies. *J. Immunol. Methods* **226,** 179–88.
5. Richardson, J.H., and Marasco, W.A. (1995) Intracellular antibodies: development and therapeutic potential. *Trends Biotechnol.* **13,** 306–310.
6. Cochet, O., Kenigsberg, M., Delumeau, I., Virone-Oddos, A., Multon, M.-C., Fridman, W.H., et al. (1998) Intracellular expression of an antibody fragment neutralizing p21ras promotes tumor regression. *Cancer Res.* **58,** 1170–1176.
7. Caron de Fromentel, C., Gruel, N., Venot, C., Debuscche, L., Conseiller, E., Dureuil, C., et al. (1999) Restoration of transcriptional activity of p53 mutants in human tumor cells by intracellular expression of anti-p53 single-chain Fv fragments. *Oncogene* **18,** 551–557.
8. Chester, K.A., Begent, R.H., Robson, L., Keep, P., Pedley, R.B., Boden, J.A., et al. (1994) Phage libraries for generation of clinically useful antibodies. *Lancet* **343,** 455–456.
9. Griffiths, A.D., Williams, S.C., Hartley, O., Tomlinson, I.M., Waterhouse, P., Crosby, W.L., et al. (1994) Isolation of high affinity human antibodies directly from large synthetic repertoires. *EMBO J.* **13,** 3245–3260.
10. Vaughan, T.J., Williams, A.J., Pritchard, K., Osbourn, J.K., Pope, A.R., Earnshaw, J.C., et al. (1996) Human antibodies with sub-nanomolar affinities isolated from a large non-immunized phage display library. *Nature Biotech.* **14,** 309–314.

11. Marasco, W.A., Haseltine, W.A., and Chen, S. (1993) Design, intracellular expression, and activity of a human anti-human immunodeficiency virus type 1 gp120 single-chain antibody. *Proc. Natl. Acad. Sci. USA* **90,** 7889–7893.
12. Hancock, J.F., Paterson, H. and Marshall, J. (1990) A polybasic domain or palmitoylation is required in addition to the CAAX motif to localize p21ras to the plasma membrane. *Cell* **63,** 133–139.
13. Cochet, O., Delumeau, I., Kenigsberg, M., Gruel, N., Schweighoffer, F., Bracco, L., et al. (1998) Intracellular targeting of oncogenes: a novel approach for cancer therapy. in: *Intrabodies, Basic Research and Clinical Gene Therapy Applications* Marasco, W.A.ed., R. G. Landes Company, Springer-Verlag, New York, pp. 129–146.
14. Lei, S.P., Lin, H.C., Wang, S.S., Callaway, J. and Wilcox, G. (1987) Characterization of the *Erwinia carotovora pelB* gene and its product pectate lyase. *J. Bacteriol.* **169,** 4379–4383.
15. Evan, G.I., Lewis, G.K., Ramsay, G., and Bishop, J.M. (1985) Isolation of monoclonal antibodies specific for human *c-myc* proto-oncogene product. *Mol. Cell. Biol.* **5,** 3610–3616.
16. Kabat, E.A., Wu, T.T., Perry, H.M., Gottesman, K.S. and Foeller, C. (1991) *Sequences of proteins of immunological interest,* 5th ed. U.S. Department of Health and Human Services, Public Health Service, National Institutes of Health, Bethesda, MD.
17. Orlandi, R., Güssow, D.H., Jones, P.T., and Winter, G. (1989) Cloning immunoglobulin variable domains for expression by the polymerase chain reaction. *Proc. Natl. Acad. Sci. USA* **86,** 3833–3837.
18. Ward, E.S., Güssow, D., Griffiths, A.D., Jones, P.T., and Winter G. (1989) Binding activities of a repertoire of single immunoglobulin variable domains secreted from *Escherichia coli. Nature* **314,** 544–546.
19. Cochet, O., Kenigsberg, M., Delumeau , I., Duchesne, M., Schweighoffer, F., Tocqué, B., et al. (1998) Intracellular expression and functional properties of an anti-p21ras scFv derived from a rat hybridoma containing specific λ and irrelevant κ light chains. *Mol. Immunol.* **35,**1097–1110.
20. Cuisinier, A.M., Fumoux, F., Moinier, D., Boubli, L., Guigou, V., Milili, M., et al. (1990) Rapid expansion of human immunoglobulin repertoire (VH, Vκ , V) expressed in early fetal bone marrow. *New Biol.* **2,** 689–699.
21. Jones, S.T., and Bending, M.M. (1991) Rapid PCR-cloning of full-length mouse immunoglobulin variable regions. *BioTechnol.* **9,** 579.
22. Cochet, O., Martin, E., Fridman, W.H., and Teillaud, J-L. (1999) Selective PCR amplification of functional immunoglobulin light chain from hybridoma containing the aberrant MOPC 21-derived Vκ by PNA-mediated PCR clamping. *BioTechniques* **26,** 818–822.
23. Egholm, M., Buchardt, O., Christensen, L., Behrens, C., Freier, S.M., Driver, D.A., et al. (1993) PNA hybridizes to complementary oligonucleotides obeying the Watson-Crick hydrogen-bonding rules. *Nature* **365,** 566–568.
24. Orum, H., Nielsen, P.E., Egholm, M., Berg, R.H., Buchardt, O., and Stanley, C.

(1993) Single base pair mutation analysis by PNA directed PCR clamping. *Nucleic Acids Res.* **21,** 5332–5336.
25. Pelegrin, M., Marin, M., Noel, D., and Piechaczyk, M. (1998) Genetically engineered antibodies in gene transfer and gene therapy. *Human Gene Ther.* **9,** 2165–2175.
26. Zhu, Q., Zeng, C., Huhalov, A., Yao, J., Turi, T.G., Danley, D., et al. (1999) Extended half-life and elevated steady-state level of a single-chain Fv intrabody are critical for specific intracellular retargeting of its antigen, caspase-7. *J. Immunol. Methods* **231,** 207–222.
27. Ramm, K., Gehrig, P., and Pluckthun, A. (1999) Removal of the conserved disulfide bridges from the scFv fragment of an antibody: effects on folding kinetics and aggregation. *J. Mol. Biol.* **290,** 535–46.
28. Visintin, M., Tse, E., Axelson, H., Rabbitts, T.H., and Cattaneo, A. (1999) Selection of antibodies for intracellular function using a two-hybrid in vivo system. *Proc. Natl Acad. Sci. USA* **96,** 11,723-11,728.
29. Rizzuto, R., Simpson A.W.M., Brini, M., and Pozzan, T. (1992) Rapid changes of mitochondrial Ca++ revealed by specifically targeted recombinant æquorin. *Nature* **358,** 325-327.
30. Harrison, J.L., Williams, S.C., Winter, G., and Nissim A. (1996) Screening of phage antibody libraries. *Methods Enzymol.* **267,** 83-109.
31. Dower, W.J., Miller, J. F., and Ragsdale C. W. (1988) High efficiency transformation of *E. coli* by high voltage electroporation. *Nucleic Acids Res.* **16,** 6127–6145.
32. Kipriyanov, S. M., Moldenhauer, G., and Little M. (1997) High level production of soluble single-chain antibodies in small-scale *Escherichia coli* cultures. *J. Immunol. Methods* **200,** 69–77.
33. Bradbury, A., Ruberti, F., Werge, T., Amati, V., Luzio, A.D., Gonfloni, S., et al. (1995) The cloning of hybridoma V regions for their ectopic expression in intracellular and intercellular immunization. in *Antibody Engineering.* Borrebaeck, C.A., ed., Oxford University Press.

24

Quantitation of mRNA Levels by RT-PCR in Cells Purified by FACS

Application to Peripheral Cannabinoid Receptors in Leukocyte Subsets

Jean Marchand and Pierre Carayon

1. Introduction

Cell-sorting and mRNA amplification techniques with RT-PCR have been combined to study the gene expression of cytokines ***(1–3)***, receptors ***(4–6)*** and differentiation antigens of purified hematopoietic cells ***(7–9)***. Nevertheless, a major drawback of these techniques is the difficulty in obtaining enough purified cells. The extended cell-sorting times necessary for this make it unsuitable for routine analysis when cell populations are poorly represented or target genes are weakly expressed. These studies are thus limited to qualitative or semi-quantitative approaches to mRNA from sorted cells. Recently, we studied the distribution patterns of central and peripheral cannabinoid receptors (CB2) in human immune tissues and leukocyte subpopulations using a quantitative reverse-transcriptase-polymerase chain reaction (RT-PCR)-based method ***(10)***. As a part of this study, we report here on a novel technique for quantitative assay of this receptor, applicable to a number of sorted cells as low as 2×10^5. This method combines quantitative RT-PCR and measurement of PCR products using an enzyme detection system in a 96-well microplate.

More precisely, this procedure comprises three steps:

1. Extraction-purification of polyA$^+$ mRNA and synthesis of first-strand cDNA are carried out with the polyATtract series 9600 commercially available from Promega. Using this system, mRNA-derived cDNA can be isolated from 96 samples in one experiment ***(11)***.

From: *Methods in Molecular Biology, vol. 193: RT-PCR Protocols*
Edited by: J. O'Connell © Humana Press Inc., Totowa, NJ

2. Quantitative PCR is performed within the exponential phase of amplification *(12)*; in this phase of amplification the amount of PCR products is correlated to the initial amount of template. The kinetic profiles of amplification are determined by analyzing the PCR products synthesized with varying numbers of cycles, using a fixed quantity of cDNA *(13)*. Relative measurements of CB2-receptor level are obtained by comparing the amplification kinetics of the target gene and the β_2-microglobulin gene (β_2-m), a relatively invariant housekeeping mRNA that is amplified in a separate reaction; this leads to relative quantitative PCR *(14)*. Consequently, variations in the amount of PCR products measured in the exponential range correspond to the same variations in the initial amount of mRNA template.
3. Measurement of PCR products is achieved using a microplate assay. This last step is based on the capture of heat-denatured PCR products by a complementary target-specific oligonucleotide probe coated onto the microplates, followed by hybridization with a second target-specific oligonucleotide linked to biotin, and finally by incubation with avidin-horseradish peroxidase (HRP) conjugate and enzymatic detection *(15)*.

2. Materials

2.1. Cell Sorting for Quantitative RT-PCR

1. Antibodies: phycoerythrin-conjugated CD4, CD8, CD11, CD56, and fluorescein-activated cell sorting (FITC)-conjugated CD3 (Becton Dickinson [BD], San Jose, CA).
2. Cells: mononuclear cells isolated from Ficoll-Hypaque density centrifugation of peripheral blood obtained from three consenting healthy donors.
3. Flow cytometer: FACS Vantage (BD).

2.2. mRNA Isolation and cDNA Synthesis

1. PolyATtract Series 9600 mRNA Isolation and cDNA Synthesis Systems (Promega, Madison, WI).
2. Phosphate-buffered saline (PBS): 0.2 g/L KH_2PO_4, 1.15 g/L Na_2HPO_4, 8 g/L NaCl, 0.2 g/L KCl, pH 7.4.

2.3. Quantitative PCR

1. Perkin-Elmer-Cetus GeneAmp PCR System 9600 (PE Applied Biosystems, Courtaboeuf, France).
2. AmpliTaq DNA polymerase (5 U/μL) with GeneAmp 10X PCR buffer (PE Applied Biosystems).
3. Deoxynucleotide 5' triphosphate (dNTPs) (20 m*M*), DNA polymerization mix (Pharmacia Biotech, Orsay, France). Make up a 1.25 m*M* solution in DNase-free water.
4. Primers (Genset, Paris, France):
   β2-m sense, 5'-CCAGCAGAGAATGGAAAGTC-3'.
   β2-m anti-sense, 5'-GATGCTGCTTACATGTCTCG-3'.

CB2 sense, 5'-TTTCCCACTGATCCCCAATG-3'.
CB2 anti-sense, 5'-AGTTGATGAGGCACAGCATG-3'.
Prepare 5 μ*M* solutions of each primer in 1X PCR buffer.

5. MicroAmp 96-well reaction tubes/tray/retainer assemblies, bases and caps (PE Applied Biosystems).

2.4. Enzymatic Microplate Assay

1. Oligonucleotide probes (Genset):
   β_2-m capture probe, 5' CAATTCTCTCTC CATTCTTCAGTAAGTCAAC-3'.
   β_2-m detection probe, 5' biotin-AGAAAGA CCAGTCCTTGCTG-3'.
   CB2 capture probe, 5'-GCCAACCTCACATCCAGCC TCATTCGGGC-3'.
   CB2 detection probe, 5' biotin-TGGGAACCAACAGATGAGGA-3'.
2. Preparation of microplate coated with capture probe: Add 200 μL of a capture oligonucleotide solution (0.5 μg/mL in phosphte-buffered saline (PBS) buffer with 0.1 *M* $MgCl_2$) in 96-well-microplates (Nunc-Immuno Module MaxiSorp F8 framed, Nunc, Roskilde, Denmark), incubate overnight at room temperature. Wash 3× with 300 μL of a 0.2 *N* NaOH, 0.5% Tween 20 solution and then 3× with deionized water. Dry the microplates for 20 min at 50°C, and immediately seal them. Under these conditions, coated microplates may be stored for several months at 4°C.
3. Hybridization buffer: 0.75 *M* NaH_2PO_4, pH 7.4, 5 *M* NaCl, 5 m*M* Ethylenediamine-tetraacetic acid (EDTA), 0.5% Tween-20, Thimerosal 0.01%.
4. Washing buffer: 0.01 *M* Tris/HCl, pH 7.4, 0.3 *M* NaCl, 0.1 % Tween-20.
5. Avidin-peroxidase solution: 1/5000-diluted Extravidin-peroxidase conjugate (Sigma, Saint Quentin Fallavier, France) in PBS/bovine serum albumin (BSA) (0.3%) buffer.
6. Ortho-phenylene diamine solution (OPD) (Sigma): 400 μg/mL in 0.05 *M* phosphate-citrate, pH 5, containing 0.03% H_2O_2.
7. Stop solution: 4 *N* H_2SO_4.

3. Methods

3.1. Preparation and Analysis of Cells by Flow Cytometry

3.1.1 Starting Leukocyte Subsets

Cell-surface phenotyping is performed by incubating peripheral-blood mononuclear cells (PBMC) with the antibodies CD4-PE/CD3-FITC (T4 cells: $CD4^+CD3^+$); CD8-PE/CD3-FITC (T8 cells: $CD8^+CD3^+$); CD56-PE (NK cells: $CD56^+$); CD14-PE (monocytes: CD14+); and CD20-PE (B cells: $CD20^+$).

1. Place 100 μL of mononuclear cells at 10^7/mL concentration in PBS in a 12 × 75 mm tube.
2. Add 10 μL of appropriate antibodies and vortex for 1 s.
3. Incubate for 20 min on ice, then centrifuge at 300*g* for 5 min.
4. Decant the supernatant, vortex for 1 s and add 2 mL of cold PBS.

3.1.2. Flow Cytometry and Sorting

1. Cytometer settings: laser power: 100 mW; wavelength: 488 nm; filters: 515 nm BP (to collect green signal for FITC) and 575 nm BP (to collect orange signal for phycoerythrin). Contamination of FITC in the phycoerythrin channel is compensated using the CaliBRITE^{TTM} 3 kit (BD).
2. Analysis and sorting. Samples are run at a flow rate of 1000 cells per second. A region of interest is drawn around the positive cells. This gate is used to discriminate T4, T8, NK, monocytes, and B cells. Cell sorting of 2×10^5 cells of each subset is performed using the Normal-C mode of the FACS Vantage Cytometer. Purification of each subset is checked by re-analyzing another sorting run. In our experience, this procedure led to cell subpopulation purities ranging from 95%– 99.5%.

3.2. Isolation of PolyA$^+$ mRNA and Synthesis of cDNA

The mRNA purification and conversion to first-strand cDNA are performed with the PolyATract mRNA Isolation System with cDNA Synthesis Reagents according to the manufacturer's protocol with minor modifications.

1. Harvest 2 to 4×10^5 counted cells (*see* **Note 1**) and wash once by centrifugation in PBS. Lyse the resulting cell pellets by adding 20 µL of extraction buffer; in these conditions total RNA is released into lysate solutions which could be stored at –20°C or –70°C.
2. Transfer the lysate solution to a well of a V-bottom 96-well plate (Costar corporation, Cambridge, MA), and mix with 40 µL of prewarmed (70°C) hybridization buffer containing synthetic biotinylated oligo(dT) probe. Incubate at room temperature for 5 min.
3. Add and mix 35 µL of polystyrene blocking beads suspension (*see* **Note 2**), centrifuge at 1500*g* for 10 min to pellet the protein; the polystyrene blocking beads cap the pellet and prevent it from contaminating mRNA in the supernatant during the rest of the procedure ***(11)***.
4. Carefully add 75 µL of washed streptavidin-coated paramagnetic particles suspension, as described in the technical manual, and incubate the cell lysate plate for 2 min at room temperature. In this step the biotinylated oligo(dT)-polyA$^+$ RNA complexes are selectively captured and then immobilized on the pins of the 96-steel-pin array magnetized with an external magnet. Incubate until the particles are cleared from the lysate (approx for 2 min).
5. Wash the collected particles as described in the manufacturer's technical manual. Briefly, the pins are withdrawn from the first microplate and lowered into a fresh one containing 180 µL of high-stringency wash buffer. By removing the magnet and replacing it for 2 min, the particles are dislodged and the mRNA complexes are alternatively released from the pins and then regathered; these washing steps allow removal of any residual contaminating proteins and DNA.
6. Finally, transfer the purified mRNA complexes to an elution plate (GeNunc Module 120) containing 20 µL of nuclease-free water (contact for 20 s). The low ionic

strength disrupts the complexes; collect the particles magnetically for 2 min, and the purified mRNA remains in solution and is ready for cDNA synthesis.

7. Convert the purified mRNA to first-strand cDNA by adding 10 µL of a reverse transcriptase mix (containing 10 U of Avian myeloblastoma virus (AMV) RT, 0.6 µg of oligo[dT]$_{15}$ primer, 0.15 m*M* dNTPs and 30 U of RNasin ribonuclease inhibitor in gold buffer) to each well of the GeNunc Module, seal the sample well and incubate at 37°C for 60 min. To stop the reaction, heat-inactivate at 95°C for 5 min and then immediately chill on ice for 5 min. The cDNA solution must be stored at –70°C.

3.3. PCR Amplification

1. To study one target, make up a 320-µL PCR master mix solution containing: 1X PCR buffer (20 m*M* Tris-HCl, pH 8.3, 50 m*M* KCl, 1.5 m*M* $MgCl_2$ and 0.01% gelatin), 0.1 m*M* dNTPs, 0.25 µ*M* each of 5' and 3' primers of the target (β_2-m for example), 8 µL of cDNA and 0.025 U/µL AmpliTaq DNA polymerase.
2. Distribute 20 µL of the master mix solution into 16 reaction tubes on the 96-well tray without the retainer set; in this way, it is possible to remove an individual well at each of 16 defined cycle numbers during the PCR, so the kinetics can be followed (*see* **Note 3**).
3. Repeat these two steps with the second pair of primers (CB2 in this case) with the same cDNA.
4. Amplify the reaction tubes with the 9600 thermal cycler; the amplification profile involves denaturation at 95°C for 20 s, primer annealing at 60°C for 30 s, and extension at 75°C for 20 s.
5. Remove one tube from the 96-well tray at cycles 16–31 for β2-m, and at cycles 27–42 for CB2, to carry out kinetic experiments.

3.4. Microplate Assay

1. Before the assay, prehybridize 16 wells of the coated-microplate (β_2-m and CB2) with 200 µL of hybridization buffer for 30 min at 37°C (*see* **Note 3**).
2. Discard the supernatant and distribute 200 µL of detection probe solution (50 ng/mL in hybridization buffer), corresponding to the target studied (β_2-m or CB2).
3. Add 7 µL of PCR products, after first heat-denaturing for 10 min at 95°C, into each well with a multichannel pipet. Then seal the microplates and incubate for 2 h at 37°C with gentle shaking (420 rpm) in a microplate shaker incubator. .
4. Wash the microplate four times with 300 µL of washing buffer using a microplate washer.
5. Incubate with 200 µL of avidin-peroxidase conjugate solution for 30 min at 37°C with shaking.
6. Wash 4× with 300 µL of washing buffer.
7. Distribute 200 µL of OPD solution and incubate at 37°C for 15 min without shaking.
8. Stop the reaction by adding 50 µL of stop solution.
9. Measure optical densities at 490 nm with a microplate reader (*see* **Notes 4** and **5**).

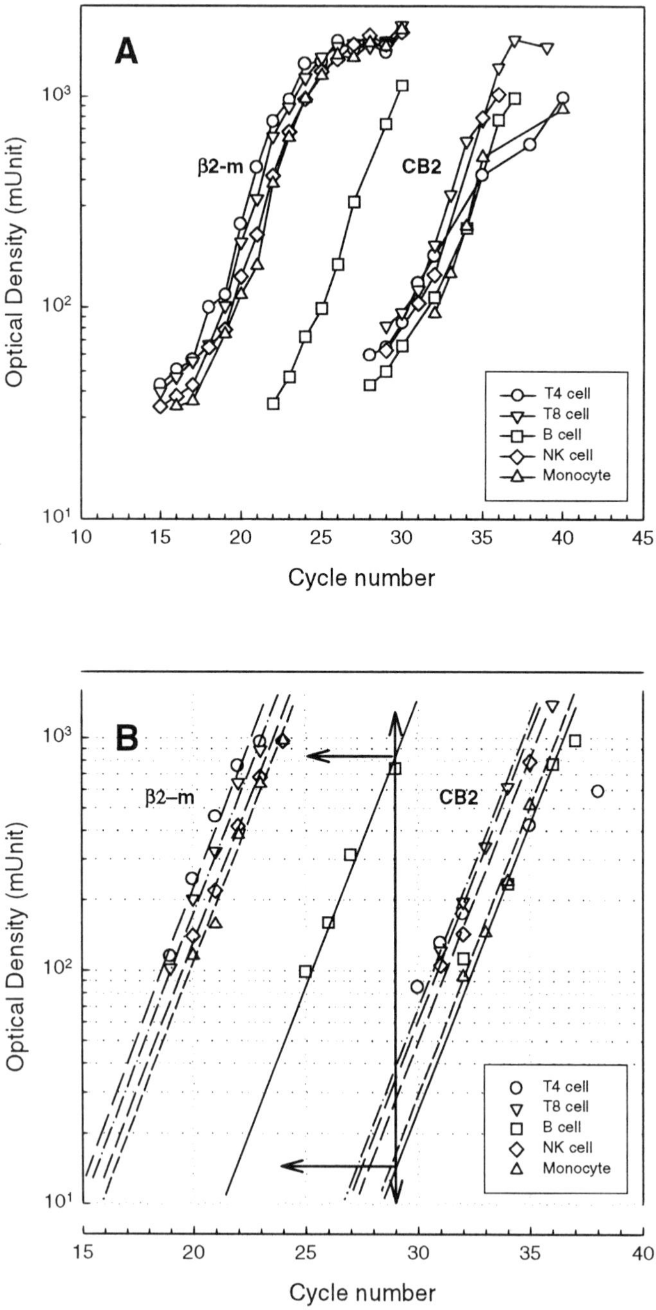

Fig. 1. Kinetic analysis of amplification of β_2-m and CB2 from leukocyte subsets from human peripheral blood. **(A)** Independent PCR amplifications with sequential cycles are performed using either β_2-m and CB2 specific primers with cDNA derived from 2×10^5 T4, T8, B, NK, and Monocyte cells. PCR reaction mixtures are analyzed by microplate assay and PCR product measurements are plotted as a function of cycle

3.5. Kinetic Analysis

To measure the amount of PCR products in the exponential range, kinetic analysis is performed over 16 cycles of PCR for each target (**Fig. 1A**). To correct for any variation in RNA content and cDNA synthesis in different preparations, each sample is normalized on the basis of its β_2-m content. The relative expression of CB2 to β_2-m was obtained by the ratio of their respective optical density (OD) values measured at a given cycle in the exponential phase that is defined as a linear part of the kinetic curve in lin-log representation (**Fig. 1B**). This normalization allowed the relative quantification of CB2 mRNA levels (**Table 1**) (*see* **Notes 6** and **7**).

4. Notes

1. Sample considerations: 10^5 cells were required to obtain a good signal for the CB2 target; conversely do not exceed the maximum sample concentration up to 4×10^5 cells per 20 μL of extraction, because the rate of purification will be decreased.
2. During the mRNA purification step procedure, after adding hybridization buffer and the blocking particle suspension, mix by pipetting up and down.
3. The analysis of cDNA obtained from 6-cell subsets requires two 96-well PCR reaction tube assemblies ($2 \times 6 \times 16$), corresponding to the amplification of the two targets studied; for the microplate hybridization assay used to measure the PCR products, for the same number of samples, one complete coated microplate is required for β_2-m and also for CB2.
4. To check for any contamination by genomic DNA, thereby distorting the quantitative results, we controlled all assayed samples by running 40 cycles of a PCR for the β2-m gene, with primers spanning one of its short introns. After analysis by agarose-gel electrophoresis, none of samples showed amplicons with the mol wt of the genomic sequence (greater than that of the cDNA), compared with a positive control run in parallel. Furthermore, the quantification of PCR products by the microplate assay applied to mRNA samples not reverse-transcribed into cDNA showed a genomic DNA contamination rate under 0.01%.
5. The sensitivity of the microplate assay was determined to be approx 10 pg/well for both targets (45 attomoles/well). The amplimer quantities synthesized during the exponential phase of PCR were between 0.15–43 ng/well. These data demonstrate that the sensitivity of the microplate assay is not the limiting step considering the assay as a whole ***(15)***.
6. The reproducibility over three separate experiments showed a CV $\leq$ 10% for the data obtained in the exponential portion of the curve used to determine the rela-

(continued) number. **(B)** The PCR products measured (OD) are plotted as a function of the number of cycles, more precisely in the exponential phase of amplification. Measured in this linear part of the curve at a given cycle, the relative amount of specific amplicon in each sample is normalized with their β_2-m contents and allowed to compare the sample with each other.

Table 1
Comparison of Relative Amounts of CB2-mRNA in Different Leukocyte Subsets Measured by RT-PCR*

Leukocyte subsets	CB2-mRNA (%)
T4 cells	9
T8 cells	7.9
B cells	100
NK cells	8
Monocyte	6.2

* In B cells the ratio of the expression of CB2 to β_2-m was found to be 2%. To simplify, the results are expressed taking the mRNA content of B cells as 100%.

tive quantities of mRNA. For the calculated final results, the reproducibility was CV $\leqslant$ 5% ***(15)***.

7. The constraints of this technique are those generally encountered with PCR reactions: defining the choice of primers, capture, and detection probes complementary to target sequences located between primers. Therefore, this method may be easily extrapolated to the assay of many targets such as receptors, cytokines, and growth factors, performed on rare cell subsets sorted by flow cytometry.

References

1. Matsue, H., Cruz, P.D., Jr., Bergstresser, P.R., and Takashima, A. (1993) Profiles of cytokine mRNA expressed by dendritic epidermal T cells in mice. *J. Investig. Dermatol.* **101,** 537–542.
2. Tollner, K.M., Scheel-Toellner, D., Sprenger, R., Duchrow, M., Trumper, L.H., Ernst, M., et al. (1995) The human germinal centre cells, follicular dendritic cells and germinal centre T cells produce B cell-stimulating cytokines. *Cytokine* **7,** 344–354.
3. Frankenberger, M., Sternsdorf, T., Pechumer, H., Pforte, A., and Ziegler-Heitbrock, L. (1996) Differential cytokine expression in human blood monocyte subpopulations: a polymerase chain reaction analysis. *Blood* **87,** 373–377.
4. Visser, J.W.M., Rozemuller, H., De Jong, M.O., and Belyavsky, A. (1993) The expression of cytokine receptors by purified hemopoietic stem cell. *Stem Cells* [Supp 2] **11,** 49–55.
5. Bani, L., David, D., Moreau, J-L., Cayota, A., Nakarai, T., Ritz, J., et al. (1996) Expression of the IL-2 receptor γ subunit in resting human CD4$^+$ lymphocytes: mRNA is constitutively transcribed and the protein stored as an intracellular component. *Int. Immunol.* **9,** 573–580.
6. Maric, D., Maric, I., Wen, X., Fritschy, J-M., Sieghart, W., Barker, J.L., et al. (1999) GABA$_A$ receptor subunit composition and functional properties of Cl- channels with differential sensitivity to zolpidem in embryonic rat hippocampal cells. *J. Neurosci.* **19,** 4921–4937.
7. Ziegler, B.L., Lamping, C.P.,Thoma, S.J., and Fliedner, T.M. (1995) Analysis of

gene expression in small numbers of purified hemopoietic progenitor cells by RT-PCR. *Stem Cells* [Supp 1] **13,** 106–116.

8. Ratech, H., Denning, S., and Kaufman, RE. (1997) An analysis of alternatively spliced CD45 mRNA transcripts during T cell maturation in humans. *Cell. Immunol.* **177,** 109–118.
9. Holyoake, T., Jiang, X., Eaves, C., and Eaves, A. (1999) Isolation of a highly quiescent subpopulation of primitive leukemic cells in chronic myeloid leukemia. *Blood* **94,** 2056–2064.
10. Galiègue, S.,Mary, S., Marchand, J., Dussossoy, D., Carrière, D., Carayon, P., et al. (1995) Expression of central and peripheral cannabinoid receptors in human immune tissues and leukocyte subpopulations. *Eur. J. Biochem.* **232,** 54–61.
11. Oaklander, A.L., and Ekenberg, S.J. (1994) Small-scale magnetic isolation of mRNA and synthesis of cDNA in 96-well plates. *Nature* **371,** 631–632.
12. Murphy, L.D., Herzog, C.E., Rudick, J.B., Fojo, A.T., and Bates, S.E. (1990) Use of the polymerase chain reaction in the quantitation of mdr-1 gene expression. *Biochemistry* **29,** 10351–10356.
13. Carayon P., Bouaboula, M., Loubet, J.F., Bourrié, B., Petiprêtre, G., Le Fur, G., et al. (1995) The sigma ligand SR 31747 prevents the development of acute graft-versus-host disease in mice by blocking IFN-γ and GM-CSF mRNA expression. *Int. J. Immunopharmacol.* **17,** 753–761.
14. Köhler, T. (1995) General aspects and chances of nucleic acid quantitation by PCR, in *Quantitation of mRNA by Polymerase Chain Reaction. Non radioactive PCR Methods* Köhler, T., ed., Springer-Verlag, Berlin Heidelberg, pp. 3–14.
15. Marchand, J., Bord, A., Pénarier, G., Lauré, F., Carayon, P., and Casellas, P. (1999) Quantitative method to determine mRNA levels by reverse transcriptase-polymerase chain reaction from leukocyte subsets purified by fluorescence-activated cell sorting: application to peripheral cannabinoid receptors. *Cytometry* **35,** 227–234.

IX

RT-PCR in Anti-Sense Technology

25

Detection of Anti-Sense RNA Transcripts by Anti-Sense RT-PCR

Michael C. Yeung and Allan S. Lau

Reverse-transcriptase-polymerase chain reaction (RT-PCR) is a powerful and sensitive tool for RNA detection. This technique is particularly valuable in cases where the amount of an RNA species present is so minute that traditional RNA analysis methodologies such as Northern blot or RNase protection assay may not be adequate. It has been estimated that RT-PCR is thousands of times more sensitive than Northern blot analysis *(1)*.

The sensitivity of RT-PCR derives from the power of PCR to exponentially amplify DNA sequences-after thirty PCR cycles, one target DNA molecule is amplified a billion-fold. Because PCR targets only DNA sequences, RNA is first converted by reverse transcriptase (RT) to single-stranded complementary DNA (cDNA) which can now be amplified by various PCR protocols. Avian myeloblastosis virus (AMV) or Moloney murine leukemia virus (MMLV) reverse transcriptases are commonly used for cDNA synthesis. For reverse transcription to occur, RNA must be primed at the 3' end (**Fig. 1**). This can be easily achieved by using primers of random sequences that will hybridize to their respective complementary RNA sequences along the RNA molecule. Thus, this random-primed method can be used for any RNA species, including ribosomal and messenger RNA (**Fig. 1A**). For mRNA, most of which ends in 3' polyadenlyation, specific 3' priming can be effected by annealing with oligo(dT) (**Fig. 1B**). If the 3' sequence of a target RNA is known, a 3' primer can be used to specifically prime the RNA, resulting in a cleaner PCR product profile (**Fig. 1C**). Priming with any of these three methods, followed by extension with reverse transcription will result in a cDNA template for the 5' sense primer in the PCR reaction, allowing the PCR to proceed. As long as an RNA molecule can be reverse-transcribed to render a target sequence for an

From: *Methods in Molecular Biology, vol. 193: RT-PCR Protocols*
Edited by: J. O'Connell © Humana Press Inc., Totowa, NJ

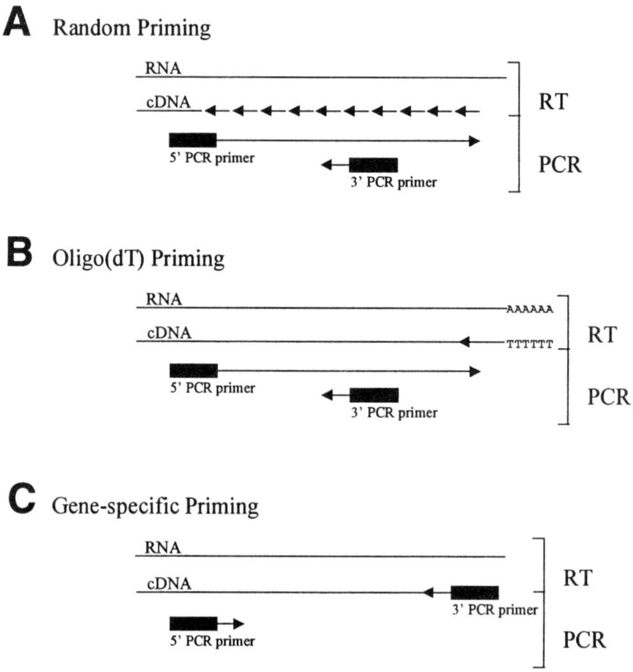

Fig. 1. cDNA priming methods. **(A)** Random priming: random sequence oligos hybridize to RNA template and are extended with RT. **(B)** Oligo(dT) priming: 3' ends of polyA-tailed mRNA molecules are annealed to oligo(dT) and extended with RT. **(C)** Gene-specific priming: gene-specific oligo, such as the 3' PCR primer, is annealed to the RNA and extended by RT.

ensuing PCR reaction, even partially degraded RNA can be amplified and be detected.

The foregoing discussion described how RT-PCR is normally employed to study sense RNA transcripts. However, there are times when anti-sense RNA transcripts are under investigation. There are at least three scenarios in which anti-sense RNA transcripts are synthesized in vivo: (1) For some RNA viruses such as picornaviruses that carry a positive-strand (sense) RNA genome, the generation of negative-strand (anti-sense) RNA templates is part of the viral replication cycle. Thus, studies on the life cycle of these viruses and the cellular virus-host interaction may necessitate identification and detection of these viral anti-sense RNA species *(2)*. (2) It is now known that anti-sense RNAs occur naturally in vivo and may be involved in gene regulation. For instance, mRNA anti-sense to the basic fibroblast growth factor (bFGF) has been identified in the rat *(3)*. The protein encoded by this anti-sense species is detectable in a tissue- and age-dependent manner. Further, this protein has been impli-

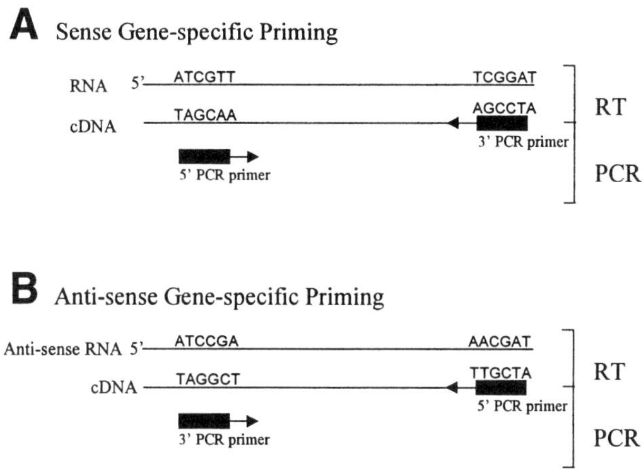

Fig. 2. Anti-sense RNA priming. (**A**) Sense RNA sequence-specific priming. This is the same as depicted in **Fig. 1C**, in which the 3' PCR primer is used to prime the sense RNA transcript, and is extended by RT. (**B**) Anti-sense RNA transcript is primed by the 5' PCR primer and is extended by RT. Note that the cDNA synthesized has exactly the same sequence as the sense RNA transcript.

cated in the regulation of the sense bFGF mRNA expression and turnover. Similarly, RNA transcripts anti-sense to the WT1 gene are readily detected by *in situ* hybridization in the human kidney, and have been associated with positive regulation of WT1 protein levels in vivo *(4)*. (3) The use of anti-sense RNA technology has been very popular in research on cancer, gene therapy, and developmental studies in recent years *(5)*. With this technology, RNA transcripts anti-sense to a gene under study are generated from an expression vector introduced by various methods into host cells. Transgenic mice harboring anti-sense expression constructs have also been made *(6)*. The anti-sense transcripts hybridize with their respective sense mRNAs, preventing the translation of the latter and thus effectively inhibit the expression of the gene in question. For these experiments, it is crucial to monitor the generation and the amount of anti-sense RNA transcripts so that the effectiveness in inhibiting gene expression may be assessed.

Because of its sensitivity, RT-PCR can be an invaluable tool for anti-sense RNA detection. However, the RT-PCR strategy described above in **Fig. 1** must be modified to accommodate the need for differentially amplifying the anti-sense but not sense RNA transcripts. This can be realized by specifically reverse-transcribing the anti-sense RNA transcripts only. **Figure 2** outlines the methodology to achieve this; note that the PCR protocol for the anti-sense RNA transcript remains the same and that no additional reagents are required.

The 5' PCR primer from an optimized PCR reaction previously designed for a given sense RNA transcript is used to prime the 3' end of the anti-sense RNA transcript. Subsequent extension from this primed site by RT generates a single-stranded cDNA strand that will serve as a template for the 3' PCR primer for the ensuing polymerase chain reaction. The final PCR product will be identical to that generated from the sense RNA transcript in every respect, except that the PCR product so generated is derived from the anti-sense RNA transcript only. With this anti-sense RT-PCR scheme, sense RNA transcripts will not be amplified.

The following example illustrates an application of the anti-sense RT-PCR protocol. The multifaceted protein kinase PKR is involved in TNF-α-induced apoptosis as well as inhibition of establishment of persistent viral infections *(2,7,8)*. To investigate these and other functions of PKR, a stable promonocytic cell line (U9K-A) harboring a PKR anti-sense expression vector was established by electroporation. To ascertain that PKR anti-sense RNA transcripts were indeed synthesized in the stable cell line, total RNA was extracted from U9K-A cells by the Chomczynski and Sacchi method *(9)* and subjected to anti-sense RT-PCR to monitor the generation of PKR anti-sense transcripts. To reverse-transcribe only the PKR anti-sense transcript, 1 μg of total RNA together with 20 pmols (1 μL from a 20-μ*M* stock) of the 5' sense PCR primer (5'-GGCACCCAGATTTGACCTTC-3') for PKR in 9 μL of nuclease-free water was heated at 90°C for 30 s. After rapidly cooling the mixture on ice, 11 μL of RT mix [4 μL of 5X RT buffer (LifeTechnology, Rockville, MD), 2 μL of 100 m*M* dithiothreitol (DTT), 1 μL of RNAGuard (50 U; APBiotech, Piscataway, NJ), 3 μL of 10 m*M* dNTPs (APBiotech, Piscataway, NJ), 1 μL of MMLV RT (200 U; LifeTechnology, Rockville, MD)] was added to the RNA/primer mix. Reverse transcription was allowed to proceed at 37°C for 2 h. Amplification of the reverse-transcribed cDNA was accomplished by subjecting up to 5 μL of the cDNA to a 100-LPCR reaction containing 50 m*M* KCl, 10 m*M* Tris-HCl, pH 9.0, at 25°C, 0.1% Triton X-100, 2 m*M* $MgCl_2$, 0.2 m*M* of each of the deoxynucleotide 5' triphosphate (dNTP), 50 pmol of each primer (PKR sense and anti-sense primers: 5'-GGCACCCAGATTTGACCTTC-3' and 5'-TCCTTGTTCGCTTTCCATCA-3', respectively; 18S ribosomal RNA sense and anti-sense primers: 5'-CGCAGCTAGGAATAATGGAA-3' and 5'- TTAT GACCCGCACTTACTGG-3', respectively), 2.5 U of Taq DNA polymerase (LifeTechnology) and 100 μL light mineral oil in a thermal cycler (Coy Corporation). PCR reactions were set up separately for PKR and 18S RNA and were allowed to proceed for 30 cycles (95°C for 1 min, 55°C for 2 min, and 72°C for 3 min) before arresting the reaction in its logarithmic phase by rapid cooling at 4°C. The PCR products were resolved on 0.7% agarose gel. **Figure 3** shows

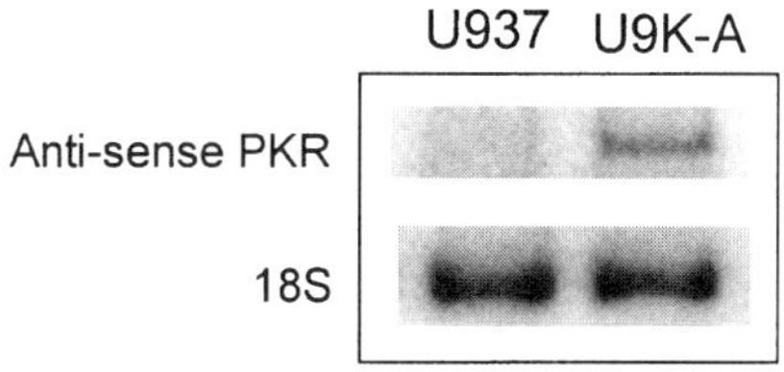

Fig. 3. Anti-sense PCR detection of PKR anti-sense RNA transcript in U9K-A cells. Total RNA was extracted from U9K-A cells harboring an integrated PKR anti-sense expression vector. PKR anti-sense RNA transcript was specifically primed with the 5' PCR primer and reverse-transcribed to yield cDNA for subsequent PCR analysis. The specificity of the anti-sense PCR is demonstrated by the absence of PKR product in the RNA extracted from parental control U937 cells. The equal intensity of the 18S RNA message from both cell lines serves as an internal control for the anti-sense RT-PCR.

that anti-sense PKR message was expressed in the U9K-A cells and not in the control parental cell line U937.

In summary, we have discussed and demonstrated how standard RT-PCR can be easily modified to study anti-sense RNA transcripts. This anti-sense RT-PCR protocol should prove useful with the increasing realization of the importance of working with anti-sense RNA in biomedical research. If a PCR protocol has been established for a gene, anti-sense RT-PCR may be utilized to screen for anti-sense RNA species that may be important for gene regulation during various biological processes including tumorigenesis, embryogenesis, and cell differentiation.

References

1. Rappolee, D.A., Wang, A., Mark, D., and Werb, Z. (1989) Novel method for studying mRNA phenotypes in single or small numbers of cells. *J. Cell. Biochem.* **39,** 1–11.
2. Yeung, M.C., Chang, D.L., Camantigue, R.E., and Lau, A.S. (1999) Inhibitory role of the host apoptogenic gene PKR in the establishment of persistent infection by encephalomyocarditis virus in U937 cells. *Proc. Nat'l Acad. of Sci. USA* **96,** 11,860–11,865.
3. Li, A.W., Seyoum, G., Shiu, R.P., and Murphy, P.R. (1996) Expression of the rat BFGF antisense RNA transcript is tissue-specific and developmentally regulated. *Mol. Cell. Endocrinol.* **118,** 113–123.
4. Moorwood, K., Charles, A.K., Salpekar, A., Wallace, J.I., Brown, K.W., and Malik, K. (1998) Antisense WT1 transcription parallels sense mRNA and protein expression in fetal kidney and can elevate protein levels in vitro. *J. Pathol.* **185,** 352–359.
5. Weiss, B., Davidkova, G., and Zhou, L.W. (1999) Antisense RNA gene therapy for studying and modulating biological processes. *Cell. Mol. Life Sci.* **55,** 334–358.

6. Imoto, S., Miki, N., Hatano, M., Aizawa, S., and Tokuhisa, T. (1993) Regulation of class II MHC gene expression by the inducible anti-sense RNA in transgenic mice. *Anal. Biochem.* **210,** 231–234.
7. Yeung, M.C., and Lau, A.S. (1998) Tumor suppressor p53 as a component of the tumor necrosis factor-induced, protein kinase PKR-mediated apoptotic pathway in human promonocytic U937 cells. *J. Biol. Chem.* **273,** 25,198–25,202.
8. Yeung, M.C., Liu, J., and Lau, A.S. (1996) An essential role for the interferon-inducible, double-stranded RNA-activated protein kinase PKR in the tumor necrosis factor-induced apoptosis in U937 cells. *Proc. Nat'l. Acad. Sci. USA* **93,** 12,451–12,455.
9. Chomczynski, P., and Sacchi, N. (1987) Single-step method of RNA isolation by acid guanidinium thiocyanate-phenol-chloroform extraction. *Anal. Biochem.* **162,** 156–159.

X

RT-PCR in cDNA Cloning

26

RT-PCR in cDNA Library Construction

Vincent Healy

1. Introduction

The construction of high-quality, full-length cDNA libraries is still the ideal method for cloning and studying mRNA populations. However, a significant problem with conventional cDNA library construction is the requirement for microgram quantities of poly A^+ RNA at the start of the procedure *(1)*. If the amount of biological material is limited and resultant poly A^+ RNA yields are low (<500 ng), the conventional approach is not an option. This problem will arise, for example, if one is attempting to construct cDNA libraries from a small number of cells, or indeed a single cell, from a tissue with a large number of cellular phenotypes such as the brain. Secondly, the investigation of gene expression at various time-points during embryonic development can be facilitated by the construction of cDNA libraries, but in many cases the amount of available embryonic tissue is minute. One obvious solution to this problem is to increase the amount of starting material, but this approach could result in higher costs and potential ethical difficulties associated with sacrificing large numbers of animals or obtaining samples from a greater number of volunteers or patients.

In the situation of a shortage of biological material the polymerase chain reaction (PCR) can be adapted to generate cDNA libraries from small amounts of RNA. PCR is used to amplify the entire population of cDNA molecules reverse-transcribed from poly A^+ RNA generating sufficient quantities of cDNA for the later steps of library construction. Using this approach, it is now possible to create a representative single-cell cDNA library starting with as little as 0.1 pg of mRNA *(2)*.

To permit amplification of the cDNA molecules, they must contain known DNA sequences at their 5' and 3' termini to act as primer-binding sites during PCR. Although a modified oligo (dT) primer can bind the poly A tail of mRNAs

From: *Methods in Molecular Biology, vol. 193: RT-PCR Protocols*
Edited by: J. O'Connell © Humana Press Inc., Totowa, NJ

at one end, a second primer-binding site must be attached to the other end for exponential amplification to occur. The two most commonly used techniques to achieve this are homopolymeric tailing —i.e., the addition of a poly (dG) *(3)* or poly (dA) *(4,5)* tail to single-stranded cDNA with terminal transferase, and secondly, ligation of adapters to double-stranded cDNA with DNA ligase *(6)*.

A further refinement that greatly simplifies cDNA library construction from small quantities of RNA is to utilize solid-phase technology by attaching the RNA to magnetic Dynabeads™. Oligo $(dT)_{25}$ Dynabeads are added to the lysed tissue or cells to bind and purify the mRNA population and to prime the synthesis of the first-strand cDNA molecules, which then become covalently attached to the beads. All further manipulations, including PCR amplification, are then performed on the immobilized cDNAs. Buffer changing and removal of excess primer/dNTPs is accomplished by simply placing the reaction tube in a magnetic particle concentrator that rapidly pellets the beads out of solution. This eliminates the need for precipitating the cDNA, and thus minimizes losses that can be critical when working with very small quantities of material *(7)*.

Two reverse-transcriptase-polymerase chain reaction (RT-PCR) protocols for cDNA library construction from small amounts of RNA are presented here. The first describes the synthesis of double-stranded cDNA from purified mRNA followed by adapter ligation to generate primer-binding sites prior to PCR amplification. The second protocol employs Dynabeads to purify the mRNA from the source material and prime first-strand cDNA synthesis. Terminal transferase is then used to tail the 3' end of the immobilized cDNA, creating a primer-binding site for PCR amplification.

2. Materials

2.1. Standard mRNA Isolation

1. Micro-FastTrack™ mRNA Isolation Kit (Invitrogen BV, the Netherlands).

2.2. cDNA Synthesis

2.2.1. First-Strand cDNA Synthesis

1. Superscript II reverse transcriptase (RT) (200 U/μL) (Gibco-BRL, Grand Island, NY).
2. 5X Superscript II RT buffer: 250 m*M* Tris-HCl, pH 8.3, 375 m*M* KCl, 15 m*M* $MgCl_2$ (Gibco-BRL).
3. 10 m*M* deoxynucleotide 5' triphosphate (dNTP) mix: 10 m*M* each of dGTP, dATP, dTTP, and dCTP (Gibco-BRL).
4. RNasin ribonuclease inhibitor (Gibco-BRL).
5. Oligo $(dT)_{18}$ primer (200 ng/μL) (**Table 1**) (Sigma-Genosys, Cambridge, UK).
6. Trehalose (Sigma Chemical Co., St. Louis, MO); prepare an 80% (w/v) solution, autoclave, and store at –20°C.
7. Glycerol (Sigma): prepare an 80% (w/v) solution and autoclave.
8. PTC-200 Peltier thermal cycler (MJ Research).

Table 1
Sequences of Oligonucleotides Used for cDNA Library Construction

Name	Sequence
T18-oligo	5'-TTTTTTTTTTTTTTTTTT-3'
T-primer	5'-GCTGTACTTAGCGGCCGCGTAATCTTTTTTTTTTT-TTTTTTT-3'
Not I	5'-GATTACGCGGCCGCTAAGTACAGC-3'
Anti-*Not* I	5'-CTTAGCGGCCGCGTAATC-3'
PCR-*Not* I	5'-GCTGTACTTAGCGGCCGCGTAATC-3'

*The underlined sequence indicates the Not I restriction enzyme cutting site.

2.2.2. Second-Strand cDNA Synthesis

1. *E. coli* RNase H (0.5 U/µL) (Promega, Madison, WI).
2. *E. coli* DNA polymerase I (10 U/µL) (Promega).
3. *E. coli* DNA ligase (10 U/µL) (Gibco-BRL).
4. T4 DNA polymerase (8 U/µL) (Gibco-BRL).
5. 5X second-strand buffer: 100 m*M* Tris-HCl, pH 6.9, 23 m*M* $MgCl_2$, 450 m*M* KCl, 0.75 m*M* β-NAD^+, 50 m*M* $(NH4)_2SO_4$ (Gibco-BRL).
6. Phenol:chloroform:isoamyl alcohol, 25:24:1 (v/v) (Gibco-BRL).

2.3. Adapter Ligation and Amplification of cDNA

1. Prepare 10X annealing buffer: 100 m*M* Tris-HCl, pH 8.0, 10 m*M* ethylenediaminetetraacetic acid (EDTA), pH 8.0, 1 *M* NaCl; autoclave and store at 4°C.
2. *Not* I, anti-*Not* I (200 ng/µL) and PCR-*Not* 1 (100 ng/µL) oligonucleotides (**Table 1**) (Sigma Genosys).
3. T4 DNA ligase (400 U/µL) (New England Biolabs, Beverly, MA).
4. 10X T4 DNA ligase buffer: 500 m*M* Tris-HCl, pH 7.8, 100 m*M* $MgCl_2$, 100 m*M* DTT, 10 m*M* adenosine triphosphate (ATP), 250 µg/mL bovine serum albumin (BSA) (New England Biolabs).
5. Expand™ High Fidelity PCR System (Roche Molecular Biochemicals, Indianapolis, IN).
6. 10 m*M* PCR nucleotide mix (Roche).
7. Dimethyl sulfoxide (DMSO) (Sigma).
8. Thin-walled PCR tubes (Biogene).
9. PTC-200 Peltier thermal cycler (MJ Research).
10. Advantage PCR-Pure Kit (Clontech, Palo Alto, CA).

2.4. Isolation of mRNA and Synthesis of First-Strand cDNA using Dynabeads™

1. Dynabeads mRNA Purification Kit (Dynal, Wirral, UK).
2. Magnetic particle concentrator (MPC-E) (Dynal).

2.5. Oligo (dA) Tailing and cDNA Amplification

1. T4 DNA polymerase (8 U/µL) (Gibco-BRL).
2. 5X T4 DNA polymerase buffer: 165 m*M* Tris-acetate, pH 7.9, 330 m*M* sodium acetate, 50 m*M* magnesium acetate, 500 µg/mL BSA, 2.5 m*M* DTT (Gibco-BRL).
3. *E. coli* RNase H (0.5 U/µL) (Promega).
4. Prepare 1X *E. coli* RNase H buffer: 20 m*M* HEPES-KOH, pH 7.8, 50 m*M* KCl, 10 m*M* $MgCl_2$, 1 m*M* DTT; filter-sterilize and store at –20°C.
5. Terminal deoxynucleotidyl transferase (30 U/µL) (Promega).
6. 5X Terminal deoxynucleotidyl transferase buffer: 500 m*M* potassium cacodylate, pH 6.8, 5 m*M* CoCl2, 0.5 m*M* Dithiothreitol (DTT) (Promega).
7. T-primer [an oligo $(dT)_{18}$ primer with an additional 24-bp sequence at its 5' end, which includes a *Not* I restriction site, (100 ng/µL and 10 ng/µL) and PCR-*Not* I (100 ng/µL) oligonucleotides (**Table I**) (Sigma Genosys)].

3. Methods

3.1. Standard mRNA Isolation

This section describes the purification of poly A^+ RNA for the RT-PCR protocol using adapter ligation that does not utilize the magnetic Dynabeads.

1. It is vital to ensure that all reagents and tubes are free from ribonuclease contamination while working with RNA (*see* **Note 1**). Many kits are commercially available for the isolation of both total RNA and mRNA; the Micro-FastTrack™ mRNA Isolation Kit (Invitrogen) has been satisfactory in our studies for the purification of poly A^+ RNA from a variety of sources, and it is used according to the manufacturer's instructions.
2. Following its elution from the oligo (dT) spun-column, precipitate the mRNA overnight at –20°C by adding 20 µg of glycogen (to act as a carrier to aid maximum precipitation), 0.1 vol of 2 *M* sodium acetate, pH 5.0, and 2.5 vol of prechilled (–20°C) 95% ethanol. Collect the RNA pellet by centrifugation in a microfuge at maximum speed for 20 min, wash it once with 70% ethanol, and resuspend in 5 µL of RNase-free H_2O (*see* **Note 2**).

3.2. cDNA Synthesis (see *Fig. 1*)

3.2.1. First-Strand cDNA Synthesis

The Superscript II reverse transcriptase (RT) (Gibco-BRL) used here converts the mRNA to first-strand cDNA primed by an oligo $(dT)_{18}$ primer which binds the poly A tail of the mRNAs (*see* **Note 3**). The reaction is performed at 55°C to weaken secondary structure in the mRNA, assisting in the synthesis of longer cDNA. The addition of trehalose acts to stabilize the RT enzyme at a higher than normal temperature (*see* **Note 4**).

1. Mix the isolated mRNA (5 µL) with 1 µL of oligo $(dT)_{18}$ primer (200 ng/µL) in a 0.5-mL microfuge tube. Heat the tube at 65°C for 10 min to denature mRNA

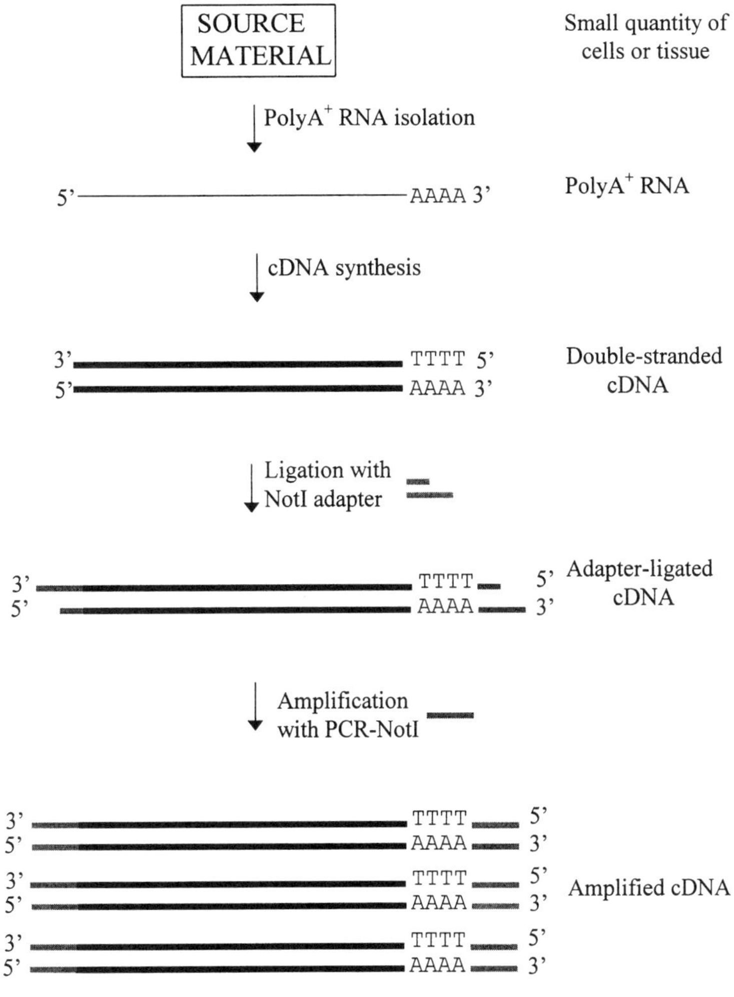

Fig. 1. RT-PCR of polyA$^+$ RNA using adapter ligation. The thin line represents RNA, and the thicker line depicts DNA. The gray lines represent the oligonucleotides that make up the adapter and the complementary PCR-*Not* I primer.

secondary structure and immediately place it on ice for an additional 5 min to complete the proper annealing of the primer to the mRNA. Briefly centrifuge the tube to collect condensation.

2. Immediately add the following to the primer-annealed mRNA: 8 µL of 5X RT buffer, 2 µL of 10 m*M* dNTPs, 4 µL of 100 m*M* dithiothreitol (DTT), 2 µL of the RNase inhibitor, RNasin (10 U/µL), 12 µL of trehalose, 80% (w/v) and 4 µL of glycerol, 80% (v/v). Mix the contents with the pipet tip. Place the reaction tube on a thermal cycler tube and preheat at 35°C for 5 min Add 2 µL of Superscript II RT (200 U/µL);

the final reaction vol is 40 µL. Incubate at 45°C for 10 min, ramp the temperature to 55°C over 2 min and continue incubation at 55°C for 1 h (*see* **Note 4**).

3.2.2. Second-Strand cDNA Synthesis

In this reaction, the RNase H begins the degradation of the RNA in the RNA/cDNA duplex, creating RNA oligonucleotides which the DNA polymerase uses as primers to synthesize the second cDNA strand. The *E. coli* DNA ligase, which then seals the phosphodiester bonds on the second strand, has been shown to have the effect of increasing the length of double-stranded cDNAs obtained from long RNAs *(8)*. Finally, T4 DNA polymerase is added to blunt the ends of the cDNA molecules in preparation for adapter ligation.

1. Following the 1-h incubation at 55°C, centrifuge briefly to collect any condensation and place the first-strand cDNA synthesis reaction on ice. Add the following to the tube: 30 µL of the 5X second-strand buffer, 3 µL of 10 m*M* dNTPs, 67 µL of nuclease-free H_2O, 4 µL of *E. coli* RNase H (0.5 U/µL), 4 µL of *E. coli* DNA polymerase (10 U/µL) and 1 µL of *E. coli* DNA ligase (10 U/µL). Mix the contents with the pipet tip and incubate the reaction (final vol, 150 µL) at 16°C for 2 h.
2. Place the reaction tube back on ice and add 1 µL of T4 DNA polymerase (5 U/µL). Mix again and continue the incubation at 16°C for another 5 min. Add 10 µL of 0.5 *M* EDTA (pH 8.0) to stop the reaction and extract it once with an equal vol (150 µL) of phenol:chloroform:isoamyl alcohol (25:24:1 v/v). Transfer the aqueous phase to a fresh tube and extract once more with an equal volume of chloroform.
3. Transfer the aqueous phase to a fresh tube and precipitate the double-stranded cDNA overnight at –20°C by adding 25 µL of 2 *M* sodium acetate, 2 µg of glycogen, to act as carrier (from the Micro-FastTrack™ mRNA Isolation Kit), and 600 µL of prechilled (–20°C) 95% ethanol. Centrifuge at top speed in a microfuge for 20 min. Wash the cDNA pellet with 70% ethanol and resuspend in 10 µL of RNase-free H_20.

3.3. Adapter Ligation and Amplification of cDNA

3.3.1. Ligation of Adapter to cDNA (see **Note 5**)

The annealing of the complementary *Not* I (24mer) and anti-*Not* I (18 mer) oligonucleotides (**Table 1**) creates the adapter used here. The design of these oligonucleotides ensures that the adapter ligates only to the blunt-ended cDNA molecules and not to itself (*see* **Note 6**).

1. Add 2 µL of 10X annealing buffer, 5 µL of *Not* I oligonucleotide (200 ng/µL), 5 µL of anti-*Not* I oligonucleotide (200 ng/µL), and 8 µL of H_2O to a tube, mix, and incubate at 65°C for 10 min. Cool to room temperature and store at –20°C until required.
2. Add the following to a 0.5-mL microfuge tube: 10 µL of double-stranded cDNA (from **Subheading 3.2.2.**), 1 µL (100 ng) of annealed *Not* I adapter, 2 µL of 10X T4 DNA ligase buffer, 5 µL of H_2O and 2 µL of T4 DNA ligase (400 U/µL). Mix with a pipet tip and incubate at 16°C overnight.

3.3.2. cDNA Amplification

The primer PCR-*Not* I binds to the adapter sequence that is ligated to both ends of the cDNA molecules and exponentially amplifies the entire length of the cDNA strands **(Fig.1)**.

1. Dilute the adapter-ligated cDNA to 200 µL (10-fold) with H_2O. Prepare the PCR by adding the following with gentle mixing to a thin-walled PCR tube on ice: 1 µL of 10 m*M* dNTPs, 1 µL of PCR-*Not* I primer (100 ng/µL), 2 µL of diluted adapter-ligated cDNA, 2.5 µL of 99.9% DMSO, 5 µL of 10X Expand™ High Fidelity buffer with 15 m*M* $MgCl_2$ and 37 .75 µL of H_2O (*see* **Note 7**).
2. Overlay the reaction with 30 µL of mineral oil. Place the sample on a thermal cycler and heat at 95°C for 2 min to denature the template. Maintain the temperature at 80°C and add 0.75 µL of Expand™ High Fidelity PCR system enzyme mix (3.5 U/µL) (*see* **Note 8**) to the reaction (final vol, 50 µL). For the first 10 cycles, use the following conditions: denaturation at 95°C for 15 s, annealing at 55°C for 30 s, elongation at 68°C for 3 min. For the final 15 cycles use: denaturation at 95°C for 15 s, annealing at 55°C for 30 s and elongation at 68°C for 5 min. Complete the PCR with a prolonged elongation of 72°C for 8 min (*see* **Note 9**).
3. Analyze the size range of the amplified cDNAs by resolving a 5-µL aliquot of the PCR reaction on a 1% agarose gel with ethidium bromide staining (*see* **Note 10**). If possible, test the amplified cDNA population for the presence of representative messages by PCR using gene-specific primers (*see* **Note 11**). Purify the amplified cDNA away from unused primers and dNTPs using the Advantage PCR-Pure Kit according to the manufacturer's instructions.

3.4. Isolation of mRNA and Synthesis of First-Strand cDNA Using Dynabeads (see *Fig. 2*)

1. Ensure all reagents and surfaces are free of ribonuclease contamination (*see* **Note 1**). Purify poly A^+ RNA from the tissue or cells of interest exactly as described in the manual accompanying the Dynabeads mRNA Purification Kit. Add approx 10–20 µL of Dynabeads when working with very small quantities ($<10^4$ cells or <5 mg tissue) of material.
2. Place the sample tube on ice and wash the Dynabeads 3× with 50 µL of 1X RT buffer. To remove the supernatant after each wash, place the tube in the magnetic particle concentrator (MPC-E) and pellet the beads. Add 8 µL of 5X reverse transcriptase buffer, 2 µL of 10 m*M* dNTPs, 4 µL of 100 m*M* DTT, 2 µL of the RNase inhibitor, RNasin (10 U/µL), 12 µL of trehalose (80% w/v), 4 µL of glycerol (80% v/v) and 6 µL of H_2O. Mix the contents with the pipet tip. Place the reaction tube on a thermal cycler and preheat at 35°C for 5 min. Add 2 µL of Superscript II RT (200 U/µL) (final reaction vol, 40 µL). Incubate at 45°C for 10 min, ramp the temperature to 55°C over 2 min, and continue incubation at 55°C for 1 h (*see* **Note 4**). Mix the reaction every 15 min during the first-strand synthesis to keep the beads in suspension.

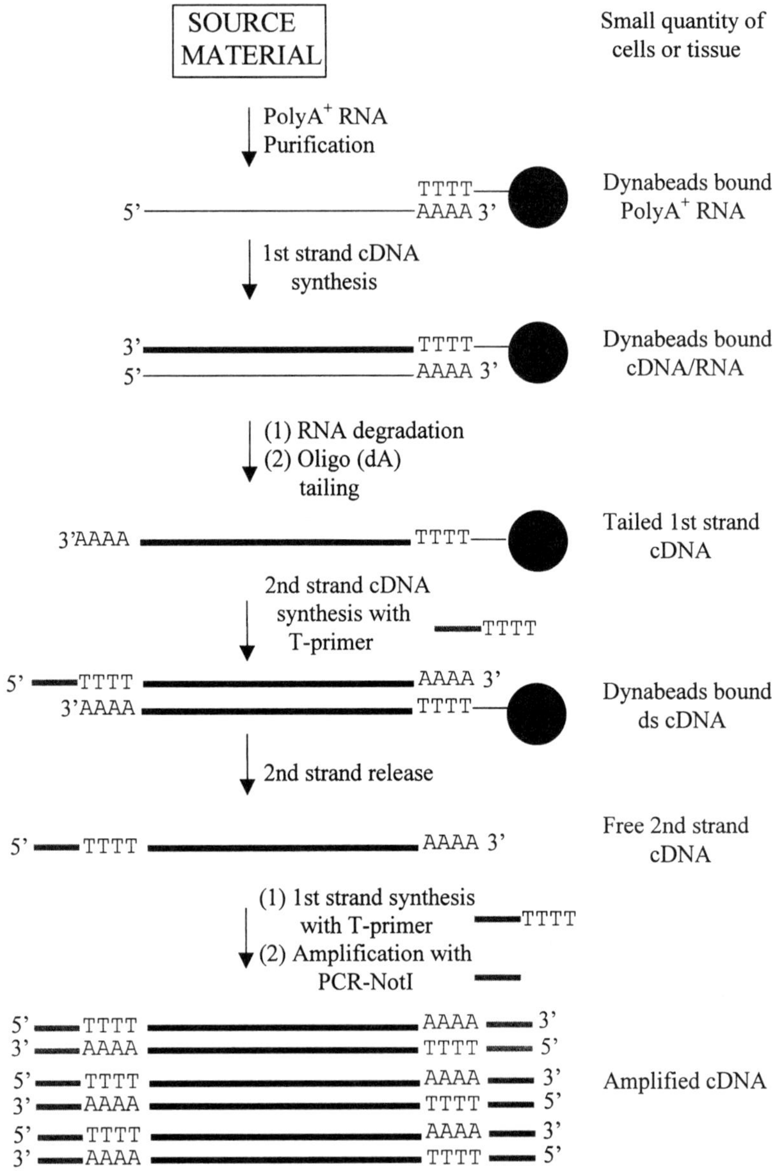

Fig. 2. RT-PCR of polyA$^+$ RNA using oligo (dA) tailing. The thin line represents RNA, and the thicker line depicts DNA. The gray lines represent both the 5' linker on the T-primer and the PCR-*Not* I primer.

3.5. Oligo dA Tailing and cDNA Amplification (see Note 12)

3.5.1. Terminal Transferase Reaction

In this reaction, the enzyme terminal transferase adds an oligo (dA) tail onto the 3' end of the first-strand cDNA (**Fig. 2**). Before this is done, however, the Dynabeads' unbound oligo (dT)s are hydrolyzed by the 3' exonuclease activity of T4 DNA polymerase *(4),* and the mRNA hybridized to the cDNA is degraded by RNaseH.

1. Following completion of first-strand cDNA synthesis, inactivate the Superscript RT by heating at 80°C for 20 min. Pellet the beads with the MPC-E magnetic concentrator and remove the buffer. Add 14.5 µL of H_2O, 4 µL of 5X T4 DNA polymerase buffer, 1.0 µL of 10 m*M* dNTPs, and 0.5 µL of T4 DNA polymerase (8 U/µL). Following incubation at 16°C for 1 h, inactivate the enzyme by heating at 75°C for 10 min and remove the solution from the beads with the MPC-E.
2. Add 19 µL of 1X *E. coli* RNase H buffer and 1 µL of *E. coli* RNase H (0.5 U/µL) to the beads and incubate at 37°C for 1 h. Stop the reaction by adding 10 µL of 5 m*M* EDTA (pH 8.0), and remove the solution using the MPC-E.
3. Wash the beads once with 50 µL of 1X terminal deoxynucleotidyl transferase buffer. Add 14.5 µL of H_2O, 4 µL of 5X terminal deoxynucleotidyl transferase buffer, 0.5 µL of 10 m*M* dATP, and 1 µL of terminal deoxynucleotidyl transferase (30 U/µL). Incubate at 37°C for 15 min and stop the reaction by adding 2 µL of 0.5 *M* EDTA (pH 8.0). Concentrate the beads and remove the buffer.

3.5.2. cDNA Amplification

At this stage, the single-stranded cDNA molecules are covalently attached to the Dynabeads and have poly A tails (approx 500 nucleotides) on their 3' ends. The thermostable DNA polymerase next performs second-strand cDNA synthesis, and heat denaturation then releases this strand from the Dynabeads (**Fig. 2**). The free strand is again made double-stranded, and then serves as template for a PCR in which the higher annealing temperature ensures that only the PCR-*Not* I primer directs exponential amplification (**Fig. 2**) (*see* **Note 13**).

1. Resuspend the Dynabeads from **Subheading 3.5.1.** in 40.5 µL of H_2O. Transfer the suspension to a thin-walled PCR tube, and place on ice. Add the following with gentle mixing: 1 µL of 10 m*M* dNTPs, 3 µL of T-primer (100 ng/µL), and 5 µL of 10X Expand™ High Fidelity buffer with 15 m*M* $MgCl_2$. Place the tube on the thermal cycler and maintain the temperature at 30°C. Add 0.5 µL of Expand™ High Fidelity PCR system enzyme mix (3.5 U/µL), and heat the reaction at 30°C for 3 min Ramp the temperature gradually to 72°C over 10 min and incubate at 72°C for an additional 10 min. Concentrate the beads and remove the supernatant.
2. Add the following to the beads with mixing: 38.75 µL of H_2O, 1 µL of 10 m*M* dNTPs, 1 µL of T-primer (10 ng/µL), 1 µL of PCR-*Not* I primer (100 ng/µL), 2.5 µL of 99.9% DMSO (*see* **Note 7**), and 5 µL of 10X Expand™ High Fidelity

buffer. Heat the sample at 95°C for 3 min on the thermal cycler to release the second cDNA strand. Pellet the beads with the MPC-E and transfer the supernatant to a fresh thin-walled PCR tube. Overlay with 30 µL of mineral oil, place the tube on the thermal cycler, and maintain the temperature at 30°C. Add 0.75 µL of Expand™ High Fidelity PCR system enzyme mix (3.5 U/µL) (*see* **Note 8**), and once more heat the reaction at 30°C for 3 min, ramp to 72°C over 10 min, and heat at 72°C for 10 min to render the cDNA in solution double-stranded.

3. Immediately perform PCR amplification. For the first 10 cycles, use the following conditions: denaturation at 95°C for 15 s, annealing at 55°C for 30 s and elongation at 68°C for 3 min. For the final 15 cycles use: denaturation at 95°C for 15 s, annealing at 55°C for 30 s and elongation at 68°C for 5 min. Complete the PCR with a prolonged elongation of 72°C for 8 min.
4. Analyze the result of the PCR amplification as described in **Subheading 3.3.2., step 3** (*see* **Notes 10** and **11**). If satisfactory, purify the amplified cDNA with the Advantage PCR-Pure Kit, following the instructions provided by the manufacturer.

3.6. Cloning the Amplified cDNA

This chapter does not include a detailed description of the procedure for cloning the amplified cDNA into a vector, as I am concentrating on the RT-PCR techniques that are employed prior to cloning. Bacteriophage λ is considered the vector of choice for cDNA library construction, as in vitro packaging of λ generally has a higher efficiency than plasmid transformation. The double-stranded cDNA molecules resulting from both protocols presented here have adapters with *Not* I restriction-enzyme cutting sites at both termini. A *Not*I digestion cleaves these sites, resulting in cohesive termini that can be ligated into a *Not*I digested vector such as the Lambda ZAP™ II bacteriophage λ cloning vector provided by Stratagene (*see* **Note 14**).

The RT-PCR protocols described in this chapter should generate an ample source of amplified cDNA to permit efficient construction of a large, representative cDNA library.

4. Notes

1. When working with RNA samples, always wear gloves. It is best to use reagents that are kept specifically for RNA work and to work on a laboratory bench dedicated to RNA work. Bake all glassware at 250°C overnight before use. Autoclave double-distilled water (MilliQ) to prepare RNase-free H_2O. If high-quality water is used, it is generally not necessary to treat it with diethylpyrocarbonate (DEPC). However, if this is not available, then add DEPC to distilled water at 0.1% concentration and stir for 12 h. Following this step, autoclave for 20 min to degrade the DEPC.
2. Although it is usually desirable to quantitate the purified mRNA spectrophotometrically and to analyze its quality by electrophoresis, the low amounts that are isolated from small quantities of starting material will make it difficult to do this

without incurring significant losses of mRNA. For this reason, it is advisable to avoid any analysis and to convert all of the purified mRNA to cDNA.

3. The T-primer (**Table 1**) can also be used to prime cDNA here. Although this primer has a 5'-linker to facilitate PCR amplification and cDNA cloning with the oligo (dA) tailing method also described in this chapter, it will not interfere with the subsequent steps in the adapter ligation protocol.
4. Some mRNAs contain elements of a complex secondary structure that can interfere with the RT as it moves along the mRNA, causing premature termination of the reverse transcription reaction ***(9)***. This decreases the amount of full-length cDNAs in the library. Carninci et al. ***(10)*** have developed a protocol to make the reverse transcription reaction more efficient by performing it at a higher temperature (55–60°C) in the presence of the disaccharide trehalose. Under these conditions the strength of the secondary structures is weakened and the enzyme can reach the 5' end of the longer (>5 kb) mRNAs. Normally, the Superscript II RT would be denatured at these temperatures (its optimum temperature under standard conditions is 45°C), but the addition of trehalose to the reaction has a thermostabilizing effect, enabling the enzyme to remain active ***(10)***.
5. Because the *Not* I adapter does not have a 5'-phosphate at its blunt end, the cDNA must be phosphorylated at both termini for ligation to occur. Previously, the cDNA was treated with T4 polynucleotide kinase at this stage to ensure its phosphorylation ***(11)***. However, if high-quality reagents and enzymes are used during first- and second-strand synthesis, the cDNA will be adequately phosphorylated after the blunting reaction, and the kinase treatment is not necessary.
6. The *Not* I adapter (**Table 1**) consists of the *Not* I and anti-*Not* I oligonucleotides which are complementary to each other. Because the adapter's blunt end has no 5'-phosphate group, and its overhang is not palandromic, the adapters can ligate only to the cDNA and not to themselves.
7. The PCR contains dimethyl sulfoxide (DMSO) to maximize PCR amplification of GC-rich fragments by decreasing their melting temperatures. Such regions of DNA are difficult for even thermostable DNA polymerases to replicate, and without DMSO it is possible that some GC-rich cDNAs may be lost from the amplified library. Baskaran and colleagues have used a combination of DMSO and another additive, trimethylglycine (betaine), to successfully amplify a CGG repeat region ***(12)*** and to construct a cDNA library from small amounts of ear tissue ***(13)***, but for most purposes DMSO alone is sufficient.
8. The Expand™ High Fidelity PCR system enzyme mix contains two enzymes, the thermostable Taq DNA polymerase and a thermostable 3'–5' proofreading exonuclease. This combination significantly decreases the mutation rate of Taq DNA polymerase and the amplified library, and therefore, contains a lower number of sequence errors.
9. The elongation temperature of 68°C and the long elongation times are recommended for templates greater then 3 kb. Using these PCR conditions here increases the proportion of full-length cDNAs in the amplified library.
10. On an ethidium bromide-stained agarose gel, the cDNAs are visible as a smear

ranging in size from approx 0.5–5 kb, with the greatest intensity at about 1.5 kb. If nothing is visible, then reamplify 5 µL of the reaction for an additional 15 cycles using the exact same conditions.

11. A significant problem that can arise when amplifying entire cDNA populations is a distortion in the representation of various cDNAs. Some transcripts are over-represented, and others are under-represented or indeed lost because of bias during PCR amplification *(2)*. It is advisable to test this by PCR analysis using gene-specific primers for several transcripts that are expressed in the source material. If no major distortions have occurred, then all PCRs should generate detectable products. In addition, PCR analysis using primer pairs located at the 5' ends, central sections, and 3' ends of known long transcripts (>5kb) can test for the presence of full-length cDNA in the amplified library.
12. Adapters can also be ligated to the Dynabeads-bound cDNA instead of the tailing reaction described here. The first-strand cDNA attached to the Dynabeads is made double-stranded exactly as described in **Subheading 3.2.2.** and adapter ligation is then performed. In addition, if a T-primer with a linker sequence containing a restriction-enzyme cutting site other than *Not* I is used instead of the one described here (**Table 1**), the amplified cDNA molecules can be cloned into a vector in a specific orientation, allowing the construction of a directional library.
13. In the two steps prior to PCR amplification, the low annealing temperature (30°C) allows the $(dT)_{18}$ section of the T-primer to bind to the poly A tail of the single-stranded cDNA. Slowly increasing the temperature to 72°C, the optimum temperature of Taq DNA polymerase, allows the synthesis of the second cDNA strand. Although the T-primer is present in the subsequent amplification reaction, the higher annealing temperature (55°C) prevents it from binding to its target sequence, and it does not interfere with the PCR.
14. The restriction enzyme *Not* I cuts at the 8-bp recognition site (5'-GCGGCCGC-3'). As this sequence appears randomly only once in every 65.5 kb of DNA, performing a *Not* I digestion to generate cohesive termini during cloning minimizes internal cutting of the cDNA strands.

References

1. Carninci P., and Hayashizaki Y. (1999) High-efficiency full-length cDNA cloning. *Methods Enzymol.* **303,** 19–44.
2. Karrer E. E., Lincoln J. E., Hogenhout S., Bennett A. B., Bostock R. M., Martineau B., et al. (1995) *In situ* isolation of mRNA from individual plant cells: creation of cell-specific cDNA libraries. *Proc. Natl. Acad. Sci. USA* **92,** 3814–3818.
3. Kim M. G., Chen C., Flomerfelt F. A., Germain R. N., and Schwartz R. H. (1998) A subtractive PCR-based cDNA library made from fetal thymic stromal cells. *J. Immunol. Methods* **213,** 169–182.
4. Lambert K. N., and Williamson V. M. (1993) cDNA library construction from small amounts of RNA using paramagnetic beads and PCR. *Nucleic Acids Res.* **21,** 775–776.

5. Ying S. Y., Lui H. M., Lin S. L., and Chuong C. M. (1999) Generation of full-length cDNA library from single human prostate cancer cells. *Biotechniques* **27,** 410–414.
6. Lukyanov K., Diatchenko L., Chenchik A., Nanisetti A., Siebert P., Usman N., et al. (1997) Construction of cDNA libraries from small amounts of total RNA using the suppression PCR effect. *Biochem. Biophys. Res. Commun.* **230,** 285–288.
7. Augenstein S. (1994) Superparamagnetic beads: applications of solid-phase RT-PCR. *Am. Biotechnol. Lab.* **12,** 12–14.
8. D'Alessio J. M., and Gerard G. F. (1988) Second-strand cDNA synthesis with *E. coli* DNA polymerase I and RNase H: the fate of information at the mRNA 5' terminus and the effect of *E. coli* DNA ligase. *Nucleic Acids Res.* **16,** 1999–2014.
9. Payvar F., and Schimke R. T. (1979) Methylmercury hydroxide enhancement of translation and transcription of ovalbumin and conalbumin mRNA's. *J. Biol. Chem.* **254,** 7636–7642.
10. Carninci P., Nishiyama Y., Westover A., Itoh M., Nagaoka S., Sasaki N., et al. (1998) Thermostabilization and thermoactivation of thermolabile enzymes by trehalose and its application for the synthesis of full length cDNA. *Proc. Natl. Acad. Sci. USA* **95,** 520–524.
11. Bhat G. J., Lodes M. J., Myler P. J., and Stuart K. D. (1991) A simple method for cloning blunt ended DNA fragments. *Nucleic Acids Res.* **19,** 398.
12. Baskaran N., Kandpal R. P., Bhargava A. K., Glynn M. W., Bale A., and Weissman, S. M. (1996) Uniform amplification of a mixture of deoxyribonucleic acids with varying GC content. *Genome Res.* **6,** 633–638.
13. Jacob A. N., Baskaran N., Kandpal, G., Narayan D., Bhargava A. K., and Kandpal R. P. (1997) Isolation of human ear specific cDNAs and construction of cDNA libraries from surgically removed small amounts of inner ear tissues. *Somat. Cell Mol. Genet.* **23,** 83–95.

27

An RT-PCR-Based Protocol for the Rapid Generation of Large, Representative cDNA Libraries for Expression Screening

Joe O'Connell

1. Introduction

1.1. cDNA Cloning

The ability to isolate genes that encode specific proteins has been fundamental to the revolution in molecular and cell biology over the past two decades. The protein-encoding versions of genes are isolated from cDNA libraries. A cDNA library is constructed by isolating mRNA from the cell or tissue that expresses the target protein, converting all the mRNA into cDNA copies via the enzyme RT, and cloning all the cDNAs into bacterial cells via a plasmid or bacteriophage vector. The library is then screened with a probe to detect a clone expressing the protein of interest. The probe is usually either a mixed oligonucleotide probe based on partial amino-acid sequence data for the protein, or alternatively, an antibody specific for the protein of interest. For the latter approach, an expression vector must be used to enable the cloned cDNA to be expressed at the protein level in *E. coli*. Advances in the methodologies involved in cDNA cloning during the 1980's enabled the identification and characterization of the cDNAs encoding thousands of important human proteins. With the advent of PCR, an alternative approach to cDNA isolation has also been employed. By using degenerate primers based on partial amino-acid sequence information from the protein of interest, its cDNA may be amplified from total cDNA without the need for cDNA library construction and screening. cDNA libraries are still valuable tools when full-length clones are sought, and in particular, when amino-acid sequence information is unavailable.

From: *Methods in Molecular Biology, vol. 193: RT-PCR Protocols*
Edited by: J. O'Connell © Humana Press Inc., Totowa, NJ

A particular advantage of cDNA libraries is that if an expression cloning vector is used, cDNA clones can be identified on the basis of reactivity of the protein product to specific antibodies. This is particularly relevant when clones are specifically sought for genes that encode antigens. Genes encoding tumor-specific antigens have recently been isolated by screening tumor cDNA libraries with antibodies derived from the sera of cancer patients—a strategy termed SEREX *(1)*.

1.2. Problems Associated with cDNA Cloning

cDNA cloning, by its nature, is an inefficient process consisting of many variable steps, all with inherent limitations in efficiency. To begin with, 0.25–1.0 mg of RNA are obtainable per g of tissue, of which only about 1% is recovered as polyA$^+$ mRNA following oligo(dT) cellulose chromatography. Conversion of mRNA into first-strand cDNA using RT occurs, at best, with 15–20% efficiency. Second-strand synthesis is usually 80–90% efficient. Ligation of *Eco*R1-ended adaptors onto blunt-ended cDNA is efficient, but ligation of the *Eco*R1-adapted cDNA into a λ cloning vector works at best with 40–60% efficiency. In vitro packaging of recombinant λ molecules into viable phage particles (to form the phage cDNA library) yields at most 10^6–10^7 recombinants per μg of recombinant λ DNA (as little as 0.001–0.1 of the recombinant molecules are actually packaged). λ in vitro packaging extracts are notoriously unstable. Added to this, loss of material at the several purification steps and trace exonuclease damage of the *Eco*R1-ended clonable ends of the cDNA render the overall process inefficient.

Often, the yield of recombinant clones obtained at the end of the cDNA cloning procedure is disappointingly low—frequently more than an order of magnitude below the number of clones required to ensure a representative cDNA library. Based on theoretical estimations and practical considerations, $>10^6$ clones are believed to be required to ensure complete representation of all the mRNAs, including low-abundance mRNAs (many mRNAs occur as a single copy per cell), expressed in a higher eukaryotic cell. We have isolated the usual cause of this to be an insufficient quantity of clonable insert cDNA to generate enough recombinant λ molecules to drive a large yield of recombinant phage particles. We have developed a simple protocol using polymerase chain reaction (PCR) to amplify cDNA to permit large cDNA library construction from limited amounts of tissue.

1.3. PCR-Based cDNA Libraries and Bias in the Representation of Clones

A number of PCR-based protocols for generating cDNA libraries have been developed. However, a limitation inherent in many protocols is that the

PCR-amplified cDNA may not contain the same distribution of cDNA molecules as that present in the starting cDNA template.

Different templates amplify at different rates during PCR. Although the size of template affects amplification such that large templates amplify less efficiently than small ones, within a limited size range, size effects are minimal, and differences in amplification rates are probably caused by other factors such as GC content, occurrence of GC-rich regions, or regions leading to stable secondary structures during the denaturation and annealing steps. Because total cDNA consists of an enormously heterogeneous population of template molecules, with approx 10,000 different cDNAs derived from a higher eukaryotic cell, we reason that bias is likely to occur during PCR amplification whereby templates which amplify efficiently will outcompete those that amplify only poorly. There is also a vast difference in the level of abundance of different cDNAs—some will be present from mRNAs which account for more than 1% of the total cellular mRNA molecules, such as β-actin, while low-abundance cDNAs may be present at the level of a single copy per cell. It therefore seems inevitable that in the course of amplification, particularly if inefficiently amplifiable templates are initially present at low abundance, such cDNAs may be under-represented or lost completely from the ultimate PCR-based cDNA library. Thus, the distribution of clones in a PCR-based cDNA library may be radically biased so that some sequences are over-represented, under-represented, or absent.

1.4. Formamide May Help to Suppress Competitive Bias in Mixed-Template PCR

Formamide is a denaturant frequently used in nucleic acid hybridization experiments to increase the stringency of hybridization by reducing the overall stability of hybridization, so that imperfectly matched hybrids are destabilized. It has also been used in PCR to minimize non-specific priming. We reasoned that the denaturing effect of formamide on nucleic acids may help to "smooth out" secondary structures or suppress their formation during PCR, and also help in efficient denaturation of GC-rich regions in template DNA. We further reasoned that such effects should have an equalizing influence on the PCR amplification efficiencies of different templates, which should lead to equal competition during mixed template PCR and result in an unbiased yield of PCR products. In a control experiment with two DNA fragments that had different PCR amplification efficiencies, we were able to verify that 5% formamide tended to equalize their amplification rates when co-amplified. When co-amplified in the absence of formamide, there was bias in the PCR products toward the efficiently amplifiable fragment, whereas equal products were obtained from both fragments when formamide was included in the PCR at 5%. Others have shown that another DNA denaturant, DMSO, can also help to minimize bias during the PCR amplification of cDNA.

1.5. Experimental Protocol for Representative PCR-Based cDNA Library Construction

We specifically developed this protocol to generate cDNA-expression libraries suitable for antibody detection of clones: thus, we use random primers during cDNA synthesis. By priming randomly, this increases the likelihood that cDNA fragments will be generated and cloned into the vector in all possible reading frames; also, better representation of coding regions of mRNA molecules is achieved than with oligo(dT) primers, which result in a bias of clones toward the 3'-untranslated regions of mRNAs (which can be several kb in length). Random priming also results in a more homogenous size range of cDNAs, favoring the generation of short fragments (usually with an average size of about 500 bp), which favor more efficient expression in *E. coli*. The size range is further narrowed during purification of the synthesized cDNA through a Sephacryl S300 gel-filtration spun column, which eliminates short cDNA fragments of less than about 300 bp. For the reasons already described, we include 5% formamide during PCR amplification of the total cDNA to minimize competitive bias. An adapter-primer strategy facilitates PCR amplification of the cDNA. Restriction digestion of the amplified cDNA is not required prior to cloning. Instead, *Eco*R1-sticky ends are generated by using the 3'-exonuclease activity of T4 DNA polymerase to "chew back" the appropriate nucleotides from the adaptor, which contains a partial *Eco*R1 site at its outer end. This chew-back generates *Eco*R1 sticky ends on the amplified cDNA. The sequence of the adaptor is also designed to favor efficient translation in *E. coli* of the mRNA sequence encoded by the cDNA insert (*see* **Subheading 2.3.** and **Fig. 1** for a full description of the adaptor-primer design). After PCR amplification, gel-filtration chromatography through Sephacryl S-300 resin is included to eliminate primer artifacts or truncated cDNAs less than 300 bp from the PCR-amplified cDNA prior to cloning. This precludes the construction of a cDNA library containing a high proportion of artifactual or short, uninformative clones. This is the basis for our strategy for generating representative PCR-based cDNA libraries.

2. Materials

1. RNA isolation: guanidine isothiocyanate (Sigma Chemical Co., St. Louis, MO); molecular-biology-grade phenol (Sigma); ethanol (Sigma). Oligo(dT)-cellulose chromatography resin (Pharmacia Biotech, Uppsala, Sweden). Several kits are commercially available for direct isolation of mRNA from tissue (e.g., the Micro-FastTrack™ mRNA isolation kit from Invitrogen BV, the Netherlands).

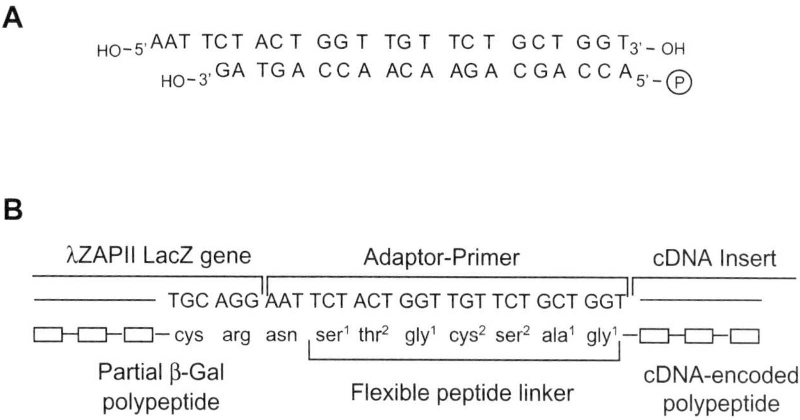

[1]most frequently used codon choice in *E.coli*

[2]second most frequently used codon choice in *E.coli*

Fig. 1. Multifunctional Adaptor-Primer. (**A**) The complementary 24-mer and 20-mer oligonucleotides are annealed to form an adaptor with one blunt end and one *Eco*R 1 sticky end. The 20-mer, but not the 24-mer, is phosphorylated at its 5' end. This allows blunt-ended ligation of the adaptor onto cDNA, but the only possible self-ligation products of the adaptor itself are "back-to-back" dimers, which are easily removed from the adapted cDNA by spun-column chromatography through Sephacryl S300. (**B**) The sequence of the adaptor-primer is in-frame with the λZAPII β-gal coding sequence, with no stop/start codons. It exhibits minimal self-complimentarity to avoid self-priming PCR artifacts. All amino-acids encoded by the adaptor have small neutral (gly, ala) or polar (ser, thr, cys, asn) side chains, providing a flexible peptide arm joining both parts of the fusion protein. This permits independent folding of both moieties of the fusion protein, favoring better antibody detection of epitopes encoded by the cDNA. The use of abundant (frequently used) *E. coli* codons ensures efficient translation from the β-gal region through the adaptor sequence into the cDNA insert.

2. cDNA synthesis: use a cDNA synthesis kit such as the cDNA Plus kit from Pharmacia. This kit includes all the enzymes/reagents necessary for first- and second-strand cDNA synthesis and the appropriate buffers, as well as prepoured Sephacryl S-300 spun columns for cDNA purification.
3. An adaptor-primer for efficient cloning and expression of PCR amplified cDNA. We designed a multifunctional adaptor/primer to permit PCR amplification of cDNA, efficient cloning of the amplified cDNA into a λ expression vector, and efficient expression of the cloned cDNA (**Fig. 1**). By embracing a number of important considerations and constraints, our adaptor-primer has the following useful features:

a. Minimal self-complementarity to avoid self-priming artifacts during PCR.
b. Codes for a short open-reading frame, which keeps the *Eco*R1-inserted cDNA sequence in-frame with the β-galactosidase coding sequence of the λZAP cloning vector.
c. Codes for a flexible arm of amino-acids. In many expression vectors, including λ ZAP, cDNA inserts are inserted into the *Lac*Z gene of the vector and are expressed as stable, recombinant β-galactosidase fusion proteins. This evidently stabilizes expression of foreign polypeptides in *E. coli*. All the amino-acids encoded by the adaptor have polar (ser, thr, cys, asn) or small, neutral (gly, ala) side chains. This provides a flexible peptide arm, joining both parts of the recombinant *Lac*Z-cDNA encoded fusion protein. This should favor independent folding of both moieties, resulting in greater availability of epitopes of the cloned polypeptide for detection by antibodies.
d. Adaptor sequence consists of frequently used *E. coli* codon choices to ensure efficient translation from the β-galactosidase region through the adaptor sequence into the cDNA insert.

4. Access to automated DNA synthesis for primer synthesis.
5. PCR: dNTPs (Promega); UlTma DNA polymerase and buffer (Perkin-Elmer, Norwalk, CT).
6. Electrophoresis: molecular-biology-grade agarose (Promega); ethidium bromide (Sigma); *Hae* III-digested ϕX174 DNA size markers (Promega).
7. T4 DNA polymerase and buffer (Promega).
8. T4 DNA ligase and buffer (Promega).
9. Bacteriophage λ ZAPII cloning vector, *Eco*R1 arms (Stratagene).
10. λ in vitro packaging kit (Promega).

3. Methods

1. Isolate total RNA from the tissue or cells of interest by lysis in guanidine isothiocyanate, followed by phenol extraction and ethanol precipitation. Polyadenylated mRNA is purified from the total RNA by oligo(dT)-cellulose chromatography. A number of kits are commercially available for direct isolation of mRNA from tissue (e.g., the Micro-FastTrack™ mRNA isolation kit from Invitrogen).
2. Synthesize double-stranded cDNA from the mRNA. Use commercially available kits (e.g., Pharmacia) to do this. The Pharmacia kit includes all the enzymes necessary for double-stranded cDNA synthesis and the appropriate buffers, as well as prepoured Sephacryl S-300 spun columns for cDNA purification. A detailed protocol is also provided.
3. Ligate the adaptor-primer (described in **Subheading 2.3** and in **Fig. 1**) onto the cDNA (**Fig. 2**). The adaptor consists of two complementary oligonucleotides, a 24-mer and a 20-mer, which anneal together to form a double-stranded duplex with one blunt end and one *Eco*R1 sticky end (**Fig. 1**). The hemi-phosphorylated form of the adaptor is used—i.e., only the 20-mer but not the 24-mer oligonucleotide is phosphorylated at its 5' end (using polynucleotide kinase). This precludes

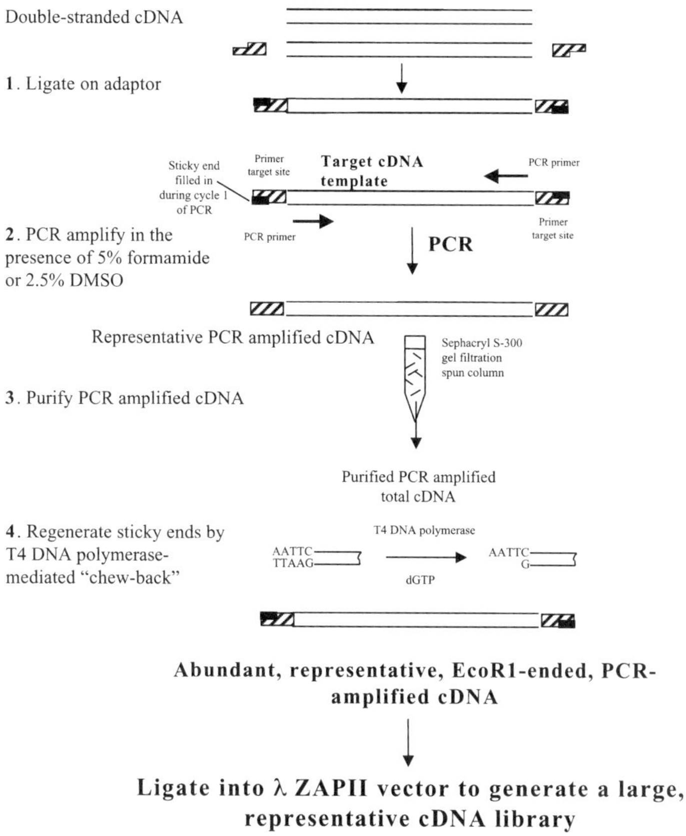

Fig. 2. Schematic diagram outlining the principal steps in the RT-PCR-based cDNA library construction. **(1)** The adaptor is ligated onto the cDNA. A high molar ratio of adaptor:cDNA drives efficient ligation. **(2)**After purification through a Sephacryl S300 spun column, which removes unligated adaptors, the adapted cDNA is amplified by PCR. The 24-mer oligonucleotide of the adaptor serves as a single PCR primer, which primes from both ends of the adapted cDNA. The sticky ends of the adaptor are filled in during the first cycle of the PCR. Inclusion of 5% formamide or 2.5% DMSO during this PCR helps to obtain unbiased amplification of all cDNAs. A proofreading thermostable polymerase is used in the PCR, to minimize misincorporations. **(3)** The adapted, amplified cDNA is purified through a Sephacryl S300 spun column, which eliminates truncated cDNAs less than about 300 bp, as well as any primer dimers/PCR artifacts. **(4)** The *Eco*R1 sticky ends of the amplified cDNA are regenerated via T4 DNA polymerase "chew-back". In the presence of a single nucleotide (dGTP), the 3' exonuclease activity of T4 DNA polymerase removes the T and A residues from the adapted cDNA, until a position is reached where the dGTP can be incorporated. The polymerizing activity of T4 DNA polymerase then predominates, to produce *Eco*R1 sticky ends. The *Eco*R1-ended, amplified cDNA is then ligated into the λ ZAPII vector. The recombinant λ-cDNA molecules are then packaged in vitro to yield a large, representative cDNA library.

autoligation of adaptor units to form large multimers, which complicate cloning and cannot be readily removed from the adapted cDNA post-ligation. Only one unit of the hemi-phosphorylated adaptor can be ligated onto each cDNA end, and the only autoligation product of the adaptor itself is a "back-to-back" dimer, which, along with free adaptor molecules, is easily removed from the adapted cDNA by gel filtration through a Sephacryl S300 spun column (*see* **step 4**). For this reason also, a large molar excess of adaptor to cDNA (about 100:1) may be employed to drive efficient ligation of the adaptor onto all of the cDNA ends (*see* **Note 1**).

4. Purify the adapted cDNA by spun-column chromatography through Sephacryl S-300 gel-filtration resin equilibrated with Tris-EDTA (TE) (the columns and a detailed protocol are provided with the cDNA synthesis kit).
5. PCR-amplify the purified, adapted cDNA using the 24-mer oligonucleotide of the adaptor as a primer. This single primer will prime from both ends of the adapted cDNA (*see* **Note 2**). Include a denaturant such as formamide at 5% or dimethyl sulfoxide (DMSO) at 2.5% in this PCR to favor representative, unbiased amplification of the cDNA. Use a thermostable polymerase with a 3' exonuclease "proofreading" activity, such as the UlTma DNA polymerase (Perkin-Elmer), to minimize misincorporation errors in the amplified cDNA. This PCR should result in a large yield of amplified cDNA for cloning, even from minute quantities of adapted cDNA template.
6. Remove the polymerase and other proteins from the amplified cDNA by phenol extraction (*see* **Note 3**). Efficient removal of the thermostable DNA polymerase is important, as it is difficult to inactivate and would interfere with the ligatable sticky ends of the DNA during the subsequent ligation step.
7. Remove unincorporated primer and nucleotides, and traces of phenol, by gel filtration chromatography through a Sephacryl S300 spun column equilibrated in TE (this replaces the PCR buffer with TE). This step eliminates truncated cDNAs less than about 300 bp, as well as any short PCR artifacts. Such short fragments would otherwise constitute an unacceptable proportion of the cDNA clones, making library screening problematic, and possibly out-competing the occurrence of informative clones in the library (*see* **Note 4**).
8. Generate *Eco*R1 ends in the amplified, adapted cDNA by the "chew-back" activity of T4 DNA polymerase (**Fig. 1**). "Chew-back" utilizes the 3'-exonuclease activity of T4 DNA polymerase to generate *Eco*R1 sticky ends in the PCR-amplified adapted cDNA. **Figure 2** shows how this 3'-exonucleolytic activity of T4 DNA polymerase is exploited to generate *Eco*R1 sticky ends in the amplified cDNA. Incubate the amplified, adapted cDNA at 37°C for 30 min with 1 U of T4 DNA polymerase in its buffer (Promega) in the presence of 2.5 m*M* dGTP (Promega) (*see* **Note 5**).
9. Purify the *Eco*R1-ended, PCR-amplified cDNA by phenol extraction and ethanol precipitation. The *Eco*R1-ended, PCR-amplified cDNA is now ready for cloning.
10. Ligate the *Eco*R1-ended, PCR-amplified cDNA into the *Eco*R1 site of the selected λ cloning vector (i.e. λ ZAP-II, Stratagene). Since λ DNA is linear, restriction cleavage at the unique *Eco*R1 site in the insertion site produces two

"arms" between which the cDNA is ligated. The vector and cDNA insert are mixed and ligated with T4 DNA ligase (Promega), in a ligase buffer (Promega) that contains adenosine triphosphate (ATP). The volume of the ligation is usually kept small (10 µL or less) (*see* **Note 6**). For a particular cDNA substrate, it is difficult to predict the best ratio of cDNA insert:vector needed to result in optimal ligation efficiency. In practice, a few different molar ratios are tested, using a fixed amount of vector DNA (the vector manufacturers provide guidelines). Since the amplified cDNA is unphosphorylated, having been generated by PCR with an unphosphorylated primer, the cDNA cannot ligate to itself. Thus, a molar excess of cDNA to vector arms will favor insertion of cDNA rather than religation of vector arms without a cDNA insert (*see* **Notes 7** and **8**).

11. Perform in vitro packaging of the recombinant molecules into viable phage particles (*see* **Note 9**). A number of packaging kits are commercially available (e.g., Promega), and are used according to the manufacturers' instructions. A pilot experiment is performed initially, where an aliquot of each of the test ligations is packaged. The number of recombinant cDNA clones generated is determined by ascertaining the "titer" or number of phage particles. Essentially, a dilution series of the packaging reaction is mixed with the appropriate *E. coli* host strain for the particular λ phage used. The cell-phage mixture is incubated at 37°C for 30 min to allow infection of the cells, and the mixture is added to 4 mL of molten "top-agar"—(LB agar containing 0.7% agar) at 48°C. The top agar is vortexed and plated onto an agar plate. After a few minutes to allow the top-agar to solidify, the plates are then incubated at 37°C overnight. After incubation, the top agar will contain an opaque "lawn" of cultured *E. coli*, with small pin-point zones of bacterial lysis called "plaques." Each plaque represents a clonal expansion of phage from a single initial phage-infected *E. coli* cell. After lysis of the initial cell, the released bacteriophage infect and lyse surrounding cells in the bacterial lawn, resulting in a plaque. The number of viable bacteriophage generated in the packaging reaction can be determined by counting the plaques and multiplying by the dilution factor. The proportion of recombinant plaques is determined by adding X-gal (40 µg mL^{-1}) and IPTG (0.5 m*M*) to the top agar prior to plating. Since cDNA is inserted into an *Eco*R1 site within the LacZ gene in the λ vector, recombinant plaques will be uncolored, compared to plaques formed by the parent vector minus insert, which will be colored blue. Detailed protocols for plating λ phage with *E. coli* to obtain bacteriophage plaques are provided by the manufacturers of the λ vector (Stratagene) and elsewhere *(2)*. Having ascertained which packaging reaction yielded the biggest number and highest percentage of recombinant clones, all of the corresponding ligation is then packaged in a full-size packaging reaction. The resultant recombinant phage particles constitute the cDNA library. The final titer of the library is determined as already described. The abundance of the PCR-amplified cDNA insert should ensure high efficiency in these steps, resulting in a large cDNA library.
12. Amplify the cDNA library to generate multiple copies of the library. This involves plating the entire library on *E. coli* at a high plaque density. For

amplification and screening of the library, use large Petri dishes (140-mm diameter) that enable 10,000–25,000 plaques to be obtained per plate. After overnight incubation, the plates are overlaid with "phage buffer" *(2)*. Bacteriophage from all the clones or plaques will be eluted into the phage buffer. Since each plaque will contain approx 10^5 bacteriophage particles, the combined eluates will contain approx 10^5 copies of the cDNA library. The titer of the amplified library is determined, and the number of copies of the library is simply the titer of the amplified library divided by the titer of the library prior to amplification. The amplified cDNA library is then aliquoted and stored at –70°C after adding DMSO (7%). For screening of an amplified library to isolate clones, it is recommended that two copies of the library are plated for screening to allow for sampling error or clone bias during amplification of the library (*see* **Note 10**). Techniques for screening cDNA libraries with antibody or DNA probes are detailed elsewhere *(2)*.

4. Notes

1. Because the adapter contains an *Eco*R1 sticky end, the adapted cDNA can be directly ligated into the *Eco*R1 site of a cloning vector. If it seems that there is sufficient adapted cDNA, this should be attempted, because this may result in a cDNA library containing the requisite number of recombinant clones without the need for PCR amplification. If not, the adapted cDNA can be amplified by PCR to generate a large library using the remainder of the protocol in this chapter.
2. The *Eco*R1 sticky ends of the adapted cDNA will be filled in during the first cycle of PCR, so that the PCR-amplified cDNA products will be blunt-ended.
3. A single phenol extraction will suffice. There is no need to perform a chloroform extraction, since the subsequent Sephacryl S300 gel-filtration column will remove traces of phenol from the DNA.
4. It is vitally important to eliminate short DNA products at this stage, for several reasons. First, even if very little low-mol-wt DNA is found in the PCR products upon agarose gel analysis, remember that very short fragments of DNA bind ethidium bromide less efficiently than longer fragments. Secondly, very short pieces of DNA such as primer dimers will contribute a much higher molar quantity of DNA ends than the same weight of larger fragments. Thus, contaminant short DNA pieces can contribute a high proportion of the subsequent cDNA clones. Finally, recombinant clones containing very short inserts will probably replicate more efficiently than those containing larger inserts, leading to potential bias and better survival of clones of the short inserts.
5. Like all DNA polymerases, T4 DNA polymerase synthesizes DNA in a template-dependent fashion in the presence of the four deoxynucleotide 5' triphosphate (dNTP) substrates. It is frequently used in cloning methodologies to "fill in" sticky ends of restriction digested DNA to provide blunt-ended DNA: the 5' overhang of the sticky end serves as a template for extension of the recessed 3' end. Most polymerases (Taq polymerase being a notable exception) also possess a 3' exonucleolytic "proofreading" activity, which, in vivo, degrades growing strands

of DNA from their 3' ends (thus the term "chew-back") to excise mis-incorporated nucleotides. When dNTPs are limiting or absent, the potent 3'–5' exonuclease activity of T4 DNA polymerase predominates. In the presence of a single dNTP, exonuclease activity will begin at the 3' ends of DNA and proceed until a position is reached where this nucleotide can be incorporated by polymerase activity. Here, polymerization activity predominates, the nucleotide is replaced at a greater rate than that of its nucleolytic excision, and exonuclease degradation is halted.

6. Keeping the volume of the ligation to a minimum helps to maximize the concentration of DNA ends, thus favoring contact and ligation.
7. There is really no alternative to determining the best ratio of insert-to-vector empirically. It is worth testing a number of ratios, as this will result in an optimal cDNA library with a high proportion of recombinant clones (90% or greater should be achievable). Remember that unlike some plasmid vectors, there is no way of selecting for the growth only of λ ZAP bacteriophage containing a cloned insert. Thus, generating a library with a low percentage of recombinants will increase the amount of phage required to ensure good representation of the cloned cDNA, and will increase the volume of phage needed to be screened to identify the required clones.
8. Although sometimes the extent of ligation can be visible by agarose gel analysis, this is not always clear, and is often misleading. The only reliable way to determine the success of the ligation is to package the ligation in vitro and determine the bacteriophage titer.
9. In vitro packaging extracts are prepared from *E. Coli* cells infected with specially modified strains of bacteriophage λ. Packaging extracts contain all the structural and bacteriophage "coat" proteins, enzymes, and other factors necessary to assemble biologically viable bacteriophage particles. The extracts do not contain "packagable" λ DNA. Thus, any extraneous recombinant λ DNA added to the extract will be packaged into mature bacteriophage particles.
10. Amplification of the λ library by generating plaques on agar plates is performed specifically to minimize bias among the clones during amplification. If the library was amplified by simply infecting a liquid *E. coli* culture with the entire library, and incubating to generate bacteriophage progeny, there would be direct competition between the clones so that efficiently replicating clones would outcompete poorly replicating ones. The resultant recombinant phage harvested at the end of the infection would be biased. In plaque format, there is no competition between growth of the different plaques, although there will be some slight variation in plaque size. Glucose (1%) should be added to the top agar used for library amplification. The presence of abundant glucose suppresses any use by the *E. coli* host cells of the LacZ promoter in the λ vector, so that the β-galactosidase-cDNA fusion protein is not expressed during library amplification. This helps to minimize variation in the plaque size, since some fusion proteins can be somewhat toxic to *E. coli* cells, and the variable rate of expression of the different fusion proteins may contribute to variation in the rate of phage replication.

Acknowledgments

This work was supported by grants from the Health Research Board of Ireland (HRB) and a "New Blood" fellowship award from the Wellcome Trust-HRB.

References

1. Yureci, O., Sahin, U., and Pfreundschuh, M. (1997). Serological analysis of human tumor antigens: molecular definition and implications. *Mol. Med. Today* **3,** 342–349.
2. Sambrook, J., Fritsch, E.F., and Maniatis, T. (1989). *Molecular Cloning: A Laboratory Manual,* 2nd ed., Cold Spring Harbor Press, Cold Spring Harbor, NY, pp. 12.16–12.20.

Index

R

S

T

V

DATE DUE

NOV 06 2002
DEC 06 2002
JAN 22 2003
MAR 25 2003
MAY 21 2003
SEP 30 2003
DEC 10 2003
JUN 7 2004
JUL 31 2004
SEP 16 2004
APR 11 2005
OCT 28 2005
MAR 15 2006
MAY 10 2006
NOV 13 2008
MAY 26 2010
AUG 08 2011

DEMCO 38 297